Gold Science: Concepts, Technology and Applications

Gold Science: Concepts, Technology and Applications

Edited by **Darren Wang**

WILLFORD PRESS

New York

Published by Willford Press,
118-35 Queens Blvd., Suite 400,
Forest Hills, NY 11375, USA
www.willfordpress.com

Gold Science: Concepts, Technology and Applications
Edited by Darren Wang

International Standard Book Number: 978-1-68285-116-6 (Hardback)

The publisher's policy is to use permanent paper from mills that operate a sustainable forestry policy. Furthermore, the publisher ensures that the text paper and cover boards used have met acceptable environmental accreditation standards.

Trademark Notice: Registered trademark of products or corporate names are used only for explanation and identification without intent to infringe.

Printed in the United States of America.

Contents

Preface

It is often said that books are a boon to mankind. They document every progress and pass on the knowledge from one generation to the other. They play a crucial role in our lives. Thus I was both excited and nervous while editing this book. I was pleased by the thought of being able to make a mark but I was also nervous to do it right because the future of students depends upon it. Hence, I took a few months to research further into the discipline, revise my knowledge and also explore some more aspects. Post this process, I began with the editing of this book.

Gold is a noble metal with various useful physical and chemical properties and a diverse array of applications in a myriad of fields like in electronics, medicine, dentistry, aerospace, etc. Its properties make gold one of the most widely used elements. This book will acquaint the readers with the latest researches in the varied areas of gold science. The students of chemistry, metallurgy and associated disciplines would find this book extremely useful.

I thank my publisher with all my heart for considering me worthy of this unparalleled opportunity and for showing unwavering faith in my skills. I would also like to thank the editorial team who worked closely with me at every step and contributed immensely towards the successful completion of this book. Last but not the least, I wish to thank my friends and colleagues for their support.

Editor

Enzyme electrodes stabilized by monolayer-modified nanoporous Au for biofuel cells

**Masataka Hakamada · Masaki Takahashi ·
Mamoru Mabuchi**

Abstract Open-cell nanoporous Au (np-Au) electrodes with pore size of approximately 40 nm were fabricated by dealloying of Au–Ag, and surfaces of the electrodes were modified with a self-assembled monolayer (SAM) of 4-aminothiophenol to enhance the electrocatalytic activities of immobilized laccase and glucose oxidase. Enzyme-immobilized SAM-modified np-Au working electrodes exhibited additional reduction–oxidation peak pairs in cyclic voltammograms in buffer solution (pH=5.0). Thus, the SAM on the np-Au facilitated electron transfer between the electrode and reactants. First-principles calculations of perfect and defective Au (111) surfaces indicated that the atomic defects at nanoligament surface of np-Au are critically responsible for the electron transfer enhancement. For the utilization of these results, a glucose/O_2 biofuel cell composed of these enzyme-immobilized SAM-modified np-Au electrodes was preliminarily fabricated, and it exhibited a maximum power density of 52 μW/cm^2 at 20°C. Further optimization of nanoporous structures and kinds of SAM will improve the performance of biofuel cells.

Keywords Nanoporous gold · Laccase · Glucose oxidase · Biofuel cell · Self-assembled monolayer (SAM)

M. Hakamada (✉) · M. Takahashi · M. Mabuchi
Department of Energy Science and Technology,
Graduate School of Energy Science, Kyoto University,
Yoshida Honmachi, Sakyo,
Kyoto 606-8501, Japan
e-mail: hakamada.masataka.3x@kyoto-u.ac.jp

Introduction

Biofuel cells convert chemical energy into current-using enzymes as biocatalysts and are promising devices for harvesting bioenergy [1–4]. The essential feature of enzymatic biofuel cells is that power can be produced from a fuel (i.e., glucose) present in the human body [5, 6]. The main advantage of such fuel cells is their ability to operate under mild conditions at temperatures of 20–40°C and at approximately neutral pH values [7, 8].

For implantable and miniature biofuel cells, materials and microstructures of the electrode surfaces are important factors [9–11]. The three-dimensional (3-D) structures and high surface areas of porous materials are attractive in preparing cells to improve power output. Materials with high surface areas such as carbon felt and porous carbon are commonly used as electrodes; however, it remains difficult to control the pore size distribution and, therefore, the real surface area of these materials [12, 13]. Nanoporous metals with 3-D networks of nanoligaments and nanopores have received much recent interest as supporting matrixes for the loading of biomolecules [14–16]. Monolithic nanoporous metals can be easily prepared by dealloying and have large specific surface areas. Their pore size can be controlled across a wide range from a few nanometers to several micrometers by thermal and/or acid treatments. With these characteristics in mind, nanoporous metals are candidates for the immobilization of enzymes. Their 3-D rigid nanoporous structure increases the number of active sites, which leads to high-density loading and effective spatial dispersion of the enzyme [16–19].

In this study, the electrochemical properties of laccase and glucose oxidase (GOx) immobilized within nanoporous Au (np-Au) have been elucidated. The pore size of np-Au was adjusted to approximately 40 nm to ensure high enzyme stability [14], and the effect of a self-assembled monolayer (SAM) of 4-aminothiophenol (4-ATP) which enhances the electron transfer [20] was examined. The enhanced electrochemical properties of enzyme-immobilized np-Au with the SAM were due to the strong covalent or electrostatic bonding between SAM amino groups and enzyme lateral amino acids, which facilitated electron transfer.

Experimental

Preparation of nanoporous and smooth Au

Commercially available Au (>99.9 mass%) and Ag (>99.9 mass%) ingots were arc-melted together under an Ar atmosphere to give the precursor $Au_{0.35}Ag_{0.65}$ alloy ingot. After homogenization at 1,173 K for 24 h under an Ar atmosphere and cold rolling, nanoporous Au was synthesized by dealloying at room temperature in 70 mass % HNO_3 for 8 days. Figure 1 shows a scanning electron microscopy (SEM) image of the np-Au in which the average pore size was 40 nm. For reference, a smooth Au surface (without the nanoporous structure) was prepared by grinding rolled Au plate with 1,200 grit SiC sandpaper, sonication in ethanol and distilled water for 5 min, immersion in 70 mass% HNO_3 for 3 min, and finally washing with distilled water.

SAM modification and enzyme immobilization

The 4-ATP monolayer was self-assembled on various np-Au samples as well as the smooth Au surface by immersion in a 20 mM ethanolic solution of 4-ATP for 65 h at room temperature in ambient air. Following immersion, samples were thoroughly rinsed with ethanol and distilled water.

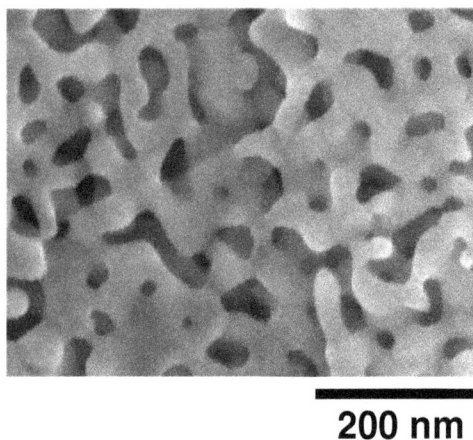

200 nm

Fig. 1 Scanning electron microscopic image of nanoporous Au with average pore size of 40 nm

Samples (both with and without SAM-modification) were kept quiescently in 2 mL of 7 mg/mL laccase (*Trametes versicolor* from Sigma Aldrich Corp. with an activity of 21 U/mg) or GOx (*Aspergillus niger* from Nacalai Tesque, Inc. with an activity of 273 U/mg) solutions (diluted with 0.1 M citrate–0.2 M phosphate buffer solution, pH=5.0) at 4°C for 24 h. Samples were then rinsed five times with 10 mL of the buffer solution to remove any weakly adsorbed enzyme on the outer surface. Assay by the Bradford method revealed that approximately 25 and 23 mg/g_{Au} of laccase were immobilized on the nanoporous Au samples with and without SAM-modification, respectively.

Electrochemical measurements

Cyclic voltammetry (CV) was carried out using a potentiostat (HZ-5000 by Hokuto Denko Corp.) at room temperature to elucidate the electrochemical properties and stability of the enzyme-immobilized samples. A three-electrode electrochemical cell with Pt wire as a counter electrode, the enzyme-immobilized np-Au electrode as a working electrode, and saturated calomel electrode (SCE) as a reference electrode was used. Potentials were documented vs standard hydrogen electrode (SHE) unless otherwise stated. For the laccase-immobilized electrode, the electrolyte of 0.1 M citrate–0.2 M phosphate buffer solution was air-saturated with dry air bubbling for 1 h. CV was conducted at 100 mV/s from −0.15 to 0.7 V. For the GOx-immobilized electrode, the electrolyte of 0.1 M citrate–0.2 M phosphate buffer solution containing 100 mM glucose was air-saturated, and CV was conducted at 100 mV/s from −0.05 to 0.85 V. In all electrochemical measurements, the apparent exposed area of the working electrode was 9 mm^2.

First-principles calculations

Geometry optimization calculations were performed using the Cambridge Serial Total Energy Package (CASTEP) [21], in which density functional theory [22, 23] was used with a plane wave basis set to calculate the electronic properties of solids from first principles. The exchange–correlation interactions were treated using the spin-polarized version of the generalized gradient approximation within the scheme due to Perdew–Burke–Ernzerhof [24]. Ultra-soft pseudo-potentials [25] represented in reciprocal space were used for all elements in the calculations. The Kohn–Sham wave functions of valence electrons were expanded to the plane wave basis set within a specified cutoff energy (=340 eV). The Brillouin zone was sampled using a Monkhorst-Pack $3 \times 3 \times 1$ k-point mesh in the defective surface models and Gaussian smearing with a width of 0.1 eV.

Calculations were performed using face-centered-cubic (111) surface slab models for the 3×3 supercells of the (1×1) unit cell in three different surface models to aid comparison with previous reports [26, 27]. The honeycomb and adatom models were generated to simulate the defective surface of np-Au by the structural relaxation calculation after removal of atoms from the (111) perfect surface model. All models contained five atomic layers where the two bottom layers were constrained while the rest were allowed to relax. The repeated slabs were separated from each other by a vacuum space of 10 Å.

Results and discussion

Enzyme electrodes

Figure 2 shows CV curves for the laccase-immobilized np-Au electrodes. As shown in Fig. 2a, the laccase-immobilized np-Au electrode without SAM exhibited one reversible broad peak pair with a midpoint potential of 0.13 V, similar to a previous study [14]. Figure 2a also shows no apparent degradation in the laccase-immobilized np-Au electrode without SAM after 100 cycles of measurement. In contrast, laccase-immobilized np-Au with SAM showed two sets of reduction–oxidation (redox) peaks in the CV curves (Fig. 2b). The first and second peak pairs had midpoint potentials of +0.26 and +0.55 V, respectively. The double peak pair indicated electron transfer via different copper centers (T2 and T1 for the lower and higher potential peaks, respectively) in *T. versicolor* laccase [28], although the peak potential is somewhat lower than the previously reported values [28, 29] perhaps because of the changed conformation of enzyme. Thus, the SAM modification of

np-Au educed the redox electron transfer at the T1 copper center of immobilized laccase.

The peak current for the laccase-immobilized np-Au electrode with SAM increased during repeated CV cycles. This may be perhaps due to the synergistic effect of electrochemical cleaning of Au electrode surface and orientation change during the CV cycles. However, laccase on the smooth Au surface (without nanoporous structure) exhibited a current decrease after 100 cycles, even when modified with the SAM (inset in Fig. 2b). Laccase could not be fully stabilized on the smooth Au surface because of the lack of nanoporous structure to geometrically confine the enzyme. Another reason for the decreased activity is that the SAM on the smooth Au surface had a weaker binding energy than that on nanostructured Au and was more readily desorbed during CV [30, 31]. The nanoporous structure and SAM were responsible for the electrochemical stability of the immobilized enzyme as np-Au geometrically captured laccase which formed strong bonding with the SAM. Furthermore, SAM itself promotes the direct electron transfer between gold and laccase [20, 32, 33]. Such high stability of enzyme and SAM may synergistically promote electron transfer through the enzyme–SAM electrode system.

CV curves of GOx-immobilized np-Au electrodes with and without SAM-modification are shown in Fig. 3. No redox peak was observed in CV curves of GOx-immobilized np-Au without a SAM (Fig. 3a); thus, this electrode exhibited no electron transfer. However, SAM-modification of np-Au resulted in an obvious redox peak pair (midpoint potential of +0.55 V) in the CV curve (Fig. 3b). This was accompanied by anodic prepeaks which have been related to fast second-order reactions [34]. The 4-ATP SAM appeared to bridge the electronic gap

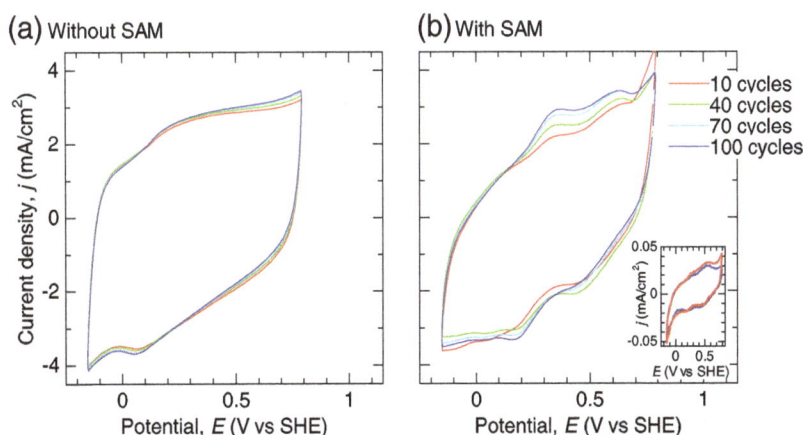

Fig. 2 Cyclic voltammetry curves for laccase-immobilized nanoporous Au (np-Au) working electrodes with (a) and without (b) a self-assembled monolayer (*SAM*) of 4-aminothiophenol. *Inset* in b shows CV curve for laccase-immobilized smooth surface of Au with the SAM. Laccases

(*T. versicolor* with an activity of 21 U/mg) were immobilized by soaking of np-Au (both with and without SAM-modification) in 2 mL of 7 mg/mL laccase solutions diluted with 0.1 M citrate–0.2 M phosphate buffer solution (pH=5.0) at 4°C for 24 h

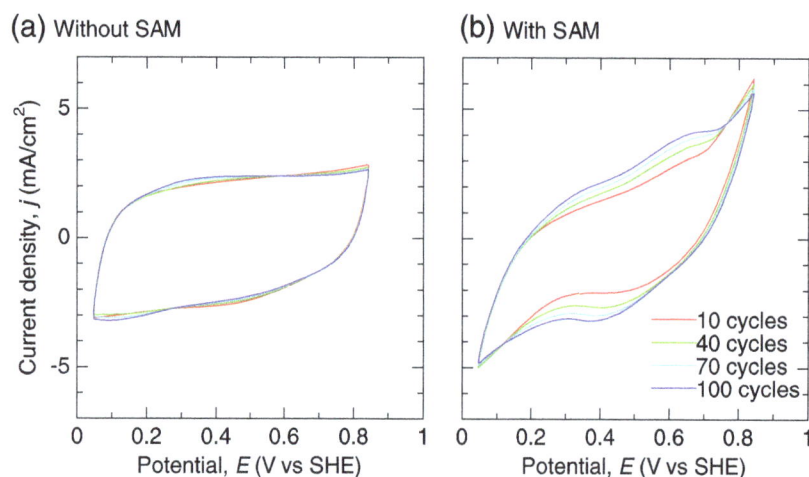

Fig. 3 Cyclic voltammetry curves for glucose oxidase (GOx)-immobilized nanoporous Au (np-Au) working electrodes with (**a**) and without (**b**) a self-assembled monolayer (*SAM*) of 4-aminothiophenol. GOx (*Aspergillus niger* with an activity of 273 U/mg) were immobilized by soaking of np-Au (both with and without SAM-modification) in 2 mL of 7 mg/mL GOx solutions diluted with 0.1 M citrate–0.2 M phosphate buffer solution (pH=5.0) at 4°C for 24 h

between the enzyme and electrode. It has been known that SAM with amine groups stabilizes GOx on Au electrode and enhances the electron transfer communication via covalent bonding [35]. Thus, the SAM on the np-Au electrodes effectively activated and stabilized GOx as well as laccase. However, the detailed nature of the strong bonding between SAM and enzymes is currently unknown. Raman and/or infrared spectroscopic elucidation may help in the understanding of the bonding state.

Electron transfer at defective surface

The effective direct electron transfer between electrolyte and electrode observed for enzyme-immobilized np-Au electrodes with SAM may have arisen from a defective surface structure [30, 36]. To investigate the effect of surface structure on electron transfer, first-principles calculations with CASTEP code were conducted. Figure 4 shows calculated local density of states (DOS) of Au atoms on the (111) surface of (a) perfect (no defect), (b) honeycomb (with defect), and (c) adatom (with defect) models. It has been reported that sharp peaks and quasi-gaps in DOS around the Fermi level indicate strong directional bonding, whereas a broad DOS around the Fermi level indicates isotropic-like metallic bonding [37, 38]. In the perfect model, sharp peaks were observed around the Fermi level, which indicated the origin of highly anisotropic electronic properties of the perfect Au (111) surface. Such localized DOS may have shortened the mean free path of surface electrons because of its constraint. The defective honeycomb and adatom models showed almost no quasi-gap around the Fermi level, which increased the mean free path of surface electrons. These properties

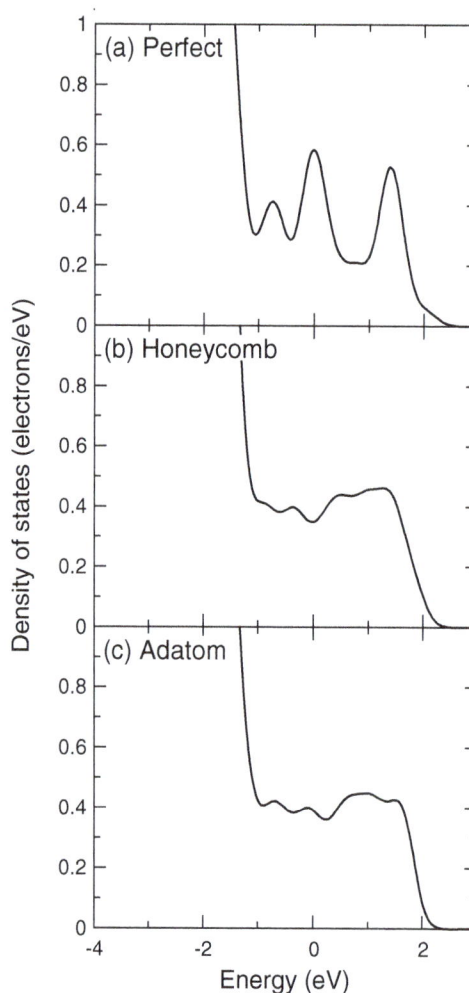

Fig. 4 Local density of states of (111) surface Au atom in **a** perfect, **b** honeycomb, and **c** adatom models calculated by first-principles calculations with CASTEP code. Fermi energy is set to 0 in the horizontal axes

of defective surfaces may have facilitated the electron transfer between SAM and metal, and consequently rendered SAM-modified nanoporous metals favorable for enzyme-immobilized electrodes. In other words, the coupling of defective surface and SAM modification might effectively enhance the electron transfer.

Preliminary performance of biofuel cells

To utilize the improved activities of the enzymes immobilized on SAM-modified np-Au, a glucose/O_2 biofuel cell consisting of GOx- and laccase-immobilized np-Au electrodes with SAM was preliminarily fabricated [39, 40]. The operating principle of the fabricated biofuel cell is shown in Fig. 5a. At the GOx-immobilized np-Au anode, glucose was oxidized to gluconic acid. At the laccase-immobilized np-Au cathode, O_2 was reduced to water by the bioelectrocatalysis of stabilized laccase. The power density vs current density plot of the glucose/O_2 biofuel cell was measured by a source measure unit (GS200 by Yokogawa Meter & Instruments Corp.) at 20°C in 250 mL of air-saturated buffer solution (pH=5.0) containing 100 mM glucose. The distance between the anode and cathode was 12 mm.

Figure 5b shows the cell voltage and power density vs current density of the fabricated biofuel cell. The maximum power density delivered by the biofuel cell was 52 μW/cm^2 at 0.21 V. The results were reproducible in two tests, although the full reproducibility should be clarified for further investigation.

Many recent studies have been conducted on glucose/O_2 system biofuel cells. For example, power densities of 64 μW/cm^2 at 23°C have been obtained using carbon fiber (2 cm×7 μm) electrodes [8]. A high power production of

442 μW/cm^2 was delivered by DNA-wrapped single-wall carbon nanotubes at room temperature [41]. Recent studies incorporating carbon nanotube electrodes shows higher maximum output of 740 μW/cm^2 [42] and 1,300 μW/cm^2 [40]. The maximum output of the present glucose/O_2 biofuel cell (52 μW/cm^2) is lower than those of other biofuel cells composed of GOx and laccase electrodes. However, there remains room for improvement. For example, the pore size should be optimized to immobilize maximum enzyme and to enhance the accessibility of cofactors and substrate. Altering the SAM to be more favorable for the respective enzymes is another possible method. Fuel cell configuration including electrode distance should be refined for the performance improvement. And also, for practical application, time variation and reproducibility of the cell performance are to be examined in more detail.

Conclusions

To conclude, the electrochemical properties of enzyme-immobilized np-Au were examined. SAM-modified np-Au had an increased stabilizing effect on enzymes compared with the smooth Au surface. The SAM was stabilized at the defective np-Au surface. Defective surfaces significantly contributed to the improved electron transfer between electrode and enzyme through SAM, resulting in increased operating current. These findings suggest that np-Au is a promising material for biofuel cells. Combined with biocompatibility and stability of np-Au, SAM-modification renders np-Au material favorable for the preparation of biofuel cells and enhancing their performance. Further research is necessary to improve enzyme stability, electronic conductivity, and consequent power production.

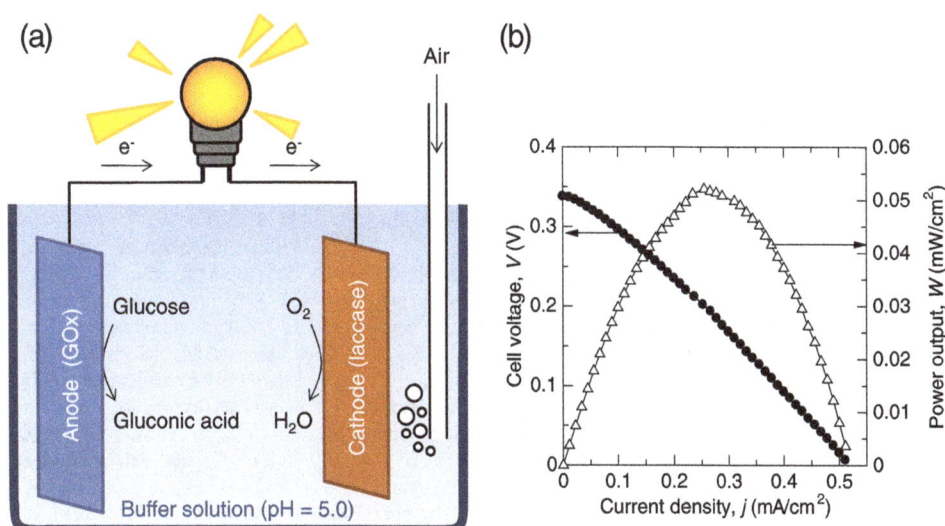

Fig. 5 a Schematic illustration of fabricated biofuel cell comprising glucose oxidase-immobilized nanoporous Au (np-Au) anode and laccase-immobilized np-Au cathode. **b** Cell voltage and power density vs current density plots generated from the fabricated biofuel cell at 20°C

Acknowledgement This study was supported by JSPS KAKENHI (22656155).

References

1. Bullen RA, Arnot TC, Lakeman JB, Walch FC (2006) Biofuel cells and their development. Biosens Bioelectron 21:2015–2045
2. Yuhashi N, Tomiyama M, Okuda J, Igarashi S, Ikebukuro K, Sode K (2005) Development of a novel glucose enzyme fuel cell system employing protein engineered PQQ glucose dehydrogenase. Biosens Bioelectron 20:2145–2150
3. Pizzariello A, Stred'ansky M, Miertuš S (2002) A glucose/hydrogen peroxide biofuel cell that uses oxidase and peroxidase as catalysts by composite bulk-modified bioelectrodes based on a solid binding matrix. Bioelectrochemistry 56:99–105
4. Katz E, Bückmann AF, Willner I (2001) Self-powered enzyme-based biosensors. J Am Chem Soc 123:10752–10753
5. Kendall K (2002) Fuel cell technology—a sweeter fuel. Nat Mater 1:211–212
6. Service RF (2002) Fuel cells: shrinking fuel cells promise power in your pocket. Science 296:1222–1224
7. Jeon SW, Lee JY, Lee JH, Kang SW, Park CH, Kim SW (2008) Optimization of cell conditions for enzymatic fuel cell using statistical analysis. J Ind Eng Chem 14:338–343
8. Chen T, Barton SC, Binyamin G, Gao Z, Zhang Y, Kim HH, Heller A (2001) A miniature biofuel cell. J Am Chem Soc 123:8630–8631
9. Barton SC, Gallaway J, Atanassov P (2004) Enzymatic biofuel cells for implantable and microscale devices. Chem Rev 104:4867–4886
10. Akers NL, Moore CM, Minteer SD (2005) Development of alcohol/O_2 biofuel cells using salt-extracted tetrabutylammonium bromide/Nafion membranes to immobilize dehydrogenase enzymes. Electrochim Acta 50:2521–2525
11. Heller A (2004) Miniature biofuel cells. Phys Chem Chem Phys 6:209–216
12. Tsujimura S, Fujita M, Tatsumi H, Kano K, Ikeda T (2001) Bioelectrocatalysis-based dihydrogen/dioxygen fuel cell operating at physiological pH. Phys Chem Chem Phys 3:1331–1335
13. Liu Y, Wang M, Zhao F, Liu B, Dong S (2005) A low-cost biofuel cell with pH-dependent power output based on porous carbon as matrix. Chem Eur J 11:4970–4974
14. Qiu H, Xu C, Huang X, Ding Y, Qu Y, Gao P (2008) Adsorption of laccase on the surface of nanoporous gold and the direct electron transfer between them. J Phys Chem C 112:14781–14785
15. Qiu H, Xu C, Huang X, Ding Y, Qu Y, Gao P (2009) Immobilization of laccase on nanoporous gold: comparative studies on the immobilization strategies and the particle size effects. J Phys Chem C 113:2521–2525
16. Shulga OV, Jefferson K, Khan AR, D'Souza VT, Liu J, Demchenko AV, Stine KJ (2007) Preparation and characterization of porous gold and its application as a platform for immobilization of acetylcholine esterase. Chem Mater 19:3902–3911
17. Xiao Y, Patolsky F, Katz E, Hainfeld JF, Willner I (2003) "Plugging into enzymes": nanowiring of redox enzymes by a gold nanoparticle. Science 299:1877–1881
18. Hudson S, Cooney J, Magner E (2008) Proteins in mesoporous silicates. Angew Chem Int Ed 47:8582–8594
19. Rekuć A, Bryjak J, Szymańska K, Jarzębski AB (2009) Laccase immobilization on mesostructured cellular foams affords preparations with ultra high activity. Process Biochem 44:191–198
20. Gupta G, Rajendran V, Atanassov P (2004) Bioelectrocatalysis of oxygen reduction reaction by laccase on gold electrodes. Electroanalysis 16:1182–1185
21. Payne MC, Teter MP, Allan DC, Arias TA, Joannopoulos JD (1992) Iterative minimization techniques for ab initio total-energy calculations—molecular-dynamics and conjugate gradients. Rev Mod Phys 64:1045–1097
22. Hohenberg P, Kohn W (1964) Inhomogeneous electron gas. Phys Rev 136:B864–B871
23. Kohn W, Sham LJ (1965) Self-consistent equations including exchange and correlation effects. Phys Rev 140:A1133–A1138
24. Perdew JP, Burke K, Ernzerhof M (1996) Generalized gradient approximation made simple. Phys Rev Lett 77:3865–3868
25. Vanderbilt D (1990) Soft self-consistent pseudopotentials in a generalized eigenvalue formalism. Phys Rev B 41:7892–7895
26. Molina LM, Hammer B (2002) Theoretical study of thiol-induced reconstructions on the Au(111) surface. Chem Phys Lett 360:264–271
27. Carro P, Salvarezza R, Torres D, Illas F (2008) On the thermodynamic stability of ($\sqrt{3} \times \sqrt{3}$) R30° methanethiolate lattice on reconstructed Au (111) surface models. J Phys Chem C 112:19121–19124
28. Frasconi M, Boer H, Koivula A, Mazzei F (2010) Electrochemical evaluation of electron transfer kinetics of high and low redox potential laccases on gold electrode surface. Electrochim Acta 56:817–827
29. Reinhammer BRM (1972) Oxidation–reduction potentials of the electron acceptors in laccases and stellacyanin. Biochim Biophys Acta 275:245–259
30. Cortés E, Rubert AA, Benitez G, Carro P, Vela ME, Salvarezza RC (2009) Enhanced stability of thiolate self-assembled monolayers (SAMs) on nanostructured gold substrates. Langmuir 25:5661–5666
31. Hakamada M, Takahashi M, Furukawa T, Tajima K, Yoshimura K, Chino Y, Mabuchi M (2011) Electrochemical stability of self-assembled monolayers on nanoporous Au. Phys Chem Chem Phys 13:12277–12284
32. Pita M, Shleev S, Ruzgas T, Fernández VM, Yaropolov AI, Gorton L (2006) Direct heterogeneous electron transfer reactions of fungal laccases at bare and thiol-modified gold electrodes. Electrochem Commun 8:747–753
33. Shleev S, Pita M, Yaropolov AI, Ruzgas T, Gorton L (2006) Direct heterogeneous electron transfer reactions of *Trametes hirsuta* laccase at bare and thiol-modified gold electrodes. Electroanalysis 18:1901–1908
34. Yokoyama K, Kayanuma Y (1998) Cyclic voltammetric simulation for electrochemically mediated enzyme reaction and determination of enzyme kinetic constants. Anal Chem 70:3368–3376
35. Willner I, Riklin A, Shoham B, Rivenzon D, Katz E (1993) Development of novel biosensor enzyme electrodes—glucose-oxidase multilayer arrays immobilized onto self-assembled monolayers electrodes. Adv Mater 5:912–915
36. Hakamada M, Nakano H, Furukawa T, Takahashi M, Mabuchi M (2010) Hydrogen storage properties of nanoporous palladium fabricated by dealloying. J Phys Chem C 114:868–873
37. Liu YL, Liu LM, Wang SQ, Ye HQ (2007) First-principles study of shear deformation in TiAl and Ti_3Al. Intermetallics 15:428–435
38. Woodward C, MacLaren JM, Rao S (1992) Electronic-structure of planar faults in TiAl. J Mater Res 7:1735–1750

39. Brunel L, Denele J, Servat K, Kokoh KB, Jolivalt C, Innocent C, Cretin M, Rolland M, Tingry S (2007) Oxygen transport through laccase biocathodes for a membrane-less glucose/O_2 biofuel cell. Electrochem Commun 9:331–336

40. Zebda A, Gondran C, Le Goff A, Holzinger M, Cinquin P, Cosnier S (2011) Mediatorless high-power glucose biofuel cells based on compressed carbon nanotube-enzyme electrodes. Nat Commun 2:370

41. Lee JY, Shin HY, Kang SW, Park C, Kim SW (2010) Use of bioelectrode containing DNA-wrapped single-walled carbon nanotubes for enzyme-based biofuel cell. J Power Sourc 195:750–755

42. Gao F, Viry L, Maugey M, Poulin P, Mano N (2010) Engineering hybrid nanotube wires for high-power biofuel cells. Nat Commun 1:2

2

Gold anion catalysis of methane to methanol

**Alfred Z. Msezane · Zineb Felfli · Kelvin Suggs ·
Aron Tesfamichael · Xiao-Qian Wang**

Abstract The oxidation of CH_4 has been investigated in the presence and absence of the atomic Au^- ion catalyst. We have employed the first principles density functional theory (DFT) and dispersion-corrected DFT calculations for the transition state on the Au^- ion and analyzed the thermodynamics properties of the reactions as well. Our results demonstrate that atomic gold anions could be used to catalyze CH_4 into valuable industrial products without the emission of CO_2, thereby making gold extremely valuable. The fundamental mechanism involves breaking the C–H bond through the formation of the anionic $Au^-(CH_4)$ molecular complex permitting the oxidation of CH_4 to methanol at the temperature of 325 K which is below that of CO_2 emission. Potentially, this could significantly impact the quality of our environment.

34.10.+x · 31.15.es · 34.50.Lf

Introduction

Nowadays, considerable efforts continue to be devoted to finding ways to reduce CO_2 emissions and atmospheric concentrations. Carbon sequestration, improving the efficiency of energy use, and reducing the carbon content of fuels are three major pathways that are currently being pursued to address the stabilization of greenhouse gas concentrations [1]. Carbon sequestration uses various approaches for CO_2 capture, storage, and reuse [1, 2]. One such process, CO_2 mineralization, uses carbonic anhydrase enzyme to convert dilute, unseparated CO_2 to HCO_3 and finally to everlasting calcium and magnesium carbonates. Biogenic methane is another of the carbon sequestrations; it involves geologic storage of CO_2 in depleting and depleted oil and gas reservoirs, with subsequent conversion of the CO_2 to CH_4 via designer microbes or biomimetic systems that operate above or below ground [1]. Common among many of these concepts is the enhancement of naturally occurring biochemical and geochemical processes through the identification and replication of natural processes for the purposes of carbon sequestration.

The catalytic partial oxidation of methane into valuable products is of great scientific importance and considerable industrial, economic, and environmental interest. However, a great challenge is that in the absence of an appropriate catalyst, methane undergoes complete combustion yielding carbon dioxide and water at approximately 340 K with minimal competition with the formation of useful products that can occur at elevated temperatures. The fundamental ideas of muon-catalyzed nuclear fusion utilizing a negative muon, a deuteron, and a triton [3] are used in the proposed oxidation of CH_4 to methanol for which we have selected the atomic gold anion as the catalyst. Here we propose the use of the atomic Au^- ion catalyst to control the temperature of the oxidation of methane to methanol around 325 K. This has the effect of lowering the transition state (TS) by 32 % compared to the case of the absence of the catalyst for the complete oxidation of methane to methanol without carbon

A. Z. Msezane · Z. Felfli (✉) · X.-Q. Wang
Department of Physics and Center for Theoretical Studies
of Physical Systems, Clark Atlanta University,
Atlanta, GA 30314, USA
e-mail: zfelfli@cau.edu

K. Suggs · A. Tesfamichael
Department of Chemistry, Clark Atlanta University,
Atlanta, GA 30314, USA

dioxide emission. We have employed the first principles density functional theory (DFT) and dispersion-corrected DFT calculations for the transition state on the Au^- ion and analyzed the thermodynamics properties of the reactions as well.

The main motivations for the investigation are: (1) the direct synthesis of H_2O_2 from H_2 and O_2 using supported Au, Pd, and Au–Pd nanoparticle catalysts [4, 5] including the theory [6, 7] that attributed the catalytic properties of Au and Pd to the formation of negative ion resonances in low-energy electron elastic total cross sections (TCSs) for Au and Pd atoms, along with their large electron affinities (EAs); (2) the recent dispersion-corrected density functional theory transition-state calculations performed on the atomic Au^- ion catalysis of water conversion to H_2O_2, revealing that the formation of the $Au^-(H_2O)_2$ anion molecular complex in the transition state provides the fundamental mechanism for breaking up the hydrogen bonding strength in the catalysis of H_2O_2 using the Au^- ion [8]. It is important to note that the Au^- ion is employed here as a prototype for negatively charged gold clusters or surfaces. The peculiar binding energy associated with the Au^- ion is of fundamental distinction as compared to that of the Au^+ ion or the neutral Au atom

Contrary to bulk gold, nanogold exhibits surprisingly high activity and/or selectivity in the combustion as well as partial oxidation of various molecules and compounds [9]. Since the publication of the paper [9], there have been considerable research activities on nanogold, particularly on its catalytic properties [9–29]. The mechanisms of charge transfer [11, 12] and relativity [13] have been advanced as possible explanations for the excellent catalytic properties of gold nanoparticles. Recently, the negative ion resonances that characterize the electron elastic scattering TCSs for atomic Au have been proposed as the fundamental mechanism driving nanoscale catalysis [6, 7]. The catalytic combustion of methane, the main component of natural gas, including its conversion to useful products, has recently received extensive experimental and theoretical attention because of the potential to reduce pollutant emissions and synthesize useful chemicals [30–36] and references therein. A recent investigation demonstrated the selective conversion of a mixture of methane and oxygen to formaldehyde at temperatures below 250 K through temperature-controlled Au^{2+} nanocatalysis [36].

Experimentally, it has been established that the Au^- anion interacts with water molecules to form the $Au^-(H_2O)_{1,2}$ complexes, causing bond breaking and with methane to form the $Au^-(CH_4)$ complex [37], thereby weakening the C–H bond. Furthermore, the strong interaction between the Au^- anion and H_2O is comparable to the hydrogen bonding in H_2O and the Au^- anion interaction with CH_4 is significant as well, but the Au^- ion does not interact with O_2 [29].

These findings [29, 37] are vital to the fundamental understanding of nanocatalysis using Au nanoparticles. To our knowledge, our proposed approach is the first to use the Au^- negative ion in the catalytic combustion of methane to useful products without the emission of CO_2.

Reactions and calculation method

The complete combustion of methane leads to the formation of carbon dioxide and water:

$$CH_4 + 2O_2 \rightarrow CO_2 + 2H_2O \tag{1}$$

Possible by-products of the partial oxidation of methane are:

$$CH_4 + \frac{1}{2}O_2 \rightarrow CO + 2H_2 \tag{2}$$

$$CH_4 + \frac{1}{2}O_2 \rightarrow CH_3OH \tag{3}$$

$$CH_4 + O_2 \rightarrow H_2CO + H_2O \tag{4}$$

$$CH_4 + O_2 \rightarrow HCO_2H + H_2 \tag{5}$$

Generally, there is little competition between the complete oxidation, reaction (1) and the selective partial oxidation (SPO), reactions (2, 3, 4, and 5), of methane. There are two reasons why the overall reaction leads to the formation of carbon dioxide and water: (1) Complete combustion of methane occurs at the lowest temperature compared to its SPO and (2) the corresponding transition state for reaction (1) is lowest compared to that of any SPO of methane to the desired products. However, the atomic Au^- negative ion activates molecular oxygen in CH_4 and increases the level of the SPO of methane to produce useful compounds.

Here the atomic Au^- catalyst is used to control the oxidation temperature of methane around 325 K to lower the transition state by 32 % compared to the case of the absence of the catalyst for the complete oxidation of methane to methanol and further oxidize methanol to formaldehyde and formic acid without CO_2 emission. We follow exactly the same procedure as in [6, 7] when applying the atomic Au^- ion catalyst to each of the reactions (1, 2, 3, 4, and 5).

The proposed mechanism of catalysis using the negative Au^- ion catalyst is as follows. When a slow electron collides elastically with a ground-state neutral gold atom, attachment can result, leading to the formation of a negative ion resonance due to the formation of compound atomic states. The energy position of this negative ion resonance corresponds to the stable bound state of the Au$^-$ negative ion formed during the collision as a resonance. The binding energy of the Au$^-$ ion defines the EA of atomic Au. Theoretically, it has been demonstrated that the EA of Au is right at the absolute minimum or the second R-T minimum (absolute) of the elastic TCS of Au [6, 7, 38, 39]. At this minimum and within the appropriate environment, the attachment of the Au$^-$ negative ion to the CH$_4$ molecule results in the formation of the Au$^-$(CH$_4$) anionic molecular complex. This complex formation results in the disruption of the stable C–H bonds in the methane molecule. The attendant change in the Gibbs energy of the system becomes negative, thereby thermodynamically favoring the formation of methanol. The Au$^-$ ion is released after the chemical reaction. We note that the dissociative energy of the Au$^-$(CH$_4$) molecular complex is within the second R-T minimum of the Au elastic TCS.

We have employed the first principles calculations based on DFT and dispersion-corrected DFT approaches for the investigation. For geometry optimization of structural molecular confirmation, we utilized the gradient-corrected Perdew–Burke–Ernzerfof parameterizations [40] of the exchange correlation rectified with the dispersion corrections [41]. The double numerical plus polarization basis set was employed as implemented in the DMol3 package [42]. The dispersion correction method, coupled to suitable density functional, has been demonstrated to account for the long-range dispersion forces with remarkable accuracy. We used a tolerance of 1.0×10^{-3} eV for energy convergence. A transition-state search employing nudged elastic bands facilitates the evaluation of energy barriers [43–45]. Finally, the energy of the transition state was calculated and the thermodynamic properties of the reaction were analyzed from the DMol3 package [42]. As the calculation of the transition barrier depends crucially on the exchange correlation scheme employed, the use of reliable dispersion-corrected approach is essential. The error in extracting the transition barrier associated with the transition pathway was estimated to be less than 0.001 eV [43–45].

Results and discussion

Figures 1, 2, 3, 4, and 5 present the optimized structures of the reactants, transition states, and products of oxidation of methane leading to the formation of CO$_2$, CO, CH$_3$OH, H$_2$CO, and HCO$_2$H, respectively. The data in (a) correspond to the absence of the Au$^-$ ion catalyst while those in (b) are data when the Au$^-$ ion catalyst is present. The red, white, gray, and gold spheres represent respectively oxygen, hydrogen, carbon, and gold atoms. The TS and EP, both in electron volts, represent respectively the calculated transition-state energy and the energy of the products. The

breaking of the stable C–H bonds in the methane molecule in the transition state resulting in the formation of methanol in the presence of O$_2$ is attributed to the formation of the anionic Au$^-$(CH$_4$) complex. The role of the Au$^-$ ion is to disrupt the stable C–H bonds in the methane molecule, allowing the formation of methanol in the presence of O$_2$. It is noted that the optimized structure corresponding to the reaction (3), namely the production of methanol, has the lowest transition-state energy (see Figs. 1b, 2b, 3b, 4b, and 5b). These results are also summarized in Table 1.

Understanding the results

Here we discuss the results of the complete oxidation of CH$_4$, reaction (1), and of the SPO of CH$_4$, reaction (3) as illustrations; the latter analysis also applies to the remaining reactions. In [6, 7] we explained the catalytic production of H$_2$O$_2$ from H$_2$O, using the atomic Au$^-$ ion catalyst, in the presence of O$_2$. Similarly, here we first apply the atomic Au$^-$ ion catalyst to the complete oxidation of CH$_4$, reaction (1), and obtain:

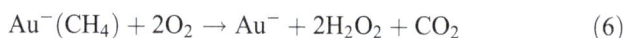

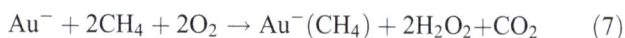

$$Au^-(CH_4) + 2O_2 \rightarrow Au^- + 2H_2O_2 + CO_2 \qquad (6)$$

$$Au^- + 2CH_4 + 2O_2 \rightarrow Au^-(CH_4) + 2H_2O_2 + CO_2 \qquad (7)$$

Adding the reactions (8) and (9), we get:

$$CH_4 + 2O_2 \rightarrow 2H_2O + CO_2 \qquad (8)$$

The Au$^-$ ion catalyst has changed nothing in the reaction, demonstrating complete combustion. The results of Table 1 (same TS values for the absence and presence of the catalyst) and Figs. 1 and 7 are illustrations of the complete combustion process. We note that the purpose of a catalyst is to decrease the reaction temperature to ambient temperature [46]. So, the Au$^-$ catalyst cannot be effective since the 340 K temperature (Table 1) is the ambient temperature for CO$_2$ production.

Next we apply the Au$^-$ ion catalyst to the reaction (3) and obtain:

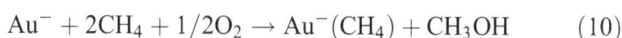

$$Au^-(CH_4) + 1/2O_2 \rightarrow Au^- + CH_3OH \qquad (9)$$

$$Au^- + 2CH_4 + 1/2O_2 \rightarrow Au^-(CH_4) + CH_3OH \qquad (10)$$

Adding the reactions (9) and (10), we have:

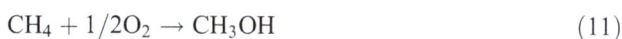

$$CH_4 + 1/2O_2 \rightarrow CH_3OH \qquad (11)$$

Contrary to the complete oxidation of methane, reaction (1), the Au$^-$ ion catalyzes the SPO of CH$_4$ to a new product, namely CH$_3$OH without CO$_2$ emission, reaction (11). As seen from comparing the TSs in column 2 and column 5 of

Fig. 1 Complete oxidation of methane to carbon dioxide and water in the absence (**a**) and presence (**b**) of the Au$^-$ negative ion catalyst. The *red, white, gray,* and *gold spheres* represent respectively oxygen, hydrogen, carbon, and gold atoms

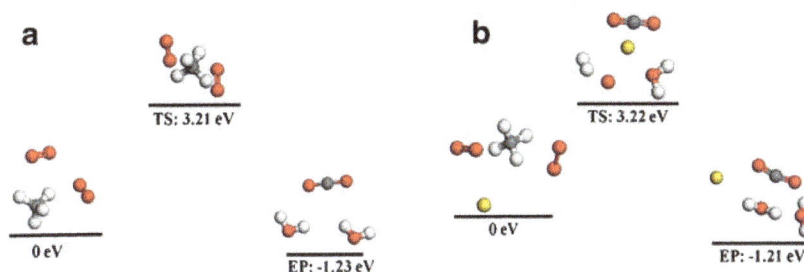

Table 1, the complete oxidation leaves the TS virtually unchanged when the Au$^-$ ion catalyst is introduced. However, for the case of the SPO of CH$_4$, reaction (3), the TSs are 4.41 and 3.01 eV in the absence and presence of the Au$^-$ ion catalyst, respectively. So, no barrier reduction is a manifestation of the complete oxidation of CH$_4$. For this case the catalyst has no effect on reaction (1). The obtained results in Table 1 and Figs. 1, 2, 3, 4, 5, 6, and 7 can be understood from three perspectives: resonance scattering theory, thermodynamics consideration, and transition-state calculations.

Resonance scattering approach

Most importantly, when a slow electron collides elastically with atomic Au, a stable negative Au$^-$ ion is formed almost exactly at the second deep R-T minimum of the electron elastic scattering TCS of atomic Au [38, 39]. The binding energy of this atomic Au$^-$ ion has been determined experimentally to be 2.309 eV [37, 47, 48]. This value also corresponds to the EA of atomic Au. If CH$_4$ is introduced at the second R-T minimum of the electron elastic TCS of atomic Au, it attaches to the Au$^-$ ion forming the anionic Au$^-$(CH$_4$) molecular complex [29, 37], with the vertical detachment energy (VDE) of 2.34 eV [29] (incidentally, the R-T minimum is used in the creation of exotic molecules such as RbCs [49, 50]). Here we observe the remarkable characteristic of atomic Au with respect to CH$_4$, namely the EA of Au and the VDE of Au$^-$(CH$_4$) are in the second R-T minimum of the Au elastic TCS. The interaction between the Au$^-$ ion and CH$_4$ is comparable to the C–H bond strength in CH$_4$ [29]. Thus the Au$^-$ ion weakens or disrupts the C–H bond in CH$_4$ permitting the formation of CH$_3$OH in the

presence of O$_2$. We note that the interaction between the Au$^-$ ion and O$_2$ is weak [29], showing the inertness of the Au$^-$ ion toward O$_2$. After the reaction the Au$^-$ ion catalyst is free to catalyze another reaction (the process is similar to the destruction of the ozone by the Cl$^-$ ion). This was the determining factor in our selecting the Au$^-$ ion as our catalyst. In [29] it has been remarked that the binding energies of the corresponding Au neutral complexes are significantly less than those of the anion species (for example, the complex Au$^-$(H$_2$O) has a binding energy that is more than an order of magnitude larger compared with that of the neutral Au(H$_2$O) complex [29]).

Thermodynamics of reactions

Low-energy chemical reaction dynamics provides the mechanism for making and breaking bonds. In the CH$_4$ catalysis to CH$_3$OH using the atomic Au$^-$ ion, the C–H bond breaking has been attributed to the formation of the anionic Au$^-$(CH$_4$) molecular complex. The C–H bonding has a direct effect on the change in the Gibbs free energy, G ($\Delta G = \Delta H - T\Delta S$) where H, T, and S represent respectively the enthalpy, temperature, and entropy. When the atomic Au$^-$ ion is introduced into the oxidation of CH$_4$, the breaking of the C–H bonding occurs. Therefore, the system changes from relative order to less order. Hence, the entropy of the system increases, whereas the enthalpy of the system decreases. The overall process results in the Gibbs free energy to be negative, resulting in the spontaneous formation of methanol. To gain a deeper understanding of the process of atomic Au$^-$ ion catalysis, the rate of the reaction was calculated using Arrhenius equation [51]. In Figs. 6 and

Fig. 2 Oxidation of methane to carbon monoxide and hydrogen gas in the absence (**a**) and presence (**b**) of the Au$^-$ negative ion catalyst. The *red, white, gray,* and *gold spheres* represent respectively oxygen, hydrogen, carbon, and gold atoms

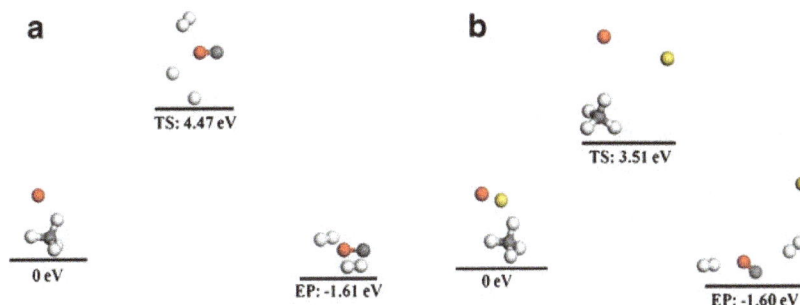

Fig. 3 Oxidation of methane to methanol in the absence (**a**) and presence (**b**) of the Au$^-$ negative ion catalyst. The *red, white, gray,* and *gold spheres* represent respectively oxygen, hydrogen, carbon, and gold atoms

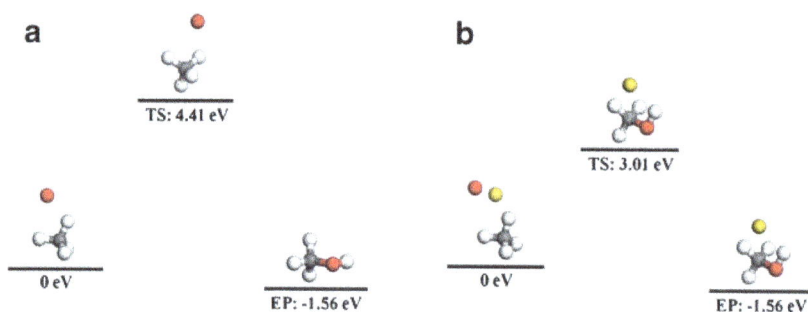

7, the ΔG versus T for all the reactions (1, 2, 3, 4, and 5) is depicted.

What is remarkable about the effect of the Au$^-$ ion catalyst on the SPO of CH_4 to CH_3OH and the complete oxidation of CH_4 is that whereas in the absence of the Au$^-$ ion catalyst, the production of methanol is at a much higher temperature (Table 1 and Fig. 7b). However, the introduction of the Au$^-$ ion catalyst into the reaction (3) dramatically impacts the rate of the reaction, lowering the temperature at which $\Delta G=0$, from 475 to 325 K (Table 1); this temperature is lower than that for the emission of CO_2 (340 K). Indeed the Au$^-$ catalyst is incredibly effective in catalyzing the conversion of CH_4 to CH_3OH without the emission of CO_2.

Transition-state calculation

Figure 1a, b presents respectively in the absence and presence of the Au$^-$ catalyst the TSs and EPs for the complete oxidation of CH_4 to CO_2. As already indicated, it is seen from both the figures that the TSs in the absence and presence of the Au$^-$ ion catalyst are virtually the same. Also the EPs differ only slightly. These results represent the signature of the complete combustion of CH_4. Henceforth, they will be used as the benchmark for assessing the SPO of the various reactions (1, 2, 3, 4, and 5).

Figure 2a, b displays the calculated TSs and EPs, in the absence and presence of the Au$^-$ ionic catalyst, respectively, for the SPO of CH_4 to $CO+2H_2$, reaction (2). Without the Au$^-$ ionic catalyst, the TS is 4.47 eV (Fig. 2a), while when the Au$^-$ ionic catalysts is present the TS drops down to 3.51 eV (Fig. 2b). This is to be expected since the role of the catalyst is to reduce the barrier.

Figure 3a, b presents respectively the data without and with the Au$^-$ ion catalyst for the SPO of CH_4 to methane, reaction (3). The introduction of the Au$^-$ ionic catalyst drops down the TS from 4.41 eV (Fig. 3a) to 3.01 eV (Fig. 3b). We note that this dramatic reduction of the TS of the reaction (3) in the presence of the Au$^-$ ion catalyst to a value below that of the complete oxidation of CH_4 is the main result of this paper. It represents a significant accomplishment in the field of catalysis using the Au$^-$ ion catalyst. The EPs are the same in both Fig. 3a, b as expected.

The results for the SPO of CH_4 to H_2CO+H_2O without and with the Au$^-$ ion catalyst are plotted, respectively in Fig. 4a, b. Just as for the reactions (2) and (3), given in Figs. 2b and 3b, the Au$^-$ ion catalyst reduces the barrier significantly. However, the TS of 3.29 eV shown in Fig. 4b is still slightly higher than that of the complete oxidation of CH_4, reaction (1). Perhaps, another atomic negative ion such as Pd$^-$ or Pt$^-$ [38] added to the Au$^-$ ion catalyst could reduce further the TS of 3.29 eV to a value significantly lower than that of the complete oxidation of CH_4. We believe that with a combination of the various atomic negative ion catalysts (see for example the various figures in [38]), all the reactions (2, 3, 4, and 5) could be catalyzed directly as in the case of the reaction (3) without CO_2 emission. This calls for further investigations.

Figure 5a, b contrasts the results for reaction (5), in the absence and presence of the Au$^-$ ion catalyst, respectively. Interestingly, for this reaction, the Au$^-$ ion catalyst reduces the TS by a small amount, 3.98 versus 3.71 eV. As expected, the EP remains unchanged in both figures. Comparing all the results presented in Figs. 1, 2, 3, 4, and 5, it is seen that the Au$^-$ ion catalyst has a dramatic effect on reaction (3). Namely, it reduces the TS of the reaction to a value below

Fig. 4 Oxidation of methane to formaldehyde and water in the absence (**a**) and presence (**b**) of the Au$^-$ negative ion catalyst. The *red, white, gray,* and *gold spheres* represent respectively oxygen, hydrogen, carbon, and gold atoms

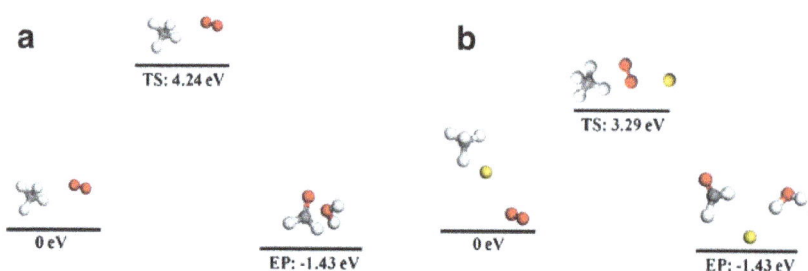

Fig. 5 Oxidation of methane to formic acid and hydrogen gas in the absence (**a**) and presence (**b**) of Au⁻ negative ion catalyst. The *red, white, gray,* and *gold spheres* represent respectively oxygen, hydrogen, carbon, and gold atoms

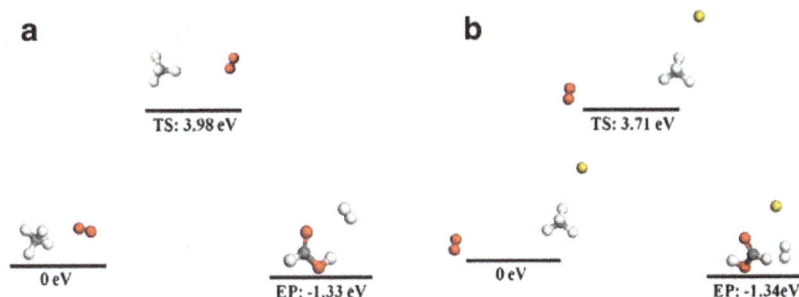

that obtained for the complete oxidation of methane. Hence, our main focus is on reaction (3). The results of these figures are summarized in Table 1.

Figure 6a, b presents the results of ΔG (in electron volts) versus T (in Kelvin) for the reactions (1, 2, 3, 4, and 5). Figure 6a represents the data in the presence of the Au⁻ ion catalyst, while Fig. 6b gives the results in the absence of the catalyst. We focus our discussion on reactions (1) and (3), namely the complete oxidation of CH_4 and the production of methanol. Note the position of the curve for the complete oxidation of CH_4, represented by the first curve in Fig. 6b, blue circles, and by the second curve in Fig. 6a, blue circles. In Fig. 6b, without the Au⁻ catalyst, the production of the methanol curve occupies the position 4, purple. However, in the presence of the Au⁻ catalyst, curve 4 jumps dramatically to position 1(Fig. 6a) ahead of the CO_2 production curve; the temperature at $\Delta G=0$ is 325 K. This can be compared with that of the CO_2 production at 340 K. Important here is that the CO_2 curve does not change its position from that it occupied in Fig. 6b. This clearly demonstrates the considerable effect the catalyst has on the methanol production. Again this represents the main result of this paper.

Figure 7a, b represents respectively the magnification of the data of Fig. 6a, b in the region around $\Delta G=0$; the colors of the curves correspond to those of Fig. 6a, b. These data exhibit clearly the extent to which a reaction has been influenced by the presence of the Au⁻ catalyst. By controlling the temperature around 325 K, methane can be completely oxidized to methanol, rather than to carbon dioxide (see Fig. 7a, first graph), and methanol can further oxidize to formaldehyde and formic acid.

Remarks on the results

As seen from Table 1, the thermodynamics properties agree excellently with the transition-state calculations of the complete and selective partial oxidation of methane. Combustion of methane to carbon dioxide and water in the presence and absence of the Au⁻ ion catalyst yields almost the same transition state. However, for the selective partial oxidation of methane, there is a significant change in the transition states when we compare the results in the presence and absence of the Au⁻ ion catalyst. The introduction of the Au⁻ ion catalyst lowers the transition states for the formation of CO, CH_3OH, H_2CO, and HCO_2H by 21, 32, 22, and 7 %, respectively. Also when we compare the transition states in the absence of a catalyst for the formation of carbon dioxide and methanol, we clearly see that the TS for the formation of CO_2 is smaller than that for the methanol formation. This elucidates why methane undergoes complete oxidation to carbon dioxide, resulting in the increased pollutant emissions. However, if the Au⁻ ion catalyst is used, the oxidation of methane favors the formation of methanol because its TS is lower than that of carbon dioxide. This is much like the separation of a mixture of alcohol and water through the temperature control.

In summary, this proposed catalytic process involving the use of the atomic Au⁻ ion catalyst promises a first and a giant step toward finding and assembling nanocatalysts atom by atom for various chemical reactions, including the direct partial oxidation of methane to useful products without CO_2 emission. This will certainly address the problem of greenhouse gas emissions, with considerable impact on the environment.

Table 1 TS, EP, and T represent, respectively, the calculated transition state, energy of the products and temperature of the reaction

	TS (eV) No catalyst	EP (eV) No catalyst	T (K) $G=0$	TS (eV) Catalyst Au⁻	EP (eV) Catalyst Au⁻	T (K) $G=0$
$CH_4 + 2O_2 \rightarrow CO_2 + 2H_2O$	3.21	−1.23	340	3.22	−1.21	340
$CH_4 + \frac{1}{2}O_2 \rightarrow CO + 2H_2$	4.47	−1.61	500	3.51	−1.60	375
$CH_4 + \frac{1}{2}O_2 \rightarrow CH_3OH$	4.41	−1.56	475	3.01	−1.56	325
$CH_4 + O_2 \rightarrow H_2CO + H_2O$	4.24	−1.42	450	3.29	−1.43	350
$CH_4 + O_2 \rightarrow HCO_2H + H_2$	3.98	−1.33	425	3.71	−1.34	400

a

b

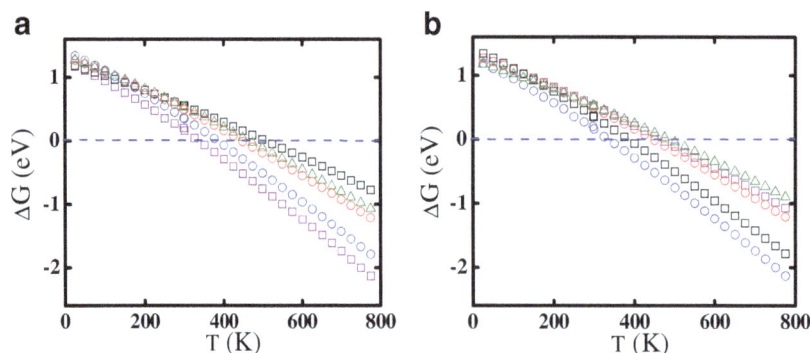

Fig. 6 **a** Change in the Gibbs free energy (in electron volts) versus temperature, T (in Kelvin), in the presence of the Au^- ion catalyst. The first (*purple squares*), second (*blue circles*), third (*red circles*), fourth (*green triangles*), and fifth (*black squares*) curves correspond respectively to the reactions leading to the production of CH_3OH, CO_2, H_2CO, CO, and HCO_2H beyond the optimum temperatures. **b** Change

in the Gibbs free energy (in electron volts) versus temperature, T (in Kelvin), in the absence of the Au^- ion catalyst. The first (*blue circles*), second (*black squares*), third (*red circles*), fourth (*purple squares*), and fifth (*green triangles*) curves correspond respectively to the reactions leading to the production of CO_2, HCO_2H, H_2CO, CH_3OH, and CO beyond the optimum temperatures

Discussion of results

Nanoparticles are essentially a small cluster of atoms; here we are dealing with a single atom (more specifically, its negative ion). The origin of the catalytic activity of supported gold nanoparticles is still not fully understood [52]. Turner et al. [52] investigated the catalytic behavior of very small size (approximately 1.4 nm) gold nanoparticles obtained from atomic gold clusters. They speculated that the remarkable catalytic behavior of the atomic nanoparticles was due partly to the strong electronic interaction between the gold and the titanium dioxide support. Here we use atomic gold and atomic gold anion, such as used in the experiment of Zheng et al. [37], which are obtained from laser-ablated gold foil. This completely avoids any complication associated with the support. In [6, 7] we have used a similar analysis to understand the experiments [4, 5] on the catalysis of H_2O_2 from H_2O using Au and Pd nanoparticles.

This investigation could also help toward understanding the issue of the support since our approach uses simply atoms and atomic anions. As pointed out in [6, 7], our approach worked for the catalysis of H_2O to H_2O_2 using the atomic Au^- catalyst for the reasons: the large EA of atomic Au, the presence of the second deep R-T minimum

in the electron elastic scattering TCS for atomic Au, and the existence of the VDE for the anionic $Au^-(H_2O)$ complex within this R-T minimum. For CH_4 catalysis the first two conditions still hold. However, the VDE (2.34 eV [37]) of the anionic $Au^-(CH_4)$ complex is still within this second R-T minimum of the Au elastic TCS. To get a sense of how the proposed mechanism might be affected when small clusters are used rather than the atoms, we recently used density functional theory to investigate the structure and dynamics of small clusters of 2, 3, 4, and 5 Pt atoms [53]; the geometric optimization was achieved using the DMol package under the generalized gradient approximation with the Perdew–Wang exchange correlation functional [42].

The electron affinities for the clusters were evaluated and compared with measurement and other theoretical calculations. Our calculated EAs were found to be closer to the measurement, demonstrating the importance of careful geometric optimization of the structures. Furthermore, the EAs for the clusters did not deviate significantly from that of the atom. This implies that the proposed mechanism would still be applicable to small clusters. However, we do not know yet how far this would hold as the cluster size is increased beyond 5. Importantly, Hakkinen et al. [54] investigated the VDE for Au_7^-; they found that the calculated VDE varied

Fig. 7 **a** Same as Fig. 6a, except that here we show the expanded region -0.50 eV $\leq \Delta G \leq 0.50$ eV. **b** Same as Fig. 6b, except that here we show the expanded region -0.50 eV $\leq \Delta G \leq 0.50$ eV. The same color scheme used for both Figures 6 and 7

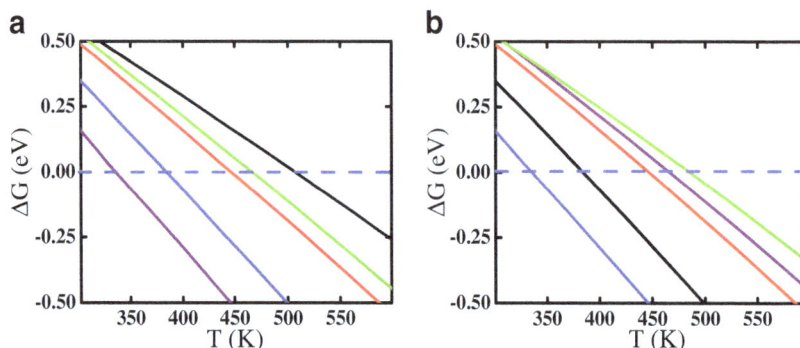

a

b

between 2.75 and 3.57 eV, with their value being 3.46 eV which agrees well with the experimental value of 3.5 eV cited in [54]. Even for a cluster of this size, our analysis would work because the VDE of Au_7^- is still within the effective range of the second R-T minimum of the Au elastic TCS [6, 7]. To firm this, we would need to calculate the electron elastic TCS for the Au_7 cluster and identify the R-T minimum and the various resonances [38]. This will certainly be one of our future research projects.

Finally, the present paper could also lead to a better understanding of the role of the noble metal particle (Au) size and the TiO_2 polymorph in the catalytic production of H_2 from ethanol [55]. Notably, Au nanoparticles of size in the range 3–12 nm were found to be particularly photoreactive.

Conclusion

The atomic Au^- ion catalyst is found to reduce the optimum temperature for the SPO of methane to about 325 K for CH_3OH production. Consequently, in the presence of the atomic Au^- ion catalyst, by controlling the temperature around 325 K, methane can be completely oxidized to methanol without the emission of the CO_2, thereby broadening considerably the scope of gold's applications. Using the Au^- ion as the catalyst essentially disrupts the C–H bonding in CH_4 oxidation through the ionic $Au^-(CH_4)$ molecular formation, thereby eliminating the competition from the carbon dioxide formation. We conclude by recommending that the negative ions of the atoms such as those in [38] be investigated individually or in combinations for possible catalytic activities in the selective partial oxidation of methane; the Pt^- negative ion will accomplish similar results as the Au^- ion catalyst.

Acknowledgments AZM is grateful to Professor Maaza for introducing him to nanogold and nanoplatinum catalysis while visiting iThemba Labs, South Africa. Research was supported by Army Research Office (Grant W911NF-11-1-0194), U.S. DOE Office of Science, AFOSR (Grants FA9550-10-1-0254 and FA9550-09-1-0672), and CAU CFNM, NSF-CREST Program.

References

1. Beecy DJ, Ferrell FM, and Carey JK (2001) In Proceedings of 1st National Conference on Carbon Sequestration, May 14–17
2. Song CS (2006) Global challenges and strategies for control, conversion and utilization of CO_2 for sustainable development involving energy, catalysis, adsorption and chemical processing. Catal Today 115:2–32
3. Armour EAG (2010) J Phys Conference Series 225:012002 and references therein
4. Edwards JK, Carley AF, Herzing AA, Kiely CJ, Hutchings GJ (2008) J Chem Soc Faraday Discuss 138:225
5. Edwards JK, Solsona B, Landon P, Carley AF, Herzing A, Watanabe M, Kiely CJ, Hutchings GJ (2005) J Mater Chem 15:4595
6. Msezane AZ, Felfli Z, Sokolovski D (2010) J Phys B 43:201001
7. Msezane AZ, Felfli Z, Sokolovski D (2010) Europhys News 41:11
8. Tesfamichael A, Suggs K, Felfli Z, Wang X-Q, Msezane AZ (2012) arXiv:1201.2191v1
9. Haruta M (1997) Catal Today 36:153
10. Dumur F, Guerlin A, Dumas E, Bertin D, Gigmes D, Mayer CR (2011) Gold Bull 44:119, and references therein
11. Sanchez A, Abbet S, Heiz U, Schneider WD, Hakkinen H, Barnett RN, Landman U (1999) J Phys Chem A 103:9573
12. Bernhardt TM, Heiz U, Landman U (2007) Chemical and catalytic properties of size-selected free and supported clusters. In: Heiz U, Landman U (eds) Nanocatalysis (nanoscience and technology). Springer, Berlin, pp 1–244
13. Gorin DJ, Toste FD (2007) Nature 446:395
14. Moshfegh AZ (2009) J Phys D 42:233001
15. Beltrán MR, Suárez Raspopov R, González G (2011) Eur Phys J D 65:411
16. Thompson DT (2007) Nano Today 2:40
17. Bond GC, Louis C, Thompson DT (2006) In: Hutchings J (ed) Catalysis by gold catalytic science series. Imperial College Press, London
18. Lim D-C, Hwang C-C, Ganteför G, Kim YD (2010) Model catalysts of supported Au nanoparticles and mass-selected clusters. Phys Chem Chem Phys 12:15172
19. van Bokhoven JA (2009) Chimia 63:25
20. Daniel M-C, Astruc D (2004) Chem Rev 104:293
21. Hashmi ASK, Hutchings GJ (2006) Gold catalysis. Angew Chem Int Ed 45:7896
22. Hashmi ASK (2007) Gold-catalyzed organic reactions. Chem Rev 107:3180
23. Jurgens B, Kubel C, Schulz C, Nowitzki T, Zielasek V, Bienrt J, Biener MM, Hamza AV, Baumer M (2007) Gold Bull 40(2):142
24. Kimble ML, Castleman AW Jr, Mitric R, Burgel C, Bonacic-Koutecky V (2004) J Am Chem Soc 126:2526
25. Liu Y-C, Lin L-H, Chiu W-H (2004) J Phys Chem B 108:19237
26. González Orive A et al (2011) Nanoscale 3:1708
27. Wong MS, Alvarez PJJ, Fang YL, Akcin N, Nutt MO, Miller JT, Heck KN (2009) J Chem Tech Biotech 84:158
28. Pretzer LA, Nguyen QX, Wong MS (2010) J Phys Chem C 114:21226
29. Gao Y, Huang W, Woodford J, Wang L-S, Zeng XC (2009) J Am Chem Soc 131:9484
30. Sorokin B, Kudrik EV, Bouchu D (2008) Chem Technol 5:T43
31. Vafajoo L, Sohrabi M and Fattahi M (2011) World Academy of Science, Engineering and Technology 73:797
32. Mohr F (ed) (2009) Gold chemistry, applications and future directions in the life sciences. Wiley, New York
33. Yuan J, Wang L, Wang Y (2011) Ind Eng Chem Res 50(10):6513
34. Chen W et al (2009) Catal Today 140:157
35. Zhang Q, He D, Zhu Q (2008) J Nat Gas Chem 17:24
36. Lang SM, Bernhardt TM, Barnett RN, Landman U (2011) J Phys Chem C 115:6788
37. Zheng W, Li X, Eustis S, Grubisic A, Thomas O, De Clercq H, Bowen K (2007) Chem Phys Lett 444:232
38. Felfli Z, Msezane AZ, Sokolovski D (2011) J Phys B 44:135204
39. Felfli Z, Eure AR, Msezane AZ, Sokolovski D (2010) NIMB 268:1370
40. Tkatchenko A, Scheffler M (2009) Phys Rev Lett 102:073005
41. Perdew JP, Burke K, Ernzerhof M (1996) Phys Rev Lett 77:3865
42. DMol3 (2011) Accelrys Software Inc., San Diego

43. Suggs K, Reuven D, Wang X-Q (2011) J Phys Chem C 115:3313
44. Suggs K, Person V, Wang X-Q (2011) Nanoscale 3:2465
45. Samarakoon D, Chen Z, Nicolas C, Wang X-Q (2011) Small 7:965
46. Nam LTH, Dat VT, Loan NTT, Radnik J, Roduner E (2010) J Chem 48:149
47. Hotop H, Lineberger WC (1985) J Phys Chem Ref Data 14:731
48. Andersen T, Haugen HK, Hotop H (1999) J Phys Chem Ref Data 28:1511
49. Simoni A, Launay JM, Soldan P (2009) arXiv:0901.3129v1
50. Balakrishnan N, Quéméner G, Dalgarno A (2009) Inelastic collisions and chemical reactions of molecules at ultracold temperatures. In: Stwalley WC, Krems RV, Friedrich B (eds) Cold molecules: theory, experiment, applications. CRC Press, Boca Raton, Florida
51. Levine RD (2005) Molecular reaction dynamics. Cambridge University Press, Cambridge
52. Turner M, Golovko VB, Vaughan OPH, Abdulkin P, Berenguer-Murcia A, Tikhov MS, Johnson BFG, Lambert RM (2008) Nature 454:981
53. Chen Z, Msezane AZ (2010) Density functional theory investigation of small Pt clusters. Bull Am Phys Soc 55(57)
54. Hakkinen H, Moseler M, Landman U (2002) Phys Rev 89:033401
55. Murdoch M, Waterhouse GIN, Nadeem MA, Metson JB, Keane MA, Howe RF, Llorca J, Idriss H (2011) Nat Chem 3:489

3

Some thoughts on bondability and strength of gold wire bonding

Muhammad Nubli Zulkifli · Shahrum Abdullah ·
Norinsan Kamil Othman · Azman Jalar

Abstract The bonding mechanisms of gold, to give the desired strength of wire bonding, still require detailed investigation, including establishing adequate and reliable testing procedures. The current practices for analysing the mechanisms of wire bonding are inadequate and do not provide a comprehensive picture. This is because the focus of the tests is not clear, which causes variation in the results obtained, changing the conclusions about the responsible mechanism. Furthermore, as the size of Au wire bonds decreases, the mechanism responsible for thermosonic Au wire bonding may change. This paper provides a comprehensive analysis of the current and possible future methods for elaborating the bonding mechanism and strength of thermosonic Au wire bonds. We discuss the testing methods, their limitations and advantages, and suggest ways in which they can be improved.

Keywords Wire bonding · Bondability · Strengthening · Bonding mechanism

Introduction

Wire bonding using gold (Au) wire is still the most preferred interconnection technique for microelectronics packaging because of its cost-effectiveness and technological maturity [1]. There are three types of wire bonding technologies that have been used in industry: thermocompression bonding, ultrasonic wedge bonding and thermosonic bonding. Thermosonic bonding is normally used in industry to weld the Au wire onto the substrate metallisation or bond pad [1] and involves combinations of several processes such as mechanical force, heat and ultrasound [2–8].

The bondability and bonding mechanism of Au wire bonding is still subject to great debate since the wire bonding process itself is affected by many factors, including capillary geometry, bond pad surface condition and environmental effects [3–31]. Furthermore, at present, there is a trend towards even smaller wires to cope with smaller bond pad pitch [21]. This will introduce many more technological challenges into the search for the bonding mechanisms responsible for wire bonding.

To date, several methods of analysis have been introduced to study the bondability and bonding mechanisms of wire bonding [1, 3–31]. The aim is to determine the phenomena responsible for forming the bond between the Au wire and the bond pad. Analyses of the lift-off or footprint morphology, intermetallic compound coverage, microstructure, frictional bonding and the strengthening reactions between Au wire and intermetallic compound have been performed. Finite element techniques and transmission electron microscopy (TEM) have also been used to evaluate the bondability and bonding mechanisms of wire bonding. However, the results obtained from the various methods suggest that different types of mechanisms control thermosonic Au wire bonding. Some of the methods, namely footprint morphology, intermetallic coverage, microstructural and frictional bonding analyses, have concluded that solid-state diffusion is the main mechanism of thermosonic Au wire bonding [1, 3–19]. However, a new possible bonding mechanism (liquid-state diffusion) has been proposed based on TEM examinations [27–30]. Each method for

M. N. Zulkifli · A. Jalar (✉)
Institute of Microengineering and Nanoelectronics (IMEN),
Universiti Kebangsaan Malaysia,
43600 UKM, Selangor, Malaysia
e-mail: azmn@ukm.my

S. Abdullah
Department of Mechanical and Materials Engineering,
Universiti Kebangsaan Malaysia,
43600 UKM, Selangor, Malaysia

N. K. Othman
School of Applied Physics, Faculty of Science and Technology,
Universiti Kebangsaan Malaysia,
43600 UKM, Selangor, Malaysia

analysing the bonding mechanisms of thermosonic Au wire bonding has its own advantages and disadvantages, but further improvement is still possible. Thus, the present paper is intended to review the general situation regarding the current and possible future methods for studying the bonding mechanisms of thermosonic Au wire bonding, to enable the formation of better quality and more reliable Au wire bonds.

Lift-off or footprint morphology analysis

Lift-off or footprint morphology analysis is one of the earliest techniques introduced to analyse the initiation and growth of the bonded region. This early analysis, conducted by Harman and Albers [1], observed that the ultrasonic parameters have a direct effect on the deformation of a wire bond, rather than working by inducing temperature changes or friction. They based their analysis on the earlier work of Langenecker [2], who studied the ultrasonic softening of metals. He showed that the stress versus elongation relationship for ultrasonic irradiation at constant temperature was equivalent to the elongation created from heat. It is known that elongation is related to the deformation of materials. Langenecker also noted that the elongation mechanism that originates from ultrasonic treatment is different to the elongation created from heat. The energy density required by the ultrasonic method to elongate metal is 10^7 times lower than the energy density required to elongate metal using heat [2]. Thus, the required compressive load to deform metals decreases when ultrasonic vibration is applied. However, after exposure to ultrasonic excitation, the metal is work hardened via acoustic hardening, in contrast to thermal excitations, which leave the metal softer or annealed. Harman and Albers [1] further reported that the oxide layer on the bond pad is shattered or pushed aside by the flow stress created from the softened or deformed wire bond. The welding process will then happen at the area of the bond pad that is free of an oxide layer. They also observed that the weld formation began from the perimeter of the wire bond and progressed inward. This finding was explored further through the wire bond lift-off patterns obtained from wire pull tests. Figure 1 shows the scanning electron microscopy (SEM) image of a wire bond lift-off pattern, as used by Harman and Albers [1] to evaluate the wire bond formation.

Several further analyses have been performed to analyse the wire bonding mechanism based on Harman and Albers' work [1]. Lum et al. [3] examined the effects of ultrasonic and bonding force parameters on the bondability of an Au wire bond and a Cu bond pad. Their analysis was based on the lift-off or footprint morphology of microwelded regions obtained from wire pull and ball shear tests. Metallurgical bonding is indicated by the presence of fractured microwelds in the

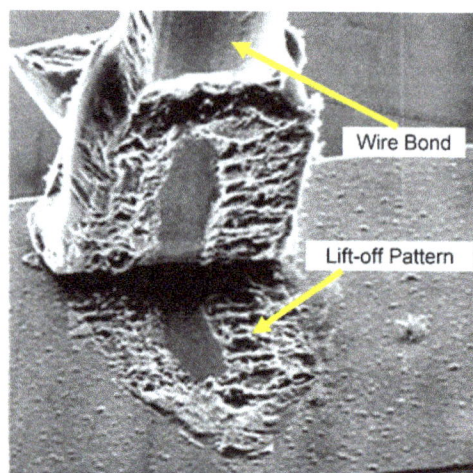

Fig. 1 Scanning electron microscopy (SEM) image of wire bond lift-off pattern [1]

footprints and is further identified by the presence of gold residues from the wire. Lum et al. [3] found that the relative motion at the bond interface varies from microslip to gross sliding when the ultrasonic energy is increased at constant bonding force. Figure 2a and b show the bond footprints obtained from a wire pull test and a ball shear test, respectively. Figure 3 shows a schematic illustration of the change in footprint morphology with increasing ultrasonic power at constant bonding force.

Lum et al. [3] also showed that an increase in bonding force will increase the contact diameter of the ball bond, and therefore, higher applied bonding power is required to obtain adequate bond strength. This is because a larger contact

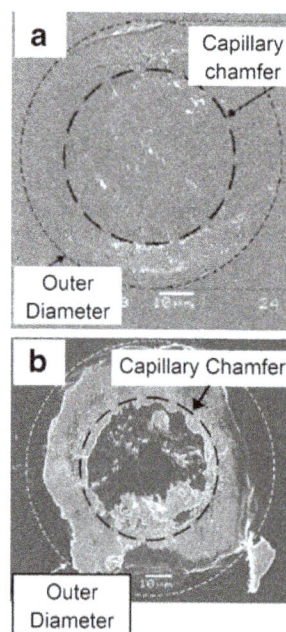

Fig. 2 Bond footprints obtained from **a** wire pull test **b** ball shear test [3]

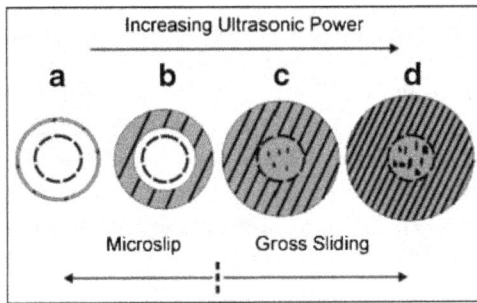

Fig. 3 Schematic illustration of the change in footprint morphology for increasing ultrasonic power at constant bonding force. The transition of footprint morphology is from microslip to gross sliding with increasing ultrasonic power. *Grey areas* indicate fretting; the *dashed circle* indicates the capillary chamfer diameter, while the *crosshatching* indicates the bonding density [3]

diameter of the ball bond requires higher ultrasonic power to form gross sliding. In addition, capillary geometry was found to play a significant role in determining the compressive stress distribution within the ball bond. Lum et al. [3] found that the highest compressive stress within the ball bond was located near the end tip of the capillary or at the end of the capillary chamfer diameter. Figure 4 illustrates the capillary, along with the compressive force versus indentation location graph, as obtained by Lum et al. [3].

Xu et al. [4, 5] extended the work done by Harman and Albers [1] and also Lum et al. [3]. They carried out experiments to analyse the effects of ultrasonic power, bonding force and bonding time parameters on the footprint morphology of the Au ball bond and Al bond pad. Again, fractured microwelds attached to the footprint area were used as an indicator of metallurgical bonding. Xu et al. [4] reported that the relative motion at the interface between the ball bond and bond pad changes from microslip to gross sliding when the ultrasonic power is increased. This finding is in agreement with the results obtained by Lum et al. [3]. Xu et al. [4] also indicated that metallurgical bonding is initiated at the periphery of the interface between the ball and pad in the direction of ultrasonic vibration and that the

bonding area grew towards the centre of the contact area with increasing ultrasonic power. They reported that the relative motion at the bonding interface also changed from gross sliding to microslip when the bonding force was increased. The unbonded area initiated in the centre of footprints and then grew outwards with increasing bonding force. Figures 5 and 6 present two models produced by Xu et al. [4], which illustrate the initiation and growth of the bonding area through the increase of ultrasonic power and bonding force.

Increasing the bonding time also increased the ball bond bonding area and subsequently increased the bonding strength, as indicated by Xu et al. [5]. They also noted that the bonding area initiated from the peripheral region of the ball bond and grew towards the centre of the contact area as the bonding time increased. This observation is in agreement with their results for increased ultrasonic power [4]. However, the full coverage of metallurgical bonding formation over the contact area occurred earlier with increasing bonding time compared with increasing ultrasonic power. Figure 7 shows a schematic of bonding initiation and growth of the bonded regions with increasing bonding time.

The findings obtained through lift-off or footprint morphology analysis only concern the initiation and growth of the bonded or contact area based on observation of the fractured microweld, as mentioned earlier. The mechanism of the bonding suggested by this technique is based on solid-state diffusion, where the formation of the bond changes from microslipping to gross sliding. However, the fractured microweld observed using this technique represents a small portion of the bonded area. This is because the methods used to fracture the microweld, namely wire pull and ball shear tests, may pull or take out the remaining microweld. Furthermore, the fracture microweld is difficult to identify, especially in the case of the ball shear test. This is because the sheared microweld may not originate from a particular bonded or contact area, but instead shearing may occur through the peripheral region of the bonded ball. Another concern is that different authors have assumed

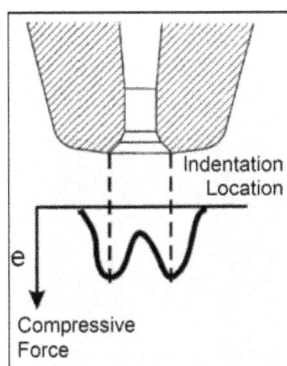

Fig. 4 Schematic illustration of capillary and compressive force versus indentation location graph [3]

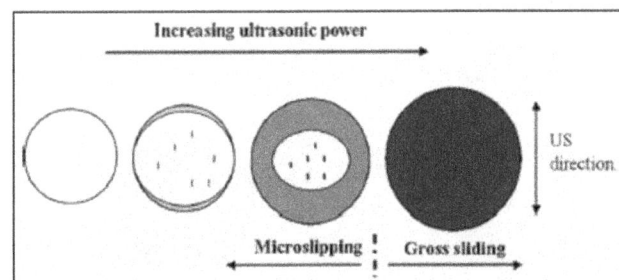

Fig. 5 Schematic illustration of initiation and growth of bonded regions with increasing ultrasonic energy. *Bright areas* are stationary regions; *grey areas* are bonded regions; *darker grey* indicates stronger bonding [4]

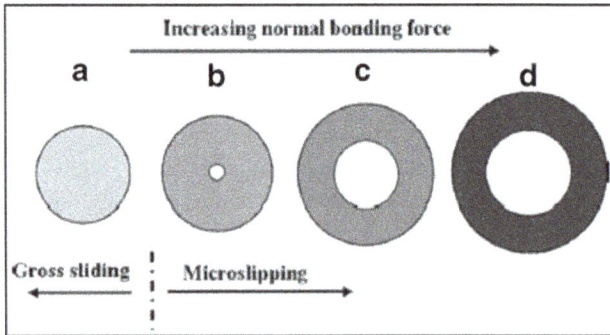

Fig. 6 Schematic illustration of evolution of the bonded regions with increasing normal bonding force. *Bright areas* are stationary regions; *grey areas* are bonded regions; *darker grey* indicates stronger bonding [4]

different shapes of growth for the contact area when interpreting their footprint morphology results. For example, Lum et al. [3] used a round shape to represent the initial contact area, while Xu et al. [4, 5] used an ellipse. This shows the difficulty in obtaining the actual contact area using the footprint morphology technique. Therefore, examination based solely on the fractured microweld is not sufficient to characterise the initiation and growth of the contact area. Determination of the actual initial contact area is crucial to make footprint morphology analysis more reliable. Wire bonding onto soft metallisation without fully bonding the wire onto a bond pad is an alternative technique that could be used to provide the bond footprint imprinted on the bond pad, without examination of the fractured microweld.

Intermetallic compound coverage analysis

Beside the usage of the footprint morphology method following wire pull and ball shear tests, several researchers have characterised the Au–Al intermetallic compound formed at the interface of the Au ball bond and Al bond pad to elucidate the wire bonding mechanism [6–8]. Qi et al. [6] used sodium hydroxide (NaOH) solution to etch away the Al bond pads and reveal the bond patterns at the

interface. They examined the bond patterns using an optical microscope so that high contrast between the bonded area with lower reflectivity and the unbonded area with higher reflectivity could be obtained. The bonded area could then be measured for different bonding conditions. Qi et al. [6] found that both the contact area and shear force increased with increasing bonding force and ultrasonic power. However, they noted that there is a certain value for the shear force at which the shear force changes only gradually with the bonding force and ultrasonic power, while the contact area keeps increasing. They inferred that a larger contact area does not necessarily increase the shear force. Figure 8 shows the optical image of the bond pattern at the underside of the gold bump. Figure 9a and b show graphs of shear force and contact area as a function of bonding power and bonding force, respectively.

Qi et al. [6] also indicated that the evolution of the bond pattern can be explained by the evolution of the plastic region and the slip area with changing bonding parameters. There are two main types of forces, normal force and tangential force, which occur during wire bonding. The normal force arises from the bonding force, while the tangential force arises from the ultrasonic vibration or ultrasonic power. Figure 10a to d are schematic illustrations of the cross-sectional ball bond describing the deformation behaviour that happens during the application of normal force and tangential force.

Wulff and Breach [7] used optical microscopy and SEM to measure the coverage of intermetallic gold aluminide compounds on the underside of the Au ball bond. The underside of the Au ball bond was accessed by etching away the Al bond pads using KOH solution. SEM examination was found to be a better tool to measure the intermetallic coverage because optical microscopy does not have the required resolution or magnification. Rosle et al. [8] used infinite focus microscopy (IFM) to characterise the two-dimensional and three-dimensional surface topography of the gold aluminide compound. Access to the underside of the Au ball bond was obtained using the same procedure as that done by Wulff and Breach [7]. Rosle et al. [8] found that the measured intermetallic coverage does not correlate with the bonding strength of the Au ball bond because the

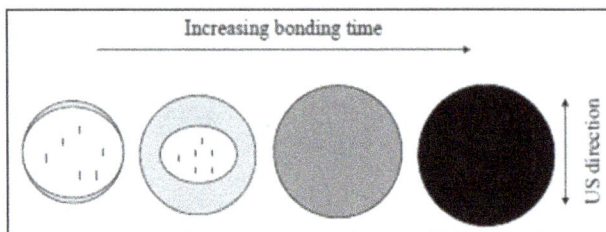

Fig. 7 Schematic illustration of bonding initiation and growth of the bonded regions with increasing bonding time. *Bright areas* are stationary regions; *grey areas* are bonded regions; *darker grey* indicates stronger bonding [5]

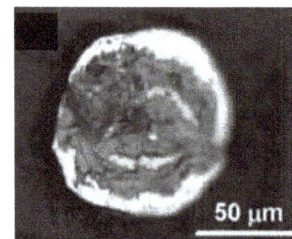

Fig. 8 Optical image of the bond pattern at the underside of the gold bump [6]

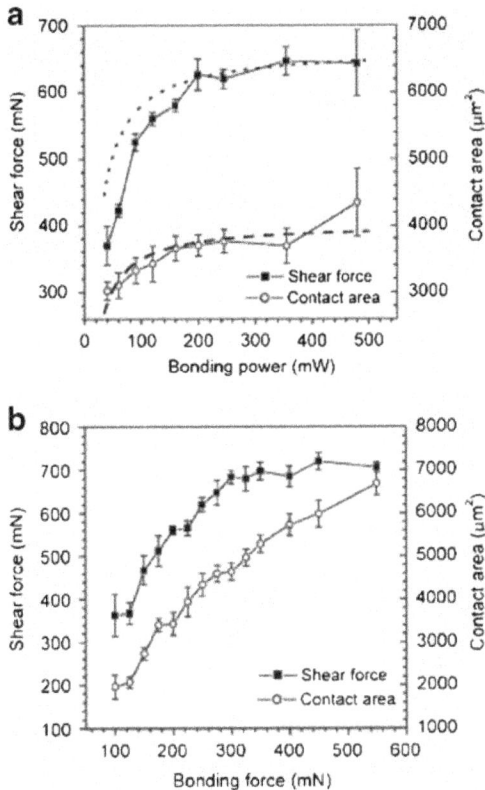

Fig. 9 *Graphs* of shear force and contact area as a function of **a** bonding power **b** bonding force [6]

higher coverage of intermetallic compound measured using IFM does not necessarily represent higher bonding strength of the Au ball bond.

Intermetallic compound coverage analysis analyses the true bond pattern at the underside of the ball bond and thus

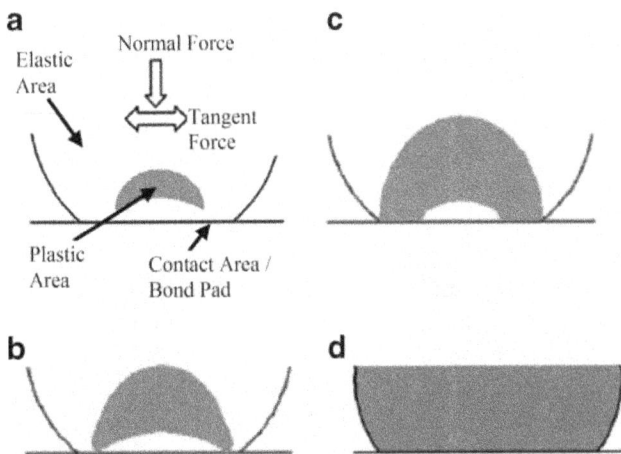

Fig. 10 Schematic illustration of the cross-section of the ball bond describing the deformation behaviour that happens during the application of normal force and tangential force. *Bright areas* are elastic regions and the *grey areas* are plastic regions. **a** The plastic region is completely surrounded by the elastic region; **b** the plastic region first reaches the contact interface; **c** the elastic core floats on the plastic region and **d** the ball is in the plastic condition [6]

provides the information that could not be obtained from lift-off or footprint morphology analysis. The coverage and also volume of the intermetallic compound is used to measure the bondability of the ball bond. The major finding of this bondability analysis of the ball bond is that no correlation between the coverage or volume of intermetallic compound and the bonding strength can be made because higher coverage of intermetallic compound does not necessarily increase the bonding strength, as noted by Qi et al. [6] and Rosle et al. [8]. Intermetallic compound coverage analysis is also quite a difficult measurement to perform, as noted by Wulff and Breach [7]. Each of the proposed coverage measurement techniques such as optical microscopy, SEM and IFM examinations have their own advantages and disadvantages in terms of resolution or magnification captured and the time required to conduct the experiment. Thus, the results obtained through this analysis are not adequate to explain the bonding mechanism of wire bonding.

An explanation for the plateau in the value of shear force obtained with increasing contact area is critical to relate the bonding parameters, the thickness of the bond pad and also the size of the wire bond to the intermetallic coverage required for the highest bonding strength. Furthermore, an understanding of the relationship between the volumes, shape and distribution of intermetallic compound might provide further insights into the wire bonding mechanism.

Microstructural analysis

Several researchers have examined the microstructural changes during the wire bonding process. The formation of large grains in the free air ball (FAB) and grain growth in the heat-affected zone of the adjacent wire is due to heat transfer during the solidification process. This solidification process happens when the Au wire tip is melted by an electrical spark (known as electro-flame off) which results in the round shape of the FAB because of the surface energy of the liquid Au. Karpel et al. [9] used TEM to show that the FAB region contains large grains and has a relatively low dislocation density as a result of the solidification process. The as-bonded Au bond has different microstructure compared with that of the FAB. Karpel et al. [9] noted that the area adjacent to the intermetallic region has a smaller grain size and the area further above the intermetallic region contains relatively large Au grains. The Au adjacent to the Au–Al interface also possesses a relatively high dislocation density, with sub-grains forming inside the Au ball bond. The high density of dislocations is a result of the plastic deformation that happens during the wire bonding process; this also leads to strain hardening of the ball bond [10, 11]. Thus, the microstructural changes across the deformed Au ball bond affect the mechanical properties and the strength of the Au ball bond.

Li et al. [12] further studied the effect of process parameters, namely bonding power, bonding force and stage temperature on the microstructure and microtexture of Au stud bumps and Au FAB using electron backscatter diffraction. They reported that an increase in bonding power will increase the deformation in a non-uniform fashion. Increasing bonding force tends to further flatten the grains. An increase in stage temperature was found to change the columnar grain structure of the Au bond into a sub-grain structure. This observation is consistent with the softening effect and stronger thermoactivation induced by ultrasonic vibration at higher temperature. Figure 11 presents several orientation maps obtained by Li et al. [12].

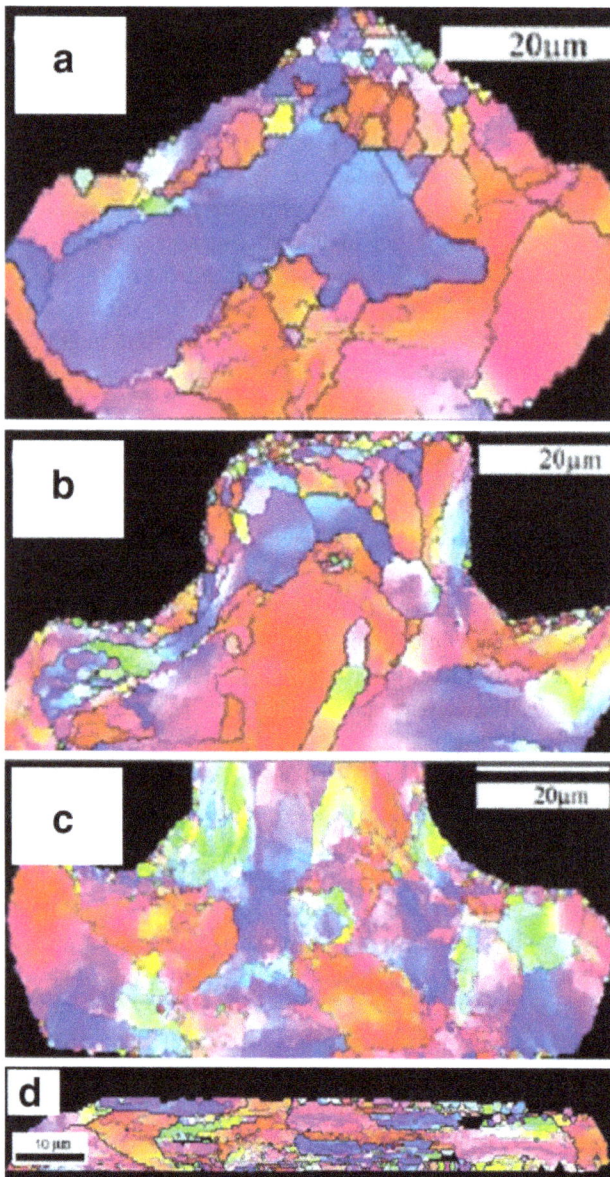

Fig. 11 Orientation mapping of the Au bumps for the case of **a** lower bonding power (20 mW), **b** higher bonding power (245 mW), **c** higher bonding force (350 mN) and **d** higher stage temperature (150 °C) [12]

Lum et al. [15] extended the work of Geißler et al. [13, 14] by studying the effect of ultrasound on softening and hardening during thermosonic Au wire bonding. It was noted that an increase in the ultrasonic amplitude will increase the acoustic residual softening effect upon the Au ball bond. They also suggested that the net result of the ultrasonic effects on the internal structure of the Au ball bond, especially the final dislocation density, determines the change in the mechanical properties (i.e. whether there is softening, hardening or no change).

Analysis of the microstructural changes in the ball bond during wire bonding has been conducted by several researchers [9–15]. This analysis has been limited to analysing the strengthening of the ball bond and has not examined the strengthening of the Au ball and the bond pad. We therefore suggest that the microstructures of both the Au ball bond and the bond pad should be analysed to examine if there are microstructural reasons for the strengthening of these regions during wire bonding. Examining the microstructural changes in both the Au ball and bond pad may also provide further insight into the diffusion that occurs during bonding and perhaps allow us to ascertain the type of diffusion state responsible for wire bonding.

Frictional bonding analysis

Several studied have examined the mechanisms of wire bonding using frictional bonding theory [16–19]. Ding et al. [16] conducted a finite element analysis of wire bonding, varying the contact pressure, real contact area and frictional energy intensity generated at the wire bond pad interface. They noted that a higher bond force does not create a higher contact pressure and that the normalised contact area was always a maximum at the periphery of the bonding region. In addition, Ding et al. [16] also found that the maximum frictional energy occurs at the periphery of the contact interface. The total frictional energy was found to increase linearly with bonding force. However, the high-energy intensity obtained at the periphery of the contact interface did not show a similar increase.

Ding et al. [17] further analysed the wire bonding mechanisms using a finite element method to evaluate the temperature rises during ultrasonic vibration or friction. They measured two different interfacial temperatures which are relevant in tribology, namely the flash and the bulk temperatures. According to Ding et al. [17], the flash temperature is the maximum temperature at the tips of interacting asperities induced by friction, while the bulk temperature is the average temperature across the frictionally heated surface—therefore the flash temperature is normally higher than the bulk temperature. Ding et al. [17] noted that the bulk temperature does not approach the melting temperature of either the Au wire or the Al bond pad and therefore is not the dominant source of bonding; in contrast, the higher flash temperature does contribute to bond formation.

The temperature changes at the wire bond interface arising from friction have also been used by several researchers in order to study mechanisms of wire bonding [18, 19]. Scheneuwly et al. [18] used a gold-nickel (Au–Ni) thermoelectric junction to measure the temperature changes of the Au ball because of friction-generated heat at the bonding interface. To realise this experiment, the Au–Ni junction is placed within the inside chamfer of the bond capillary. The thermal change or variation is used as the thermal response. The temperature at the interface rose to about 100 °C at the interface because of the friction during the wire bonding process; this temperature rise provides good bond contact quality. Ho et al. [19] used a thin film K-type thermocouple sensor to measure the contact temperature of the frictional interfaces during the ultrasonic wire bonding process. They found that the contact temperature increases monotonically with increasing preload and ultrasonic power of vibration and observed a temperature rise of 300 °C.

Frictional bonding analysis also uses solid-state diffusion to explain the wire bonding mechanism, and the frictional energy intensity or temperature rise is regarded as the source of bonding. However, the highest temperature rise measured so far using frictional bonding analysis is only 300 °C, which is low compared with the melting temperature of Au [19]. This temperature only represents the bulk temperature and not the flash temperature. Up until now, the flash temperature has been difficult to measure. Therefore, a technique to measure the flash temperature is required before the responsible mechanism for the diffusion phenomenon can be ascertained. Once the flash temperature is known, the formation of the intermetallic compound can be explained in much more detail.

Au wire and intermetallic compound strength analysis

The strength of the Au base metal and the bonded interface or intermetallic compound has been measured by several researchers to analyse the strengthening of the Au ball bond [20, 21]. Li et al. [20] carried out an experiment to analyse the tensile rupture characteristics of interface of the Au ball bond and the Al bond pad using a wire pull test. They observed that the ball bond has dimpled rupture features. The fracture occurred within the base material (Au) and not at the bonded interface (intermetallic compound). This result shows that the bonded interface or the intermetallic compound has higher bonding strength because it can withstand the deformation strain during the wire pull test. They inferred that atomic diffusion at the bond interface enhanced the microstructural strength and increased it beyond that of

the base materials. Figure 12a and b show SEM images of the bonded interface in the Au wire and the Al bond pad following the wire pull test and the dimpled rupture characteristics, respectively.

Jalar et al. [21] carried out a nanoindentation test to measure the hardness of the Au base metal and the intermetallic compound of the Au ball bond and found that the intermetallic compound has higher hardness than the Au base metal. Hardness is proportional to yield strength, based on the Tabor relationship [22]. Hence, the formation of the intermetallic compound, with its higher yield strength, increases the strength at the bonding interface.

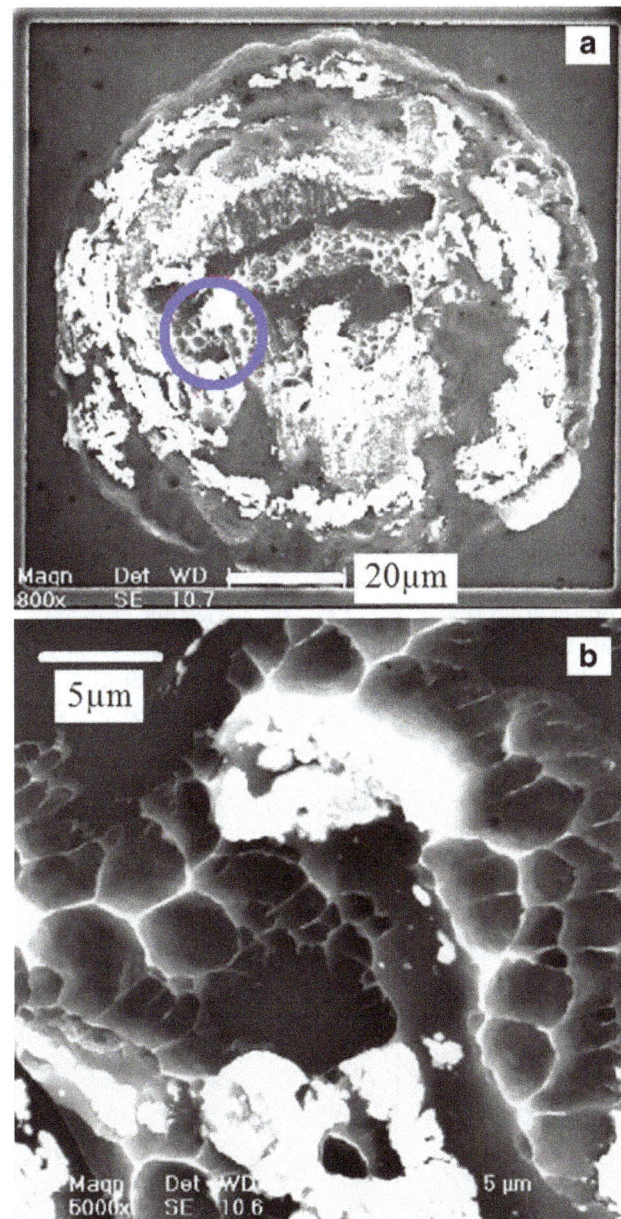

Fig. 12 SEM images of **a** the bonded interface in the Au wire and the Al bond pad following the wire pull test; **b** dimpled rupture characteristics at the *blue circle* in **a** [20]

The microstructural analysis of the Au wire and intermetallic compound provides additional information. The intermetallic compound is stronger than the Au base metal, based on the dimpled rupture features and the higher hardness value of the intermetallic. Further measurements of the strengthening behaviour of the intermetallic and its relationship with the Au base metal may be needed. The nanoindentation test used by Jalar et al. [21] might facilitate the measurement of the strength of the intermetallic and the Au base metal. However, this technique is still prone to measurement difficulties; for example, the size of the indenter can be greater than the size of the intermetallic region. To allow the use of smaller indentations that can fit within a particular phase or grain boundary, electropolishing can be used to obtain a surface free of dislocations or strain hardening. This then enables detailed analysis of the strengthening effects of each individual phase, grain and boundary using nanoindentation tests.

Finite element analysis

Finite element analysis (FEA) has been used extensively by several researchers to analyse the plastic deformation that occurs during the wire bonding process [23–26]. Saiki et al. [23] investigated the effect of capillary tip shape on ball bonding using finite element analysis. They found that the amount of sliding of the ball bond increases with decreasing capillary tip angle. In addition, bonding can be achieved when the amount of sliding is more than 0.01 μm. Figure 13 shows a schematic of the cross-section of the capillary tip, defining the capillary tip angle, θ.

Wulff et al. [24] and Hsu et al. [25] carried out FEA to examine the effect of the capillary upon the stress distribution within the ball bond. They indicated that the distribution of stress varies across the ball bond. The highest stress

is located at the periphery and is perpendicular to the deformed area of the ball bond that is created by the end of the capillary tip. Figure 14 shows the image of the FEA result for a ball bond that has been deformed by application of bonding force originating from the capillary.

Wulff et al. [24] and Stephen et al. [26] carried out experiments using microhardness testing to measure the hardness distribution across the bonded ball. They reported that the hardness values are proportional to the stress distribution across the bonded ball obtained through FEA. The highest hardness is correlated with the highest stress experienced by the bonded ball. Figure 15 shows the hardness variation across the bonded ball region of Au and Cu ball bonds, as obtained by Wulff et al. [24].

Most of the analyses that have been conducted using FEA have been more focused on the deformation behaviour of the Au ball bond during the wire bonding process. The highest stress experienced by the ball bond from the application of the bonding force is found to be located at the periphery of the ball bond. Thus, the FEA findings further confirm that the initiation of bonding at the peripheral region of the ball bond is because this is where the highest stress occurs, which is consistent with the results of footprint morphology analysis. It is suggested that the combination of finite element analysis to evaluate the effects of friction, applied temperature, bonding power, bonding force and bonding time could provide a more detailed result and assist us to understand the wire bonding mechanism. The stress distribution and temperature rise are examples of FEA results that could be used as indicators of contact initiation.

Transmission electron microscopy (TEM) analysis

Several researchers have used TEM to examine the wire bonding mechanism in more detail [27–30]. Karpel et al. [27] indicated that during the bonding stage, ultrasonic and thermal activation can result in the formation of a liquid state

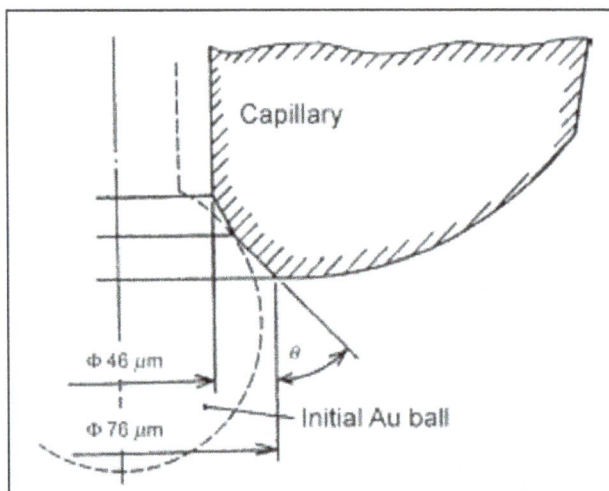

Fig. 13 Diagram of cross-section of capillary tip showing capillary tip angle, θ [23]

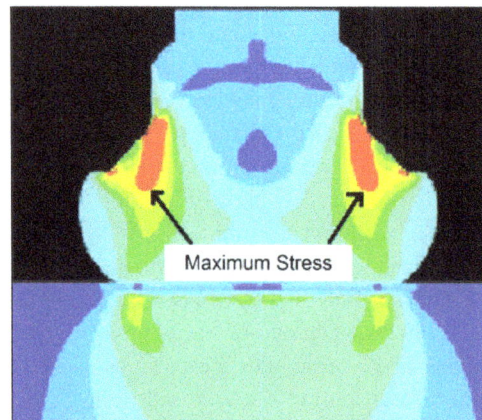

Fig. 14 FEA image showing the stresses in a bonded ball [24]

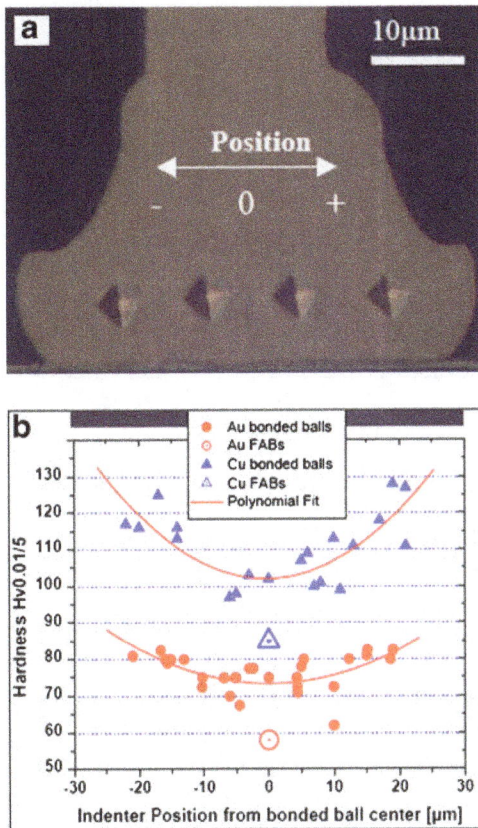

Fig. 15 **a** Location of the indentations, **b** variation of hardness across the bonded ball area [24]

Fig. 16 Graph of shear strength versus silicon die temperature [27]

multilayers with sub-interfaces inside Au_8Al_3 intermetallic phases and intrusion of the Al wire into the Au bond pad metallisation, respectively. Figure 18 shows the HRTEM image of the layered-diffusion phenomenon at the bond interface.

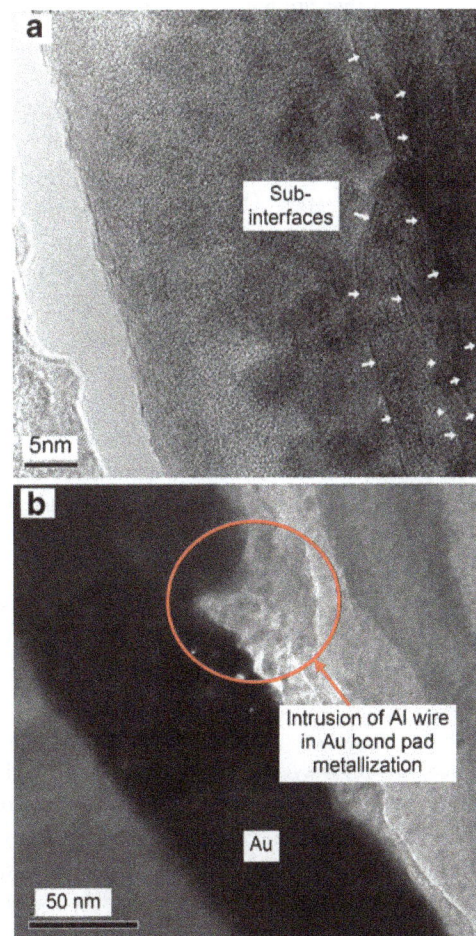

at the bond interface. This finding is proven by the presence of high temperature intermetallics, $AlAu_4$ and Al_3Au_8. They also observed that the intermetallic region that formed under conventional bonding conditions contained a high density of flaws in the intermetallic grains and at the grain boundaries. These flaws, which are known as solidification voids, and are formed inside the intermetallic region, are the result of the volume changes upon the formation of the Al–Au intermetallic. Karpel et al. [27] also noted that increasing the silicon die temperature increased the shear strength of the ball bond as a result of further intermetallic compound formation. Figure 16 shows the relationship between shear strength and silicon die temperature obtained by Karpel et al. [27].

Ji et al. [28] used high-resolution transmission electron microscopy (HRTEM) to analyse the Au wire bonding mechanism. They reported that solid-state diffusion theory cannot be used to explain the wire bonding mechanism because a thick intermetallic compound still can be formed within milliseconds at room temperature. They also showed that the cyclic ultrasonic energy created multilayers with sub-interfaces inside the Au_8Al_3 intermetallic phases and that the Al wire intruded into the Au bond pad metallisation. In addition, diffusional features, specifically layered-diffusion and an interface pattern of alternating dark and bright bars near Au, were observed because of their thermal effect. Figure 17a and b show HRTEM images of

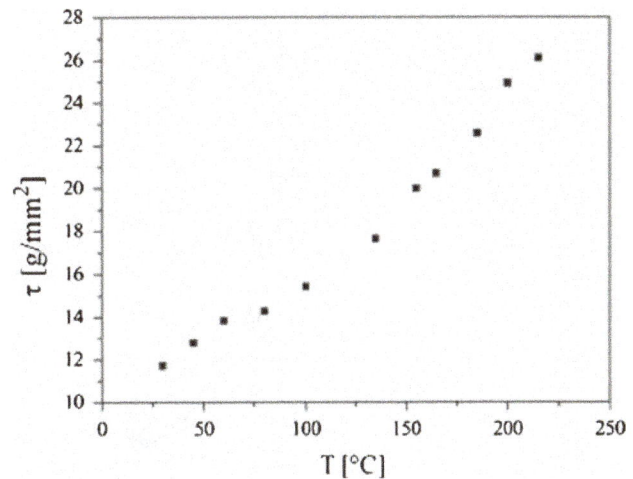

Fig. 17 HRTEM images of **a** multilayers with sub-interfaces inside the Au_8Al_3 intermetallic phases, **b** intrusion of the Al wire into Au bond pad metallisation [28]

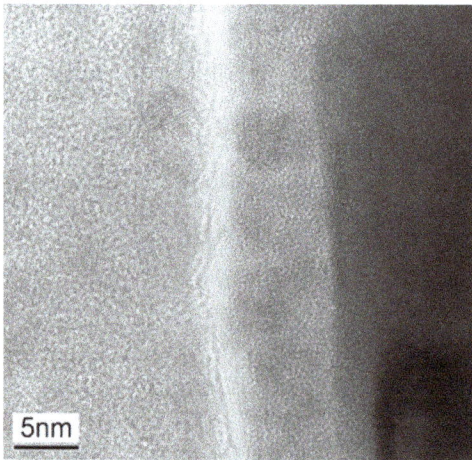

Fig. 18 HRTEM image of the layered-diffusion phenomenon at the bond interface [28]

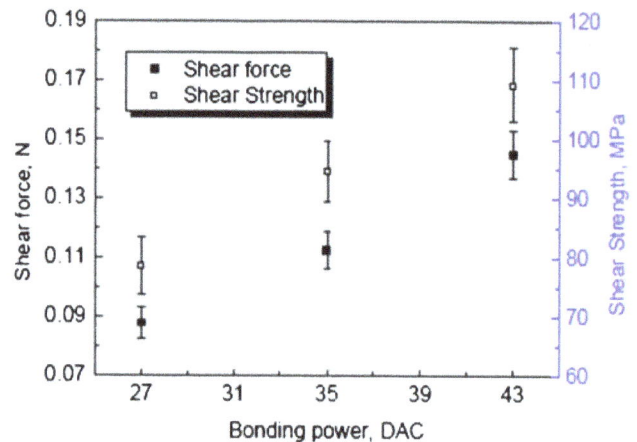

Fig. 19 Shear force and shear strength versus bonding power [30]

Li et al. [29] used HRTEM to inspect the ultrasonic bonding interface features. The application of ultrasonic vibration increased the dislocation density within the metals at the interface metals between the Au wire and the Al bond pad. They also stated that the bond is created between the Au ball bond and the Al bond pad through short-circuit diffusion. This is because the atoms located near the dislocation are easily activated or diffused. It is known that the activation energy for this dislocation-enhanced diffusion or short-circuit diffusion is about half that for body diffusion [29]. Using HRTEM, Li et al. [29] observed that, within several milliseconds, the thickness of atoms that have diffused to the bonding interface is about 100 to 300 nm, and this finding supported the idea of short-circuit diffusion.

The native aluminium oxide layer existing on the surface of the Al bond pads can act as a barrier layer that blocks the inter-diffusion of the Au wire and the Al bond pad. Thermosonic wire bonding produces ultrasonic vibration that locally and physically breaks down the native aluminium oxide layer and enables direct Au and Al contact under certain bonding pressure, as indicated by Xu et al. [30, 31]. Xu et al. [31] noted that the dominant intermetallics created during the wire bonding process are Au_4Al and $AuAl_2$. They also estimated that the heating at the interface from the ultrasonic effect is equivalent to more than 600 °C, while the overall average temperature is probably near 320 °C. Consequently, this produces numerous diffusion pathways via grain boundaries and dislocations that have lower activation energy. The formation of the Au and Al intermetallic compounds could significantly improve the bond strength [30, 31]. Figure 19 shows the shear force and shear strength versus bonding power graphs obtained by Xu et al. [30] to describe the influence of ultrasonic power on bonding strength.

The analysis of the bonding mechanism using TEM provides a more detailed explanation because of the nanometer scale of the measurement. New observations, such as

the high temperature intermetallic compound and the formation of thick intermetallic compound, are used to explain the liquid-state diffusion phenomenon. However, the value of the temperature rise that is responsible for the liquid-state diffusion can only be estimated. Thus, in situ measurement of the temperature rise due to the flash temperature might provide more valuable data to support the observation of the formation of a high temperature intermetallic compound. Correlation of the results of TEM analysis with the results obtained from the other techniques that have been used to analyse the bonding mechanism may give a more detailed understanding of the mechanisms. For example, the distribution of the diffusion pathways created after the removal of the native aluminium oxide layer can be correlated with the results obtained from the intermetallic compound coverage analysis. This in turn will enhance the understanding regarding the formation and distribution of the intermetallic compound during the wire bonding process.

Conclusion

The present paper reviews the methods that have been introduced to determine the bonding mechanisms of gold wire bonding. Each of the techniques that have been introduced leads to different explanations for the Au wire bonding mechanism. Therefore, the applicability of each technique needs to be examined, and the relationships between the techniques reviewed to provide a more comprehensive explanation of the bonding mechanism. Furthermore, new techniques should be used to improve the current practice for analysing the bonding mechanism of wire bonding. The determination of the actual initial contact area using soft metallisation, the relationship between the volumes and distribution of intermetallic compound, the interaction between the strengthening of the Au ball bond and the bond pad, and also the measurement of actual, in situ flash temperatures are some of the suggested new methods that could shed further light on the thermosonic Au wire bonding mechanism. In addition, the suggested new approaches: examining the effect of individual phase and surroundings on the strengthening produced by the Au–Al intermetallic compound; combining FEA based on friction and wire bonding parameters and correlating TEM results with results obtained from other techniques should enable a more detailed understanding of the bondability and strength of thermosonic Au wire bonds.

Acknowledgements This work was sponsored by the National University of Malaysia under research university grants (UKM-RRR1-07-FRGS0257-2010, OUP-2012-120 and ERGS/1/2011/STG/UKM/02/10).

References

1. Harman GG, Albers J (1977) The ultrasonic welding mechanism as applied to aluminum and gold-wire bonding in microelectronics. IEEE Trans Parts, Hybrids, Packag 13:406–412
2. Langenecker B (1966) Effect of ultrasound on deformation characteristics of metals. IEEE Trans Sonics Ultrason 13:1–8
3. Lum I, Jung JP, Zhou Y (2005) Bonding mechanism in ultrasonic gold ball bonds on copper substrate. Metall Mater Trans A 36:1279–1286
4. Xu H, Liu C, Silberschmidt VV, Chen Z, Wei J (2010) Initial bond formation in thermosonic gold ball bonding on aluminium metallization pads. J Mater Process Technol 210:1035–1042
5. Xu H, Liu C, Silberschmidt V (2010) The role of bonding duration in wire bond formation: a study of footprints of thermosonic gold wire on aluminium pad. Microelectron Int 27:11–16
6. Qi J, Hung NC, Li M, Liu D (2006) Effects of process parameters on bondability in ultrasonic ball bonding. Scripta Mater 54:293–297
7. Wulff F, Breach CD (2006) Measurement of gold ballbond intermetallic coverage. Gold Bull 39:175–184
8. Rosle MF, Abdullah S, Hamid MAA, Daud AR, Jalar A, Kornain Z (2009) Surface topographical characterization of gold aluminide compound for thermosonic ball bonding. J Electron Packag 132:1–6
9. Karpel A, Gur G, Atzmon Z, Kaplan WD (2007) Microstructural evolution of gold-aluminum wire-bonds. J Mater Sci 42:2347–2357
10. Murali S, Srikanth N, Vath CJ III (2003) Grains, deformation substructures, and slip bands observed in thermosonic copper ball bonding. Mater Charact 50:39–50
11. Srikanth N, Murali S, Wong YM, Vath CJ III (2004) Critical study of thermosonic copper ball bonding. Thin Solid Films 462–463:339–345
12. Li CM, Yang P, Liu DM, Hung NC, Li M (2007) A preliminary electron backscatter diffraction study of microstructures and microtextures evolution during Au stud and flip chip thermosonic bonding. J Electron Mater 36:587–592
13. Geissler U, Schneider-Ramelow M, Lang KL, Reichl H (2006) Investigation of microstructural processes during ultrasonic wedge/wedge bonding of AlSi1 wires. J Electron Mater 35:173–180
14. Geissler U, Schneider-Ramelow M, Reichl H (2009) Hardening and softening in AlSi1 bond contacts during ultrasonic wire bonding. IEEE Trans Compon Packag Technol 32:794–799
15. Lum I, Huang H, Chang BH, Mayer M, Du D, Zhou Y (2009) Effects of superimposed ultrasound on deformation of gold. J Appl Phys 105:024905
16. Ding Y, Kim JK, Tong P (2006) Numerical analysis of ultrasonic wire bonding: effect of bonding parameters on contact pressure and frictional energy. Mech Mater 38:11–24
17. Ding Y, Kim JK (2008) Numerical analysis of ultrasonic wire bonding: part 2. Effects of bonding parameters on temperature rise. Microelectron Reliab 48:149–157
18. Schneuwly A, Groning P, Schlapbach L, Muller G (1998) Bondability analysis of bond pads by thermoelectric temperature measurements. J Electron Mater 27:1254–1261
19. Ho JR, Chen CC, Wang CH (2004) Thin film thermal sensor for real time measurement of contact temperature during ultrasonic wire bonding process. Sens Actuators A 111:188–195
20. Li J, Han L, Duan J, Zhong J (2007) Microstructural characteristics of Au/Al bonded interfaces. Mater Charact 58:103–107
21. Jalar A, Zulkifli MN, Abdullah S (2011) Nanoindentation test for the strength distribution analysis of bonded Au ball bonds. Adv Mater Res 148–149:1163–1166
22. Shah M, Zeng K, Tay AAO (2002) Mechanical characterization of the heat affected zone of gold wirebonds using nanoindentation. J Electron Packag 126:87–94
23. Saiki H, Nishitake H, Yotsumoto T, Marumo Y (2007) Deformation characteristics of Au wire bonding. J Mater Process Technol 191:16–19
24. Wulff FW, Breach C, Stephen D, Saraswati DKJ (2005) Further characterisation of intermetallic growth in copper and gold ball bonds on aluminium metallization. Proceedings of 6th Electronics Packaging Technology Conference. EPTC 2004:348–353
25. Hsu HC, Chang WY, Yeh CL, Lai YS (2011) Characteristic of copper wire and transient analysis on wirebonding process. Microelectron Reliab 51:179–186
26. Stephan D, Chew YH, Goh HM, Pasamanero E, Thient EPP, Calpito DRM, Ling J (2007) A comparison study of the bondability and reliability performance of Au bonding wires with different dopant levels. Proceedings of 9th Electronics Packaging Technology Conference. EPTC 2007:737–742
27. Karpel A, Gur G, Atzmon Z, Kaplan WD (2007) TEM microstructural analysis of as-bonded Al–Au wire-bonds. J Mater Sci 42:2334–2346
28. Ji H, Li M, Kim JM, Kim DW, Wang C (2008) Nano features of Al/Au ultrasonic bond interface observed by high resolution transmission electron microscopy. Mater Charact 59:1419–1424
29. Li J, Fuliang W, Han L, Zhong J (2008) Theoretical and experimental analyses of atom diffusion characteristics on wire bonding interfaces. J Phys D: Appl Phys 41:135303
30. Xu H, Liu C, Silberschmidt VV, Chen Z, Sivakumar M (2009) TEM interfacial characteristics of thermosonic gold wire bonding on aluminium metallization. Proceedings of 11th Electronics Packaging Technology Conference. EPTC 2009:512–517
31. Xu H, Liu C, Silberschmidt VV, Pramana SS, White TJ, Chen Z, Sivakumar M, Acoff VL (2010) A micromechanism study of thermosonic gold wire bonding on aluminum pad. J Appl Phys 108:113517

4

Tin-modified gold-based bulk metallic glasses

Shuo-Hong Wang · Tsung-Shune Chin

Abstract Tin was selected as a modifying element in low-gold-content (50 at.%) bulk metallic glasses (BMGs) aiming at developing alloys with cost-effective performance. New gold-based Au–Sn–Cu–Si alloys were fabricated by injection-casting into a copper mold. The as-cast BMG $Au_{50}Sn_6Cu_{26}Si_{18}$ with 18.6-karat gold and a diameter of 1 mm possessed a lower glass transition temperature (T_g) of 82°C (355 K), a lower liquid temperature of 330°C (603 K), and a super-cooled liquid region of 31°C. The viscosity range of this BMG $Au_{50}Sn_6Cu_{26}Si_{18}$ was from 10^8 to 10^9 Pa s measured at a low applied stress of 13 kPa. To compare the viscosity with different applied stresses, its viscosity clearly increased with applied stress below T_g but not so obvious above T_g. The low viscosity of this BMG $Au_{50}Sn_6Cu_{26}Si_{18}$ at around 102°C, which is very close to the boiling temperature of water (100°C), rendered easy thermal–mechanical deformation in a boiling water-bath by hand-pressing and tweezers-bending. Such a deformation capability in boiling water is beneficial to the further applications in various fields.

Keywords Gold-based alloys · Bulk metallic glasses · Glass transition temperature · Thermal mechanical analysis · Viscosity · Deformation capability

Introduction

Pure gold is a precious metal and possesses the characteristics of luster, softness, malleability, and ductility. However, in its pure form, it is too soft to be used for monetary exchange and for producing jewelry materials. Thus, hardening of gold by alloying it with Cu, Ag, Ni, or other metals have been a common practice for a long time [1, 2]. Moreover, because of its superior thermal conductivity, excellent electrical conductivity, and high resistance to corrosion, gold has been widely applied in modern industries including IC electronics, aerospace, medicine, and dentistry [3]. Although pure gold and gold-based alloys with higher karats without toxic elements are useful and acceptable for the human body, the existence of grain boundaries in crystalline gold-based alloys is an ongoing concern with biomedical implants.

Metallic glass is amorphous in structure, containing no crystalline anisotropy, dislocation, grain boundary, or crystalline defects [4]. Bulk metallic glasses (BMGs) usually possess unique properties such as high strength and elasticity, increased hardness, good toughness, and excellent resistance to corrosion compared to their crystalline counterparts

S.-H. Wang · T.-S. Chin (✉)
Department of Materials Science and Engineering,
National Tsing Hua University,
No. 101, Section 2, Kuang-Fu Road,
Hsinchu 30013, Taiwan
e-mail: tschin@mx.nthu.edu.tw

T.-S. Chin
e-mail: tschin@fcu.edu.tw

T.-S. Chin
Center for Nanotechnology, Materials Science and Microsystems,
National Tsing Hua University,
No. 101, Sec-2, Kuang-Fu Road,
Hsinchu 30013, Taiwan

T.-S. Chin
Department of Materials Science and Engineering,
Feng Chia University,
No. 100, Wenhwa Road, Seatwen District,
Taichung 40724, Taiwan

[5, 6]. Since 1988, a large number of BMG systems have been developed notably those Mg- [7], La- [8], Zr- [9], Fe- [10], Pd- [11], Pt- [12], Ti- [13], Ni- [14], and Ca- [15] alloy systems. These BMGs exhibit large critical glass forming size and high thermal stability and hence reveal new possibilities for industrial applications.

In 1960, Klement et al. [5] synthesized the first amorphous alloy using the binary eutectic $Au_{82}Si_{18}$ composition, in thickness less than 50 μm. However, its poor glass-forming ability (GFA) has resulted in limited development over the past several decades. Schroers et al. exploited gold-based multi-component BMGs. These Au-based BMGs possessed a low glass transition temperature (T_g) of at most 128°C, good thermal stability at ambient temperature, and increased hardness, and they were easier to process. For Au–Ag–Pd–Cu–Si BMGs, the addition of Pd and Ag enhances GFA but they increase T_g [3]. On the other hand, Zhang et al. increased the gold content in Au–Ag–Cu–Si BMGs and resulted in a sharp decrease in T_g. The lowest reported T_g for an Au-based BMG is 66°C in the high-Au content composition $Au_{70}Cu_{5.5}Ag_{7.5}Si_{17}$ [16]. The above properties make Au-based BMGs useful for applications such as jewelry, micro-electromechanical systems (MEMS), electronics, nano-molding technologies [17, 18], and dentistry [19].

The purpose of this study is to explore new Au-based BMG composition with cost-effective performance. Composition design is to keep the gold content as low as possible, yet not less than 18 karats. We used tin as an inexpensive alloying element instead of palladium or silver. The performance of major concern is the thermal properties in particular the relationship between viscosity and deformation capability (processing characteristics) of such BMGs. More specifically, our ultimate goal is to develop a new Au-based BMG with low glass transition temperature which will facilitate processing capability to benefit further applications.

Experimental details

In this work, alloy ingots with nominal compositions $Au_{50}Sn_6Cu_{26}Si_{18}$, $Au_{50}Sn_6Cu_{24}Si_{20}$, and $Au_{50}Sn_9Cu_{23}Si_{18}$ were prepared by arc melting the mixtures of pure elements Au, Sn, Cu, and Si with purities of at least 99.9% under vacuum at 10^{-2} Torr. For homogeneity of the alloys, we melted the ingots six times for each composition. Bulk alloy rods with diameters of 1 mm were fabricated by conventional injection casting into a copper mold under an argon atmosphere. The structure of the as-cast rods was examined by X-ray diffractometry (XRD, Shimadzu XRD-6000) using Cu $K_{\alpha 1}$ radiation. The thermal properties were studied by a differential scanning calorimetry (DSC; PerkinElmer Diamond DSC) at a fixed heating rate of 20°C/min under

flowing argon. To study the processing ability, we used a thermal mechanical analyzer (TMA; PerkinElmer Diamond TMA). The as-cast rods were cut into pieces, each with an aspect ratio of around 2 and polished at both ends. TMA measurements were performed with applied loads of 10 and 50 mN. We raised the temperature of the as-cast rods at a fixed heating rate of 10°C/min starting from room temperature to 200°C. The microstructure was observed on cross sections of the as-cast rods using a transmission electron microscope (TEM; JOEL 2010F) operating at an accelerating potential of 200 kV. Structural identification of the phases was carried out by conventional selected area electron diffraction (SAED).

Results and discussion

Figure 1 depicts the X-ray diffraction (XRD) patterns taken from the transverse cross sections of the as-cast 1-mm-diameter rods of $Au_{50}Sn_6Cu_{26}Si_{18}$, $Au_{50}Sn_6Cu_{24}Si_{20}$, and $Au_{50}Sn_9Cu_{23}Si_{18}$ alloys. The XRD results indicate that no obvious diffraction peaks could be identified for the $Au_{50}Sn_6Cu_{26}Si_{18}$ alloy, which shows a fully amorphous structure. The $Au_{50}Sn_6Cu_{26}Si_{18}$ BMG has a GFA capable of casting into amorphous rods of at least 1 mm in diameter and is the easiest to process of all our Au–Sn–Cu–Si quaternary alloys. The new Sn-added alloys show poor GFA when the Cu content decreased from 26 at.% (by being replaced by Si or Sn). This means that the increase in Si content is harmful to the formation of the quaternary Au–Sn–Cu–Si BMG. On the other hand, as Cu is partially substituted by Sn, the GFA also obviously decreases. We thus aimed to further investigate the BMG $Au_{50}Sn_6Cu_{26}Si_{18}$.

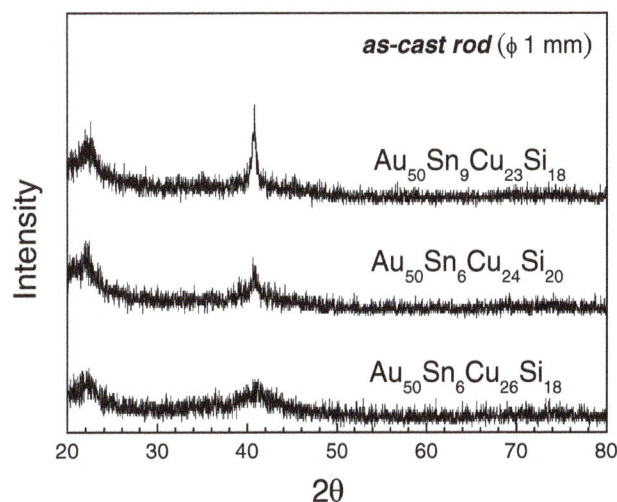

Fig. 1 X-ray diffraction patterns taken from cross sections of as-cast 1-mm diameter rods of $Au_{50}Sn_6Cu_{26}Si_{18}$, $Au_{50}Sn_6Cu_{24}Si_{20}$, and $Au_{50}Sn_9Cu_{23}Si_{18}$ alloys

Fig. 2 Typical TEM image and selected area diffraction pattern (*SADP*) of specimen taken from as-cast 1-mm diameter rod of $Au_{50}Sn_6Cu_{26}Si_{18}$ alloy

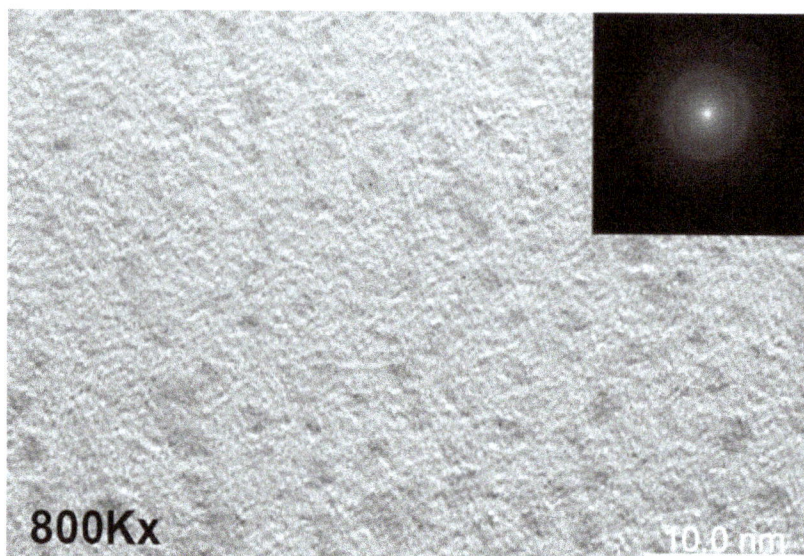

The microstructure of the $Au_{50}Sn_6Cu_{26}Si_{18}$ as-cast BMG rod was studied by TEM (Fig. 2) together with a selected area diffraction pattern (SADP). The lack of diffraction spots or rings in the SADP indicates that the structure of the BMG $Au_{50}Sn_6Cu_{26}Si_{18}$ was amorphous even though there were slight contrasts in the matrix.

A typical DSC curve of the $Au_{50}Sn_6Cu_{26}Si_{18}$ as-cast BMG rod is shown in Fig. 3. The crystallization temperature T_x and T_g are 113°C and 82°C, respectively. The liquid temperature (T_l) of $Au_{50}Sn_6Cu_{26}Si_{18}$ BMG is 330°C. The calculated thermal criteria are the super-cooled liquid region (ΔT_x), 31 K; the reduced glass transition temperature (T_{rg}), 0.59; and the γ-factor 0.403. Although the ΔT_x of the $Au_{50}Sn_6Cu_{26}Si_{18}$ BMG was not as high as those reported in the literature [16], it was

expected to be potential for easier processing below 100°C in hot water because of low T_g and moderate ΔT_x. Therefore, we carried out further study on its viscosity at elevated temperatures, as to be delineated later. Table 1 summarizes the thermal parameters of Au-based BMGs measured using the DSC. By comparing the results with those of other Au-based BMGs whose Au content 46–52 at.%, the T_g of our $Au_{50}Sn_6Cu_{26}Si_{18}$ BMG is obviously much lower than those of Au-based BMGs containing Pd and Ag. The ΔT_x (31°C) of $Au_{50}Sn_6Cu_{26}Si_{18}$ BMG is higher than that of BMG $Au_{46}Ag_5Cu_{29}Si_{20}$ (ΔT_x=25°C) but lower than that of Pd-containing BMG $Au_{49}Ag_{5.5}Pd_{2.3}Cu_{26.9}Si_{16.3}$ (ΔT_x=58°C). Among these Au-based BMGs, the most well known is the composition $Au_{49}Ag_{5.5}Pd_{2.3}Cu_{26.9}Si_{16.3}$. Despite its high T_g, this BMG has been widely used as the mold material in nano-imprint technology and in MEMS. Therefore, our BMG alloy, which has a lower T_g and does not contain Pd and Ag, is more cost-effective in similar applications.

Thermal mechanical analysis (TMA) and differential TMA (DTMA) curves of the $Au_{50}Sn_6Cu_{26}Si_{18}$ as-cast BMG rod at an applied load of 50 mN are plotted in Fig. 4. The DTMA curve shows the differentiated results of displacement related to time. To exhibit the relationship between temperature and viscosity several parameters are defined. They are the onset temperature of viscous flow for the initial state (T_{onset}); the temperature corresponding to the lowest viscosity which is the viscosity of semi-steady state (T_{vs}); and the finishing temperature of viscous flow for the full crystallization state (T_{finish}). These are marked on the DTMA curve. From the DTMA curve, T_{onset}, T_{vs}, and T_{finish} are 84°C, 108°C, and 116°C, respectively. The values of T_{onset} and T_{finish} are very close to those of T_g (82°C) and T_x (113°C), respectively. TMA data and DSC data are almost the same for these two specific temperatures. The viscosity

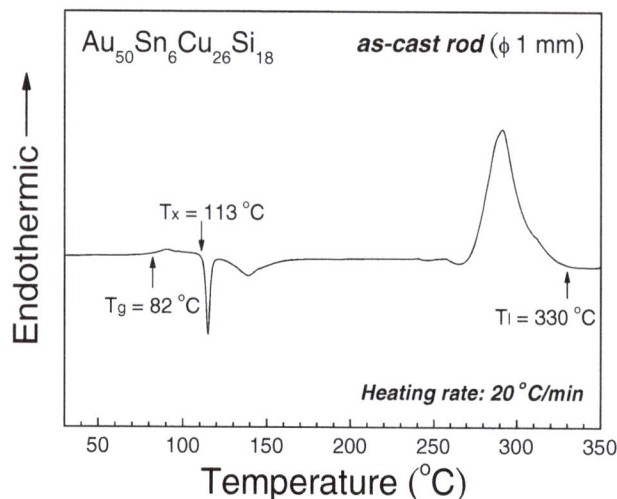

Fig. 3 Typical DSC curve of as-cast 1-mm diameter rod of $Au_{50}Sn_6Cu_{26}Si_{18}$ alloy

Table 1 Thermal parameters of several Au-based BMGs found in the literature along with the values determined from this study [3]

Composition (at.%)	T_g (K)	T_x (K)	ΔT_x	T_l (K)	T_{rg}	γ	Ref.
$Au_{50}Sn_6Cu_{26}Si_{18}$	355	386	31	603	0.59	0.403	This study
$Au_{50}Cu_{33}Si_{17}$	383	405	22	679	0.56	0.381	[3]
$Au_{46}Ag_5Cu_{29}Si_{20}$	395	420	25	664	0.59	0.397	[3]
$Au_{52}Pd_{2.3}Cu_{29.2}Si_{16.5}$	393	427	34	651	0.6	0.409	[3]
$Au_{49}Ag_{5.5}Pd_{2.3}Cu_{26.9}Si_{16.3}$	401	459	58	644	0.62	0.439	[3]

of the $Au_{50}Sn_6Cu_{26}Si_{18}$ BMG can be evaluated according to the Stefan equation [20]:

$$\eta = \left(\frac{\sigma}{3\dot{\varepsilon}}\right)\left(1 + \frac{d_0^2}{8l_0^2(1+\varepsilon_n)^2}\right)^{-1}, \qquad (1)$$

where σ is the stress (Pa), $\dot{\varepsilon}$ is the strain rate (s^{-1}), d_0 is the initial diameter of the specimen, l_0 is the initial length of the specimen, and ε_n the engineering strain (nominal strain). The stress is calculated by $\sigma = (F/A_0)(l/l_0)$, where F, A_0, and l are the applied loading force (mN), initial cross-sectional area of the specimen, and length after deformation, respectively. Calculated results are plotted in Fig. 5.

The curves in Fig. 5a and b show that the viscosity values of the $Au_{50}Sn_6Cu_{26}Si_{18}$ BMG are between 10^8 and 10^9 Pa s at an applied stress of 13 kPa, and 2×10^8 and 10^{10} Pa s at an applied stress of 65 kPa. The viscosity of the BMG $Au_{50}Sn_6Cu_{26}Si_{18}$ is in a range similar to that of the BMG $Au_{49}Ag_{5.5}Pd_{2.3}Cu_{26.9}Si_{16.3}$ (10^7–10^9 Pa s) obtained by Tang et al. [17]. However, the lower limit is relatively higher than those of BMGs $Mg_{65}Cu_{25}Gd_{10}$ and $Mg_{65}Cu_{25}Gd_{10}P_3$ (10^6–10^{10} Pa s) obtained by Chang et al. [20]. In addition, we observed that viscosity increases as increasing applied stress from 13 to 65 kPa. These tendencies are consistent with

those reported by Chang et al. [21] for the BMG $Mg_{58}Cu_{31}Y_{11}$. In the literature, Tang et al. reported that the applied stress was in the range of 40–400 kPa or even higher for the BMG $Au_{49}Ag_{5.5}Pd_{2.3}Cu_{26.9}Si_{16.3}$ [17]. Hence, the applied flow stress of 13 kPa for the BMG $Au_{50}Sn_6Cu_{26}Si_{18}$ is relatively low [21, 22].

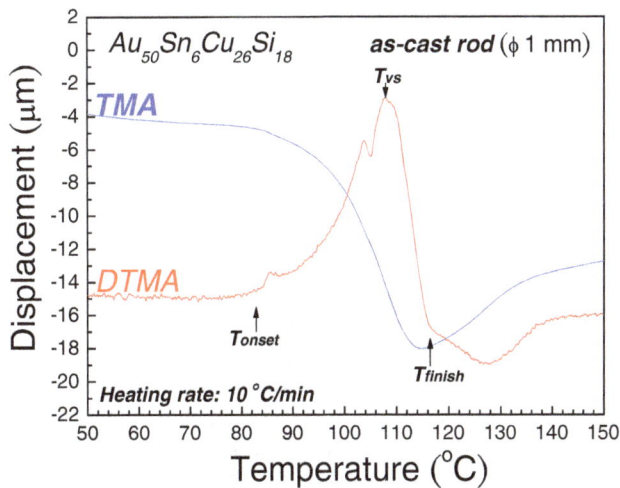

Fig. 4 TMA and DTMA curves of as-cast $Au_{50}Sn_6Cu_{26}Si_{18}$ BMG rod at the applied load of 50 mN

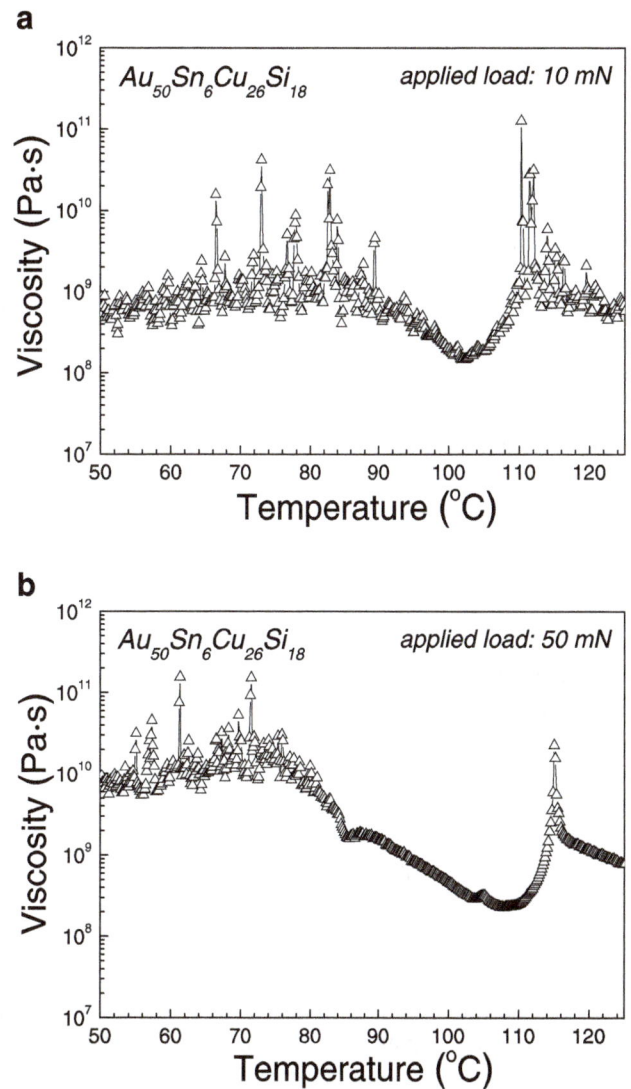

Fig. 5 Estimated viscosity of as-cast $Au_{50}Sn_6Cu_{26}Si_{18}$ BMG, measured under an applied load of a 10 mN (corresponding to 13 kPa) and b 50 mN (65 kPa)

Fig. 6 Deformation capability of as-cast $Au_{50}Sn_6Cu_{26}Si_{18}$ BMG in the boiling water-bath by **a** hand-pressing and **b** tweezers-bending

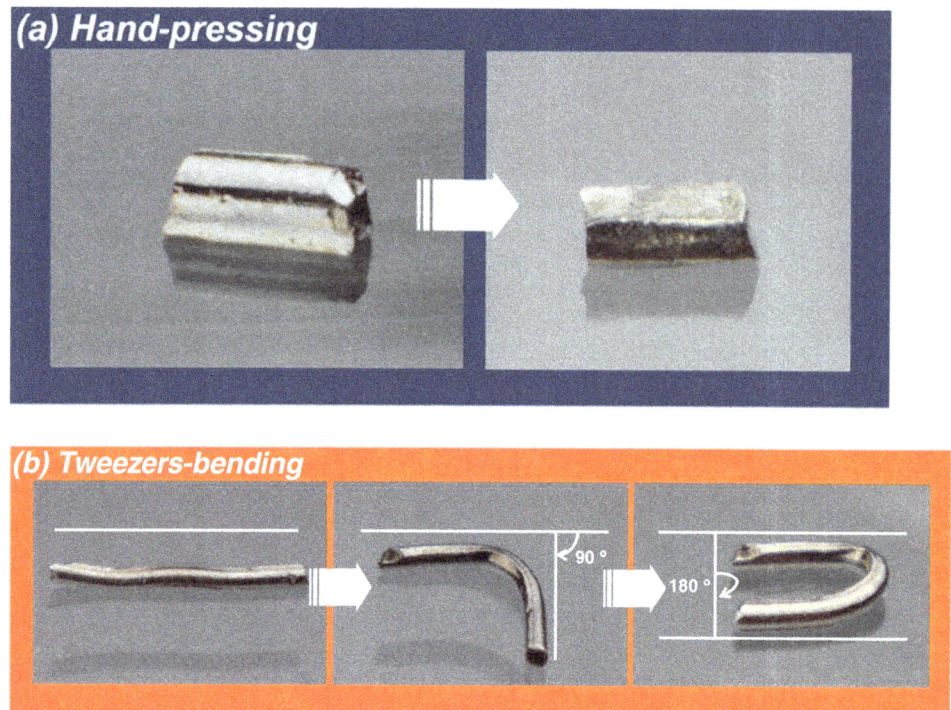

In addition, the viscosity value is highly dependent on temperature. In fact, the much higher viscosity measured at 65 kPa is at a temperature much lower than 83 °C, which is the TMA glass transition temperature. As the temperature becomes higher than 83 °C, the viscosity soon declines to its minimum value. The viscosity decreases from 10^9 to 10^8 Pa s for both applied stresses of 13 and 65 kPa. Over this temperature range, the BMG $Au_{50}Sn_6Cu_{26}Si_{18}$ exhibits a typical viscosity of a Newtonian liquid. There exists a subtle difference in minimum viscosity values, 10^8 Pa s at 13 kPa and 2×10^8 Pa s at 65 kPa, and at different temperatures of the lowest viscosity (T_{vs}), 102 °C at 13 kPa and 108 °C at 65 kPa. This indicates that the applied stress does influence the measured viscosity to some extent. Above 113 °C, the BMG no longer exhibits the viscosity that is characteristic of glassy materials, but instead shows the deformation characteristics of a rigid crystalline solid in responding to applied stress at elevated temperatures. These results indicate that the BMG $Au_{50}Sn_6Cu_{26}Si_{18}$ can have a viscous flow.

According to the result of viscosity, the proposed best temperature for thermal deformation is at 102 °C, which is very close to the boiling temperature of water (100 °C). Hence, we think it is possible to investigate the deformability of the as-cast $Au_{50}Sn_6Cu_{26}Si_{18}$ BMG, 1 mm in diameter, in boiling water-bath. Figure 6a and b exhibits the results of hand-pressed and tweezers-bent $Au_{50}Sn_6Cu_{26}Si_{18}$ BMG in boiling water-bath. It shows that rod-shaped $Au_{50}Sn_6Cu_{26}Si_{18}$ BMG can be pressed to be flattened, and can be bent to 90–180° repeatedly for several times without breaking or hardening. We did deformation out of the boiling water-bath and the BMG rod soon cracked. These are direct proofs that the BMG alloy deforms by a viscous flow mechanism in boiling water. Such a deformation capability in boiling water, which does not need a temperature controller to maintain isothermal condition, is very useful for application in various fields. These include but not limited to nano-imprint molds, dental prosthesis, jewelry, or MEMS devices.

Conclusions

In summary, we explored a cost-effective quaternary gold-based BMG $Au_{50}Sn_6Cu_{26}Si_{18}$ with a gold content of 18.6 karats in this research. This alloy retains a good GFA which renders possibility of inject-casting into BMG rod of at least 1 mm in diameter. This new Au-based BMG is characteristic of low glass transition (82 °C), crystallization (113 °C), and liquid (330 °C) temperatures and a moderate super-cooled liquid region (ΔT_x, 31 °C). Moreover, the $Au_{50}Sn_6Cu_{26}Si_{18}$ BMG exhibits a viscosity ranging from 10^8 to 10^9 Pa s measured under an applied stress as low as 13 kPa. The merit of its significant deformation capability at around 100 °C (i.e., at the boiling temperature of water) renders easy and precise deformation without the use of a temperature controller. Based on these outstanding and cost-effective characteristics, the new BMG $Au_{50}Sn_6Cu_{26}Si_{18}$ will find extensive applications in molds for nano-imprinting, dental prosthesis, jewelry, and MEMS devices in the future.

Acknowledgment The authors are grateful to the National Science Council of the Republic of China for sponsoring this research under grant NSC 97-2221-E-035 -011-MY2.

References

1. Drummond IM (1987) The gold standard and international monetary system. Macmillan, New York
2. Vilar P, White J (1991) A history of gold and money. Verso, London
3. Schroers J, Lohwongwatana B, Johnson WL, Peker A (2005) Gold based bulk metallic glass. Appl Phys Lett 87:061912.
4. Greer AL (1995) Metallic glasses. Science 267:1947–1953.
5. Klement W, Willens RH, Duwez P (1960) Non-crystalline structure in solidified gold-silicon alloys. Nature 197:869–870.
6. Loffler JF (2003) Bulk metallic glasses. Intermetallics 11:529–540.
7. Inoue A, Ohtera K, Kita K, Masumoto T (1988) New amorphous Mg–Ce–Ni alloys with high-strength and good ductility. Jpn J Appl Phys 27:2248–2251
8. Inoue A, Zhang T, Masumoto T (1989) Al–La–Ni amorphous-alloys with a wide supercooled liquid region. Mater Trans JIM 30:965–972
9. Inoue A, Zhang T, Masumoto T (1990) Zr–Al–Ni amorphous-alloys with high glass-transition temperature and significant supercooled liquid region. Mater Trans JIM 31:177–183
10. Lin CY, Tien HY, Chin TS (2005) Soft magnetic ternary iron-boron-based bulk metallic glasses. Appl Phys Lett 86:162501.
11. Inoue A, Nishiyama N, Matsuda T (1996) Preparation of bulk glassy $Pd_{40}Ni_{10}Cu_{30}P_{20}$ alloy of 40 mm in diameter by water quenching. Mater Trans JIM 37:181–184
12. Schroers J, Johnson WL (2004) Highly processable bulk metallic glass-forming alloys in the Pt-Co-Ni-Cu-P system. Appl Phys Lett 84:3666–3668.
13. Lin XH, Johnson WL (1995) Formation of Ti–Zr–Cu–Ni bulk metallic glasses. J Appl Phys 78:6514–6519.
14. Tien HY, Lin CY, Chin TS (2006) New ternary Ni-Ta-Sn bulk metallic glasses. Intermetallics 14:1075–1078.
15. Amiya K, Inoue A (2002) Formation, thermal stability and mechanical properties of Ca-based bulk glassy alloys. Mater Trans 43:81–84.
16. Zhang W, Guo H, Chen MW, Saotome Y, Qin CL, Inoue A (2009) New Au-based bulk glassy alloys with ultralow glass transition temperature. Scr Mater 61:744–747.
17. Tang TW, Chang YC, Huang JC, Gao Q, Jang JSC, Tsao CYA (2009) On thermomechanical properties of Au–Ag–Pd–Cu–Si bulk metallic glass. Mater Chem Phys 116:569–572.
18. Kumar G, Tang HX, Schroers J (2009) Nanomoulding with amorphous metals. Nature 457:868–872.
19. Lee SH, Lim IS, Cho MH, Pyo AR, Kwon YH, Seol HJ, Kim HI (2011) Age-hardening and overaging mechanisms related to the metastable phase formation by the decomposition of Ag and Cu in a dental Au–Ag–Cu–Pd–Zn alloy. Gold Bull 44:155–162.
20. Chang YC, Huang JC, Cheng YT, Lee CJ, Du XH, Nieh TG (2008) On the fragility and thermomechanical properties of Mg–Cu–Gd– (B) bulk metallic glasses. J Appl Phys 103:103521.
21. Chang YC, Huang TH, Chen HM, Huang JC, Nieh TG, Lee CJ (2007) Viscous flow behavior and thermal properties of bulk amorphous $Mg_{58}Cu_{31}Y_{11}$ alloy. Intermetallics 15:1303–1308.
22. Jang JSC, Chang CF, Huang YC, Chiang WJ, Nieh TG, Liu CT (2009) Viscous flow and microforming of a Zr-base bulk metallic glass. Intermetallics 17:200–204.

5

Electrospun polystyrene fiber-templating ultrafine gold hollow fiber production

Shinji Sakai · Shogo Kawa · Koichi Sawada · Masahito Taya

Abstract Ultrafine gold hollow fibers of about 1.5 μm in outer diameter and 1 μm in hollow core diameter, i.e., about 250 nm of gold sheath thickness, were produced in combination with electrospinning technique, photoreduction, electroless plating, and heat treatment for removing template polymer fibers. The gold sheath was fabricated through electroless plating on electrospun polystyrene fibers of about 1 μm diameter using the gold nanoparticles formed in situ in the fibers by irradiation of ultraviolet light as seeds of the gold sheath formation. Then, a hollow core structure was developed by treating the resultant fibers at 180, 350, or 800 °C. Scanning electron microscopy imaging showed the formation of a hollow core structure in the fibers by the treatment at 180 °C as well as those treated at 350 and 500 °C. These results demonstrate the possibility of fabrication of ultrafine gold hollow fibers using electrospun polystyrene fibers as a template of the hollow core structure.

Keywords Electrospinning · Electroless plating · Gold hollow fiber · Photoreduction · Polystyrene

Introduction

Electrospinning first patented in the USA in 1902 [1] has received a dramatic revival of interest in the last decade as a technology by which ultrafine fibers with diameters ranging from several microns to down to 100 nm or less can be prepared from a wide variety of materials [2]. The small diameter of individual fibers induces remarkable characteristics of the resultant nonwoven fabrics due to a very large surface area-to-volume ratio, high porosity, and very small pore size [3]. A variety of electrospun fibers have been made for applications in energy storage, healthcare, biotechnology, and environmental engineering [3]. The aim of the present study is to fabricate ultrafine gold hollow fibers using the electrospun fibers as a template of the hollow core. Fabrications of nanotubes and ultrafine hollow fibers have attracted much attention for their structural particularity because such structural modifications could further enhance material properties for the aforementioned applications [4]. In the present work, we attempted to prepare ultrafine gold hollow fibers by combining electrospinning technique with subsequent processes of electroless plating and removal of the original polymer fibers. Electroless plating is a method of depositing metal onto insulating substrates without having to use electricity. There are literatures for successful coating of gold onto electrospun polyacrylonitrile (PAN) fibers using the electroless plating process [1, 5]. In the preceding literatures, no attempts have been performed for developing a hollow core structure in the fibers. In this study, we used electrospun polystyrene (PS) fibers for the electroless plating of gold because polystyrene decomposes even around 200 °C in air [6]. The ultrafine gold hollow fibers were prepared through following four steps: (1) The PS fibers containing HAuCl$_4$ were electrospun, (2) gold ions in the PS fibers were reduced by ultraviolet (UV) irradiation, (3) a gold sheath was electrolessly deposited on the PS fibers, and (4) a hollow core structure was developed by decomposing the PS fibers by heat treatment (Fig. 1).

Experimental

Electrospinning

Polystyrene (MW 280,000, Sigma, MO, USA) was dissolved in dimethylformamide (DMF) at 15 % (w/w), followed by addition of 3 % (w/w) HAuCl$_4$ 4H$_2$O (Kanto Chemical, Tokyo, Japan) and stirring to a homogeneous yellowish color solution for about 10 min. The resultant solution was loaded into a plastic syringe

S. Sakai (✉) · S. Kawa · K. Sawada · M. Taya
Division of Chemical Engineering, Department of Materials Science and Engineering, Graduate School of Engineering Science, Osaka University, 1-3 Machikaneyama-cho, Toyonaka, Osaka 560-8531, Japan
e-mail: sakai@cheng.es.osaka-u.ac.jp

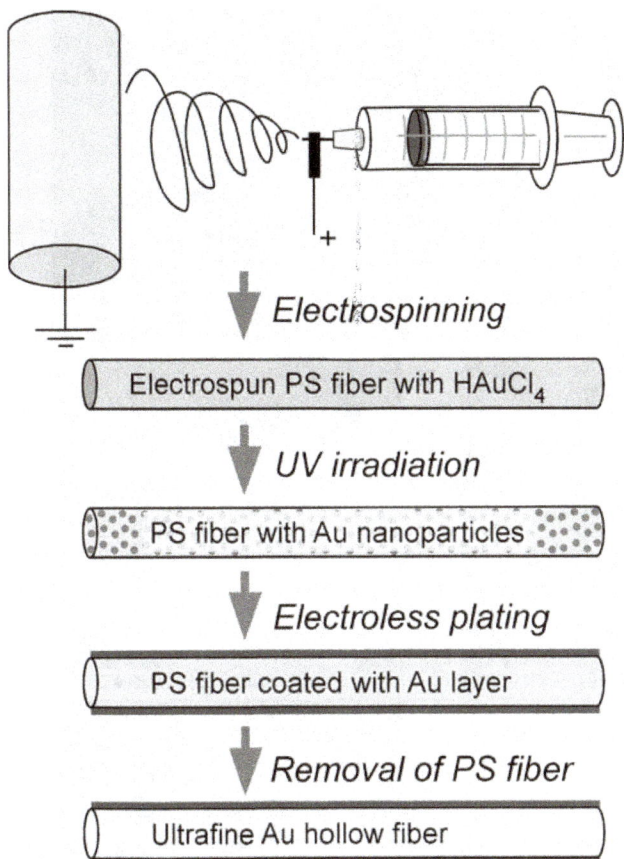

Fig. 1 Scheme of ultrafine gold hollow fiber preparation from electrospun PS fibers enclosing $HAuCl_4$

equipped with a 20-gauge stainless steel needle, and then extruded at 2 mL/h under being applied at +20 kV using a high-voltage DC generator. The needle connected to the DC supply was located 12 cm from an earthed counter electrode.

Electroless plating

The resultant PS fibrous mat containing $HAuCl_4$ was dried in vacuo. Then, the fibrous mat was irradiated UV for 1 day by putting the specimen 60 cm from a UV lamp (51 mW/m^2). Through this process, the gold salt embedded in the PS fibers was reduced to form gold nanoparticles as catalytic seed metal of the subsequent gold coating. The PS fibers containing gold nanoparticles (4 mg) were put into a membrane filtration folder of 25 mm in diameter. Then, 600 mL of 0.01 % (w/w) $HAuCl_4$ $4H_2O$ solution cooled in an ice bath was circulated at 12 mL/min through the PS fibrous mat until the electroless plating solution became colorless for 2 h.

Removal of PS fibers

The fibrous specimen treated with the electroless plating was rinsed with distilled water and then dried in vacuo.

Subsequently, the dried specimen was kept in an oven at 180 °C for 2 or 3 days, or at 350 or 500 °C for 2 h.

Scanning and transmission electron microscopies observations

Morphologies of the electrospun fibers before and after the treatment of the electroless plating were observed using a scanning electron microscope (SEM, Model S-2250 N, Hitachi Ltd., Tokyo, Japan) after drying in vacuo. Formation of gold nanoparticles in the fibers after the UV irradiation was observed using a transmission electron microscope (TEM, Model H-800, Hitachi Ltd., Tokyo, Japan) operating at 100 kV and magnification value of 100 k. Samples were dispersed in water, and a drop of the resultant suspension was poured on a carbon coated-copper grid.

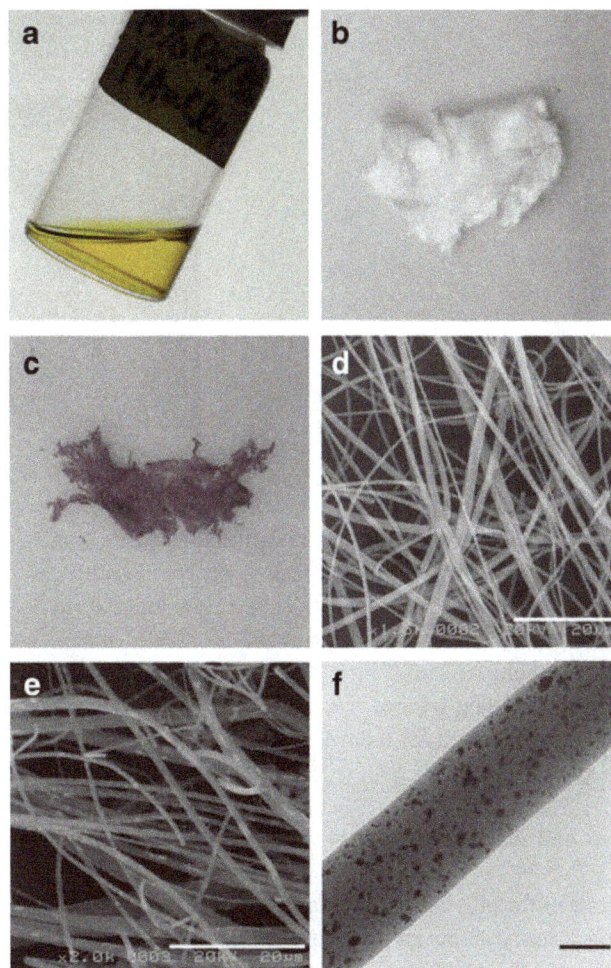

Fig. 2 Photographs of **a** homogeneous solution containing PS and $HAuCl_4$ $4H_2O$ at 15 and 3 % (w/w) before electrospinning, and fibrous mat of electrospun PS fibers **b** before and **c** after UV irradiation. SEM images of PS fibers **d** before and **e** after UV irradiation. **f** TEM image of gold nanoparticles (black dots) formed in PS fibers through UV irradiation. Bars in **d** and **e**, 20 μm; Bar in **f**, 200 nm

Fig. 3 SEM images of PS fibers after electroless plating of fibers **a** with and **b** without containing gold nanoparticles, calcined at **c** 350 °C and **d** 500 °C for 2 h, and at 180 °C for **e** 2 days and **f** 3 days. The *arrows* in **e** and **f** indicate remaining PS layer. *Bars* in **a** and **b**, 20 μm; *Bar* in **c**, 3 μm; *bars* in **d** and **e**, 2 μm; and *Bar* in **f**, 1.5 μm

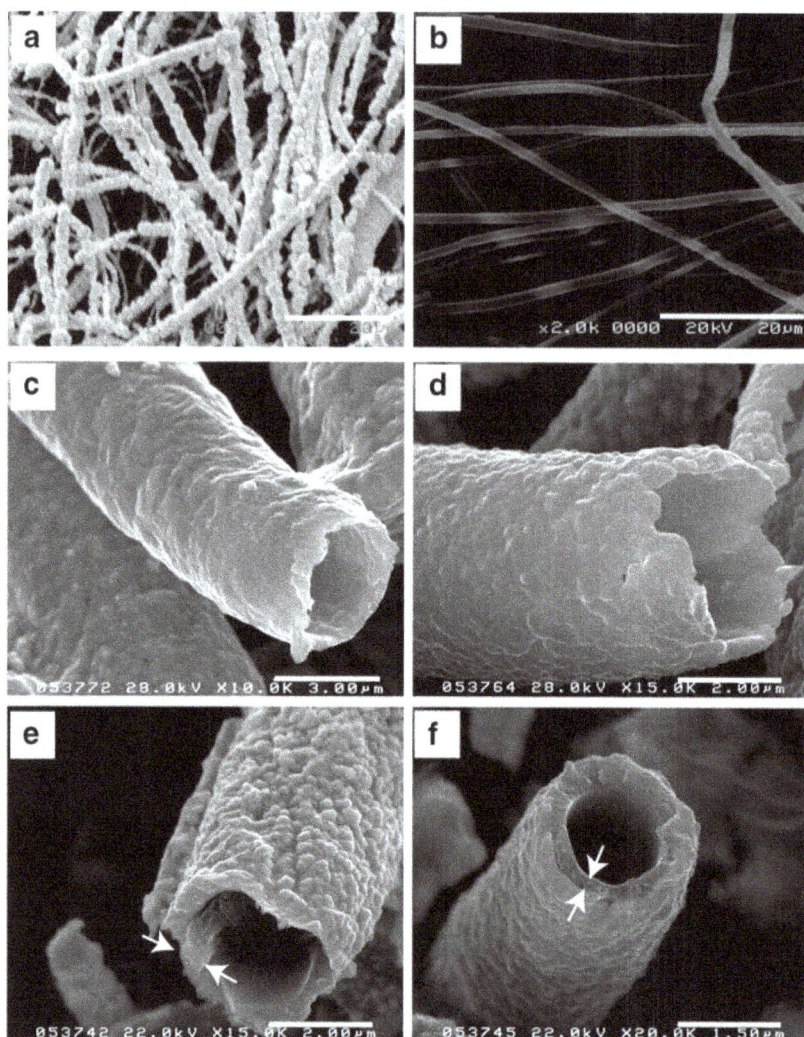

Thermal measurement

Weight change of the dried specimen with a gold sheath was determined using a thermal analyzer (STA6000, PerkinElmer, Inc, MA, USA) in the temperature range from 80 to 600 °C in air. At 180, 350, and 500 °C, these temperatures were held for 1 or 2 h. A heating rate except for the temperature holding periods was 10 °C/min. About 10 mg of specimen was used for the measurement.

Results and discussion

PS is one of the polymers which has been electrospun [7, 8]. In preliminary experiment, we attempted to dissolve $HAuCl_4$ $4H_2O$ at 15 % (*w/w*) with 15 % (*w/w*) PS in DMF. Under this condition, homogeneous solution was not obtained. Thus, we decreased the content of $HAuCl_4$ $4H_2O$ to 3 % (*w/w*), and could obtain homogeneous yellowish solution (Fig. 2a). Electrospun fibers containing $HAuCl_4$ were prepared from the homogeneous solution. The

diameter of as spun PS fibers containing $HAuCl_4$ was 0.89 ± 0.37 μm. The contents of PS and $HAuCl_4$ in the resultant fibers were 83 % (*w/w*) and 17 % (*w/w*), respectively, after removing DMF.

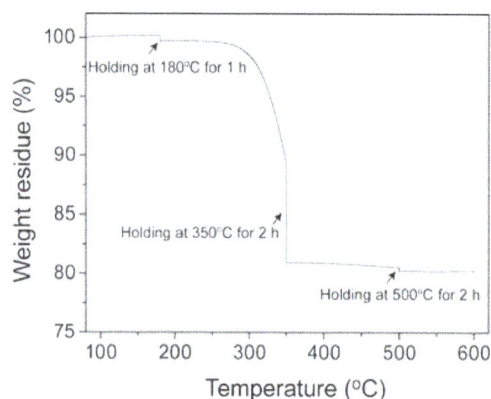

Fig. 4 Transition of weight residue of electrospun PS fibers with a gold sheath by heating. The weight of specimen at 80 °C was set as 100 %. *Arrows* show the periods at which temperatures were held for 1 h at 180 °C and for 2 h at 350 and 500 °C

Next, we attempted to prepare gold nanoparticles in/on electrospun PS fibers as seeds of a gold sheath formation on the fibers. Incorporations of silver and gold nanoparticles in electrospun PAN fibers were reported as simple but effective ways for fabricating these metal sheaths on the fibers through electroless plating [9, 10]. For the formation of the gold nanoparticles, we employed UV irradiation based on the report by Anka et al. [10] for in situ preparation of gold nanoparticles in electrospun PAN fibers. After the UV irradiation in this study, whitish as spun fibers (Fig. 2b) turned into purplish ones showing the existence of gold nanoparticles (Fig. 2c). In fact, gold nanoparticles of several to dozens nm were found by the observation of the fibers using a TEM (Fig. 2f). The diameters of the UV-irradiated fibers were not changed significantly before (0.89 ± 0.73 μm, Fig. 2d) and after the gold nanoparticle formation (0.90 ± 0.33 μm, Fig. 2e). A gold sheath formation was confirmed on the PS fibers containing gold nanoparticles by applying the electroless plating (Fig. 3a). The diameter of the resultant fibers with gold sheath was 1.49 ± 0.36 μm. It means the formation of a gold sheath of about 300 nm thickness. In contrast, the gold sheath formation was not observed on the fibers without containing gold nanoparticles after applying the same electroless plating step (Fig. 3b).

Finally, we attempted to develop a hollow core structure in the PS fibers with the gold sheath. Gold hollow fibers were obtained by the heat treatment at 500 °C for 2 h in air (Fig. 3c). The decrease in temperature to 350 °C also resulted in the fibers with hollow structure (Fig. 3d). The hollow core structure was also induced by the heat treatment at 180 °C (Fig. 3e). It is known that PS decomposes to styrene, cumene, and ethylbenzene around 200 °C [6]. In fact, the weight of PS fibers before the treatment with the electroless plating decreased to about 35 % after 2 days of standing at 180 °C. An obvious difference between the hollow fibers obtained at 350 and 180 °C treatment was the double-layered structure found only for the cross-sectional image of the latter specimen (Fig. 3e). It is intuitive that the inner layer is PS. This inner layer did not disappear after 3 days of treatment at 180 °C (Fig. 3f). Exact reason is not clear now but the remaining of the PS layer neighboring the gold sheath would be caused by the interaction between PS molecules and gold ions. Transition of the change in weight during heating in air (Fig. 4) supports the findings of the hollow core structure formation through heat treatments: During 1 h of holding at 180 °C, weight of the specimen with the gold sheath decreased slightly but surely. The residual weight of the specimen decreased to 81 % by the end of 2 h of holding at 350 °C. The weight further decreased to 80 % by the end of 2 h of holding at 500 °C. No weight change was detected during the following period of heating from 500 to 600 °C and 2 h of holding at 600 °C. These results demonstrate majority of the PS in the fibrous

specimen disappeared by the heat treatment at 350 °C for 2 h. In addition, the heat treatment at 500 °C is enough for removing the PS in the gold sheath. It was reported that gold sheathes could be formed on electrospun PAN fibers through electroless plating (but not reported on the production of gold hollow fibers) [9, 10]. Therefore, we also investigated the possibility of using electrospun PAN fibers as a template of the hollow core structure in a gold sheath. A hollow core structure could be obtained for the PAN fiber system treated at 500 °C. However, the treatment at 350 °C was not enough despite it being enough for the PS fiber system. These results demonstrate the usefulness of PS fibers as a template of gold hollow fiber preparation.

Conclusion

In this study, we aimed to develop the ultrafine gold hollow fibers. We first prepared the electrospun PS fibers containing $HAuCl_4$ and then irradiated UV light for photoreduction resulting in the formation of gold nanoparticles in/on the PS fibers. The in situ formed gold nanoparticles were used as seeds of the subsequent gold sheath formation through electroless plating. A hollow core structure was developed by the heat treatment above 180 °C. The temperature required for the decomposition of the PS fiber template was lower than that necessary for PAN fibers. From these results, we concluded that the ultrafine gold hollow fibers can be obtained in combination with electrospinning technique, photoreduction, electroless plating, and heat treatment for removing core polymer material. In addition, PS is the useful material as a fiber material on which the gold sheath forms through the electroless plating.

Acknowledgments We gratefully thank Mr. M. Kawashima, GHAS laboratory at Osaka University, for his help in the SEM measurement. In addition, we thank Prof. H. Yasuda and Dr. T. Sakata, Research Center for Ultra-High Voltage Electron Microscopy at Osaka University, for supporting the TEM measurement.

References

1. Liu X, Li M, Han G, Dong J (2010) The catalysts supported on metallized electrospun polyacrylonitrile fibrous mats for methanol oxidation. Electrochim Acta 55:2983–2990
2. Huang ZM, Zhang YZ, Kotaki M, Ramakrishna S (2003) A review on polymer nanofibers by electrospinning and their applications in nanocomposites. Compos Sci Technol 63:2223–2253
3. Ramakrishna S, Fujihara K, Teo WE, Yong T, Ma ZW, Ramaseshan R (2006) Electrospun nanofibers: solving global issues. Mater Today 9:40–50

4. Srivastava Y, Loscertales I, Marquez M, Thorsen T (2008) Electrospinning of hollow and core/sheath nanofibers using a microfluidic manifold. Microfluid Nanofluid 4:245–250

5. Marx S, Jose MV, Andersen JD, Russell AJ (2011) Electrospun gold nanofiber electrodes for biosensors. Biosens Bioelectron 26:2981–2986

6. Pfaffli P, Zitting A, Vainio H (1978) Thermal degradation products of homopolymer polystyrene in air. Scand J Work Environ Health 4(Suppl 2):22–27

7. Casper CL, Stephens JS, Tassi NG, Chase DB, Rabolt JF (2004) Controlling surface morphology of electrospun polystyrene fibers:

effect of humidity and molecular weight in the electrospinning process. Macromolecules 37:573–578

8. Wannatong L, Sirivat A, Supaphol P (2004) Effects of solvents on electrospun polymeric fibers: preliminary study on polystyrene. Polym Int 53:1851–1859

9. Song X, Lei J, Li Z, Li S, Wang C (2008) Synthesis of polyacrylonitrile/Ag core–shell nanowire by an improved electroless plating method. Mater Lett 62:2681–2684

10. Anka FH, Perera SD, Ratanatawanate C, Balkus KJ (2012) Polyacrylonitrile gold nanoparticle composite electrospun fibers prepared by in situ photoreduction. Mater Lett 75:12–15

6

The electrochemical corrosion behaviour of quaternary gold alloys when exposed to 3.5 % NaCl solution

L. P. Ward · D. Chen · A. P. O'Mullane

Abstract Lower carat gold alloys, specifically 9 carat gold alloys, containing less than 40 % gold, and alloying additions of silver, copper and zinc, are commonly used in many jewellery applications, to offset high costs and poor mechanical properties associated with pure gold. While gold is considered to be chemically inert, the presence of active alloying additions raises concerns about certain forms of corrosion, particularly selective dissolution of these alloys. The purpose of this study was to systematically study the corrosion behaviour of a series of quaternary gold–silver–copper–zinc alloys using dc potentiodynamic scanning in saline (3.5 % NaCl) environment. Full anodic/cathodic scans were conducted to determine the overall corrosion characteristics of the alloy, followed by selective anodic scans and subsequent morphological and compositional analysis of the alloy surface and corroding media to determine the extent of selective dissolution. Varying degrees of selective dissolution and associated corrosion rates were observed after anodic polarisation in 3.5 % NaCl, depending on the alloy composition. The corrosion behaviour of the alloys was determined by the extent of anodic reactions which induce (1) formation of oxide scales on the alloy surface and or (2) dissolution of Zn and Cu species. In general, the improved corrosion characteristics of alloy #3 was attributed to the composition of Zn/Cu in the alloy and thus favourable microstructure promoting the formation of protective oxide/chloride scales and reducing the extent of Cu and Zn dissolution.

Keywords Gold alloy · 3.5 % NaCl · Potentiodynamic scanning · Anodic polarisation · Characterisation · Selective dissolution

Introduction

The inherent physical properties of gold, including electrical conductivity and chemical inertness combined with attractive appearance, have established its use in a number of areas, such as the electronics, dental and jewellery industries. However, pure (24 carat) gold is expensive and mechanically weak (soft) and therefore has limited applications even in jewellery applications. Consequently, attention has focussed on the development of a range of lower carat gold alloys, particularly 9 carat gold alloys, where compositions vary from ternary gold–silver–copper alloys to quaternary alloys where zinc is added primarily to improve workability through contraction of the two-phase field of the phase diagram [1] and to a lesser extent counteract the red colour induced by copper [2].

In order to function satisfactorily in jewellery applications, certain requirements have to be met with respect to mechanical, chemical and biological properties, particularly where prolonged contact with human skin is evident. Gold itself is considered chemically inert and biologically compatible with the body. However, there is concern with respect to the corrosive nature of these gold alloys, particularly the lower carat, quaternary alloys where the presence of non-noble metals in the alloy are likely to strongly influence the corrosion resistance.

Rapson [3] has reported that the significant areas where corrosion and tarnishing of gold alloys becomes important are (1) tarnishing of gold electrical contacts and gold jewellery; (2) corrosion of gold dental alloys and gold jewellery

L. P. Ward (✉)
School of Civil, Environmental and Chemical Engineering, RMIT University, Melbourne, VIC, Australia
e-mail: liam.ward@rmit.edu.au

D. Chen
Department of Chemistry and Chemical Engineering, Foshan University, Foshan, Guangdong, China

A. P. O'Mullane
School of Applied Sciences, RMIT University, Melbourne, VIC, Australia

and (3) stress corrosion cracking of gold jewellery. Gold jewellery is known to corrode in a number of environments which include sea water and chlorinated water.

Numerous studies have been conducted on the mechanisms and consequential effects of stress corrosion cracking (SCC), a well-known form of environmental attack in low carat gold jewellery, particularly below 14 carat gold content [3–10]. While there appears to be less information available on other forms of corrosion, such as pitting and intergranular corrosion, the concepts of selective leaching of alloying elements to modify alloys for decorative purposes has been around for many years [11, 12]. Her, gold alloys are depleted intentionally of alloying elements to produce "gold of higher purity". Copper depletion with the aid of acid plant juices was practiced by the ancient South Americans [11] and Forty [12] referred to the production of a gold-rich surface from a gold–silver alloy as far back as the twelfth century through the selective corrosion of the less noble species using nitric acid. The electrochemical corrosion behaviour of a series of Au–Cu–Ag–Pd alloys and the metallurgical dependence has been studied by Nakagawa [13] who reported that single-phase alloys showed superior corrosion resistance than the dual-phase alloys. Forty [14] has studied the selective dissolution of active constituents in gold alloys and found that selective dissolution is more likely to occur in the lower carat gold alloys. Laub and Stanford [15] reported that for ternary Au–Ag–Cu alloys, both copper and silver were involved in the corrosion process; corrosion occurring primarily in the silver-rich regions and secondarily in the copper-rich regions. Fujita and co-workers [16] observed for an Ag–Au–Pt–Cu alloy, comprising of a mixture of a major phase of the matrix and minor phase of small grains embedded in the matrix, corrosion occurred preferentially in the matrix and not the precipitates. It was suggested that Ag^+ ions were released from the Ag–Au-rich matrix phase but not from the Pt–Cu-rich second phase.

The purpose of this study was to systematically study the corrosion behaviour of a series of quaternary gold–silver–copper–zinc alloys using dc potentiodynamic scanning in saline (3.5 % NaCl) environment. Phase 1 of the study consisted of conducting wide range anodic and cathodic scans in order to determine the overall characteristics of the alloy. Phase 2 of the study consisted of conducting a series of selective anodic scans and terminating the scans at different anodic potentials, in order to determine any selective dissolution occurring.

Experimental

The samples used in this study were a selection of commercial 9 carat gold quaternary alloy (Au–Ag–Cu–Zn) wires (Golden West) with a thickness of 2 mm. Alloy compositions are shown in Table 1. Samples were annealed after drawing by heating to 350 °C, in order to remove the effects of cold working, followed by quenching in flowing cold water. Sample surfaces were abraded with 1200G SiC paper to remove any residual oxide layers on the surface and to provide a relatively uniform surface finish to the alloy surface. Optical metallography on the gold alloys was performed by cold mounting the gold wires end on, grinding to 1200G finish using SiC paper and polishing to 1 μm surface finish using diamond paste. Samples were then chemically etched in a modified aqua regia electrolyte containing one part concentrated HNO_3 plus two parts concentrated HCl for 20 s to reveal the grain structure.

All electrochemical corrosion tests were carried out using a Voltalab 21 Potentiostat. A conventional three-electrode cell set up consisting of the working electrode, a saturated calomel electrode as the reference electrode and a platinum wire counter electrode was used. All tests were carried out in (3.5 % NaCl at room temperature (20 °C)). Specimens were initially stabilised at the free corroding potential prior to conducting potentiodynamic scans. Anodic/cathodic scans in phase 1 of the study were conducted in the range −1,000 to +1,000 mV at a scan rate of 1 mV/s, in order to determine the overall corrosion characteristics of the alloy. The results are presented as potentiodynamic polarisation curves in the form E vs. log I plots and Tafel Extrapolation was used to determine the corrosion rates (E_{corr} and i_{corr} values) of the system.

Phase 2 of the study consisted of conducting a series of selective anodic scans and terminating the scans at different anodic potentials. The surface morphology and composition of the alloy and corroding media were analysed using scanning electron microscopy (SEM), energy dispersive X-ray spectroscopy (EDX), X-ray photoelectron spectroscopy (XPS) and atomic absorption spectroscopy (AAS) in order to determine the corrosion mechanisms and the occurrence of any selective dissolution.

Results

Potentiodynamic scans for phase 1 of the study

An overlay of the potentiodynamic cathodic/anodic scans for the gold alloy samples #1, #2 and #3 in 3.5 % NaCl, are

Table 1 Composition of gold alloys

Element	Alloy 1	Alloy 2	Alloy 3
Au	37.60	37.70	37.60
Ag	10.57	5.90	6.00
Cu	42.86	46.70	51.00
Zn	9.07	9.94	5.40

shown in Fig. 1. Analysis of the potentiodynamic scans indicate similar cathodic trends being observed on the forward scan from −1,000 mV to the free corroding potential. Here, E_{corr} values were typically in the region of −830 to −860 mV for all alloys considered in this study. E_{corr} results as shown in Table 2 indicate little variation in potential for alloys #1 and #2, giving values of −864.3 and −860.0 mV respectively. However, a slightly more positive value of −831.6 mV was observed for alloy #3.

Analysis of the I_{corr} data from Table 2 shows that similar values, typically 0.4 and 0.8 mA/cm^2 were displayed by alloys #1 and #2, respectively. In contrast the I_{corr} value for alloy #3 was significantly lower by a factor between 6X and 13X lower than the other samples, yielding a value of 0.06 mA/cm^2.

Further potentiodynamic scanning from the free corroding potential to +1,000 mV for all the gold alloys revealed the presence of a discontinuous curve in the anodic region (a series of distinct transitions showing a rise in the current density, followed by a plateau/lowering of the current density), suggesting the presence of a number of competing anodic reactions, possibly associated with the selective dissolution of the alloying elements. This is looked at in more detail in phase 2 of this study.

Further detailed analysis of the anodic portions of the curves reveal the first transition showing a pronounced rise and subsequent lowering of the current density in the E_{corr} to −100 mV range was observed for all the alloy samples. In contrast, while two further transitions were observed in the ranges −100 to +200 mV and +200 to +400 mV respectively for alloys #2 and #3, only one transition was observed for alloy #1 in the voltage range −100 to +400 mV, as shown in Fig. 1. Further, this transition was more pronounced than the two transitions observed for alloys #2 and #3 in the same regions.

Table 2 E_{corr} and I_{corr} values for gold alloys in 3.5 % NaCl	Alloy	Ecorr (mV)	Icorr (mA/cm^2)
	Alloy #1	−864.3	0.40
	Alloy #2	−860.0	0.80
	Alloy #3	−831.6	0.06

Structural and compositional analysis at selected anodic potentials (phase 2)

Scanning electron microscopy analysis of gold alloys

In this phase of the study, the gold alloy was subject to potentiodynamic scans to selected anodic potentials, corresponding to the various positions show in Fig. 2. Anodic voltages chosen for this study were selected at values (1) prior to and after the onset of the transition regions (positions A and D) and (2) at each position in the curve where a plateau/lowering in the anodic current density was observed (positions B and C). Figure 3a to d show typical SEM images of alloy #1 and alloy #3 polarised at positions A and D, respectively.

For alloy #1, the surface morphology of the micrographs indicate the presence of areas where possible dissolution of selected regions of the alloy may have occurred after polarising to position D (Fig. 3b), compared with position A (Fig. 3a). Similarly, for alloy #3, the presence of a region where corrosion, possibly associated with pitting and/or selective dissolution, resulting in the formation of porous regions is evident (Fig. 3d).

Energy dispersive X-ray analysis of gold alloys

Tables 3, 4 and 5 show the relative compositions of the alloys #1, #2 and #3 respectively taken at the various anodic potentials.

The results from Table 3 indicate that as the anodic potential is increased the Cu and Zn concentrations are reduced significantly, particularly after polarising to position B of the anodic curve for alloy #1 and position D for alloys #2 and #3 (Tables 4 and 5, respectively). Further, there is a significant increase in the Au and Ag content corresponding with the reduction in Cu and Zn at the various positions on the curve. The presence of oxygen was observed for all samples after polarising to all positions showing increased levels at locations corresponding with decreased Cu/Zn and increased Ag/Au concentrations. This suggests that metal oxides are formed during the polarisation process. In summary, these results would suggest that selective dissolution of the Cu and Zn has initiated at position B for alloy #1 and position D for alloys #2 and #3, evidenced by the reduced levels of Cu/Zn and increased levels of Ag/Au and oxygen.

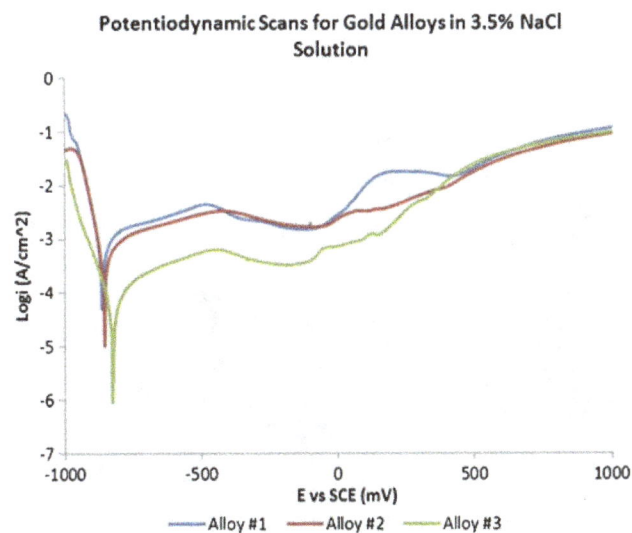

Fig. 1 Potentiodynamic scans for gold alloys in 3.5 % NaCl

Fig. 2 Four anodic potentials
chosen for further analysis

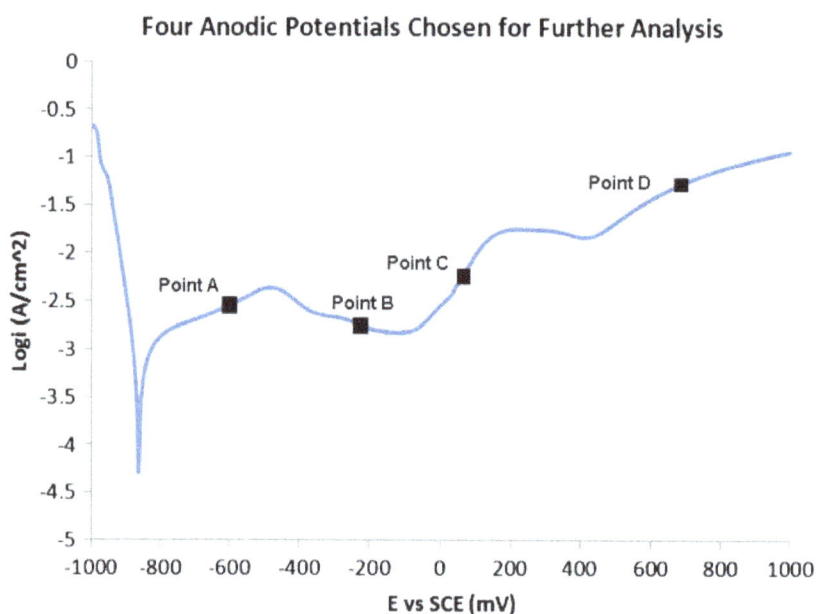

Four Anodic Potentials Chosen for Further Analysis

In addition, Table 6 shows the composition variations for alloy #2 after exposure to the environment for 20 min duration, while polarised to each of the four different anodic potentials for a constant time duration of 20 min. The results show reduced Cu and Zn levels at position D yielding compositional changes similar to those observed in Table 4.

X-ray photoelectron spectroscopy analysis of gold alloys

XPS spectra and subsequent elemental analysis from XPS of the surface of gold alloy #3 at each of the various stages of anodic polarisation are shown in Fig. 4 and Table 7, respectively. The results show that an increase in the Zn

Fig. 3 a SEM image of alloy
#1 polarised to position A. **b**
SEM image of alloy #1
polarised to position D. **c** SEM
image of alloy #3 polarised to
position A. **d** SEM image of
alloy #3 polarised to position D

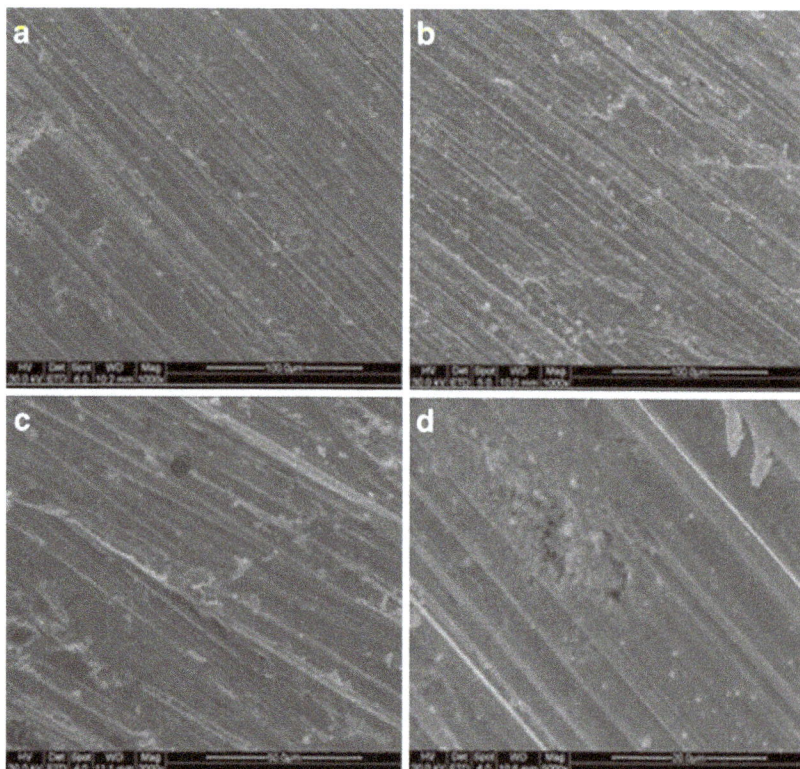

Table 3 Elemental composition (wt.%) of alloy #1 at the four anodic potentials

	Original	Position A	Position B	Position C	Position D
O		1.71	3.46	3.79	5.58
Ag	10.57	15.03	21.4	19.43	22.14
Cu	42.86	39.02	18.67	26.39	24.52
Zn	9.07	8.08	3.44	5.29	4.69
Au	37.60	36.15	53.03	45.1	43.07

Table 5 Elemental composition (wt.%) of alloy #31 at the four anodic potentials

	Original	Position A	Position B	Position C	Position D
O		1.82	3.07	2.88	2.99
Ag	6.00	7.81	7.18	8.75	12.82
Cu	51.00	44.25	33.86	42.7	16.42
Zn	5.40	8.43	7.44	7.83	2.57
Au	37.60	37.69	48.45	37.08	65.21

content at point C corresponding with increased O and Cl content may be attributed to localised diffusion of Zn and the formation of oxides and/or chlorides of Zn at the surface. Further, at point D, the increase in Cu concentration accompanied by further increase in Cl and O content may be associated with (1) localised Cu diffusion and the formation of oxides and/or chlorides of Cu or (2) the dissolution and subsequent re-formation of Cu or Cu chlorides/oxides due to reduction of Cu ions. The reduction of copper ions to metallic copper will be facilitated by the presence of Zn at the surface through a galvanic replacement process. In addition, the Zn concentration decreased at position D, supporting the dissolution of Zn from the alloy at this point.

The Cu spectra of gold alloy #3 at positions A and D of the cyclic polarization curve, as shown in Fig. 5, reveal that the peak binding energy (BE) of Cu2p increases from 932.72 at position A (scan A) to 935.39 at position D (Scan B), which correspond to Cu and $Cu(OH)_2$, respectively. These results support the theory that oxidised copper species are formed as part of the corrosion/dissolution process. The presence of shake up satellite peaks also support that the formation of $Cu(OH)_2$ occurs.

Atomic absorption spectroscopy analysis of gold alloys

The results of the atomic absorption spectroscopy analysis for Cu and Zn ions from all three gold alloys after being anodically polarised to position D are shown in Table 8. These results further confirm the dissolution of Cu and Zn into the solution. Further, the highest concentration of Cu

and Zn cations (78.8 mg/L of Cu^{2+} and 15.6 mg/L of Zn^{2+} present in the final solution after completion of testing) was observed in the gold alloy #1 solution and the least in gold alloy #3, suggesting that increased selective dissolution was observed for alloy #1 compared with alloys #2 and #3

Corrosion behaviour of gold alloys under non-accelerated (free corroding) conditions

Corrosion studies under non-accelerated conditions were also carried out, whereby all three alloys were immersed for 240 h in the environment under free corroding conditions. Analysis of the samples using SEM imaging and EDX analysis revealed dense corrosive products were formed on the surface of alloys #1 and #3 (Fig. 6) in contrast to more severe corrosion/selective dissolution being observed on sample #2 (Fig. 7). Table 9 is a summary of the surface EDX analysis of all three alloys after 240 h immersion confirming some of the results and observations from the accelerated tests as indicated in section "Structural and compositional analysis at selected anodic potentials (phase 2)". The higher levels of oxygen and chloride associated with alloys #1 and #3 compared with alloy #2 would indicate the dense corrosion products on the surface are oxides/chlorides of Zn and Cu, possibly in the form of ZnO, $CuCl_2$ and CuCl. In contrast, dissolution of gold alloy #2 may be responsible for the lower amounts of O and Cl being detected (Fig. 8).

Table 4 Elemental composition (wt.%) of alloy #2 at the four anodic potentials

	Original	Position A	Position B	Position C	Position D
O		1.19	2.36	1.47	2.87
Ag	5.90	12.93	8.76	6.66	8.08
Cu	46.70	14.02	41.55	42.42	16.23
Zn	9.94	2.49	7.96	8.41	3.13
Au	37.70	68.47	39.01	40.76	69.69

Table 6 Elemental composition (wt.%) of alloy #2 after 20 min exposure to the environment at the four anodic potentials

	Original	Position A	Position B	Position C	Position D
O		1.99	3.14	2.78	9.1
Ag	5.90	6.26	8.49	7.88	9.56
Cu	46.70	45.02	43.69	42.19	23.37
Zn	9.94	8.97	7.65	8.35	3.47
Au	37.70	37.75	37.03	38.8	54.5

Fig. 4 XPS spectra for gold alloy #3 at the various anodic polarisation potentials

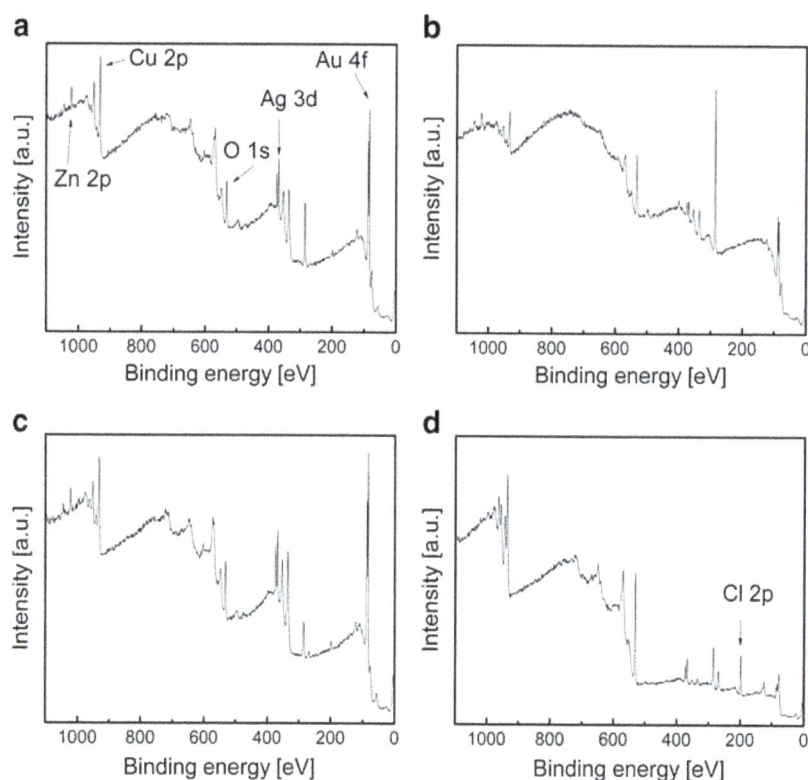

General Discussion on the Selective Dissolution of the Gold Alloys

Influence of gold chemistry and metallurgy on selective dissolution

Another important factor in the dissolution mechanism of gold alloys is the metallurgy of the alloy, specifically if the alloy contains a homogeneous or heterogeneous structure. Selective dissolution of homogeneous binary Au–Ag and Au–Cu alloys has been studied in detail [3]. Mechanisms for attack of the selective dissolution of the more active atoms (alloying species) is influenced by a number of factors to include (1) the activation energy or overpotential to remove these species; (2) presence of these atoms within the lattice,

Table 7 Elemental composition (at%) from XPS of gold alloy #3 at the four anodic polarisation potentials

Element	Position A (at%)	Position B (at.%)	Position C (at.%)	Position D (at%)
Au	6.5	1.5	7.8	0.3
C	63.4	81.1	47.4	36.4
Cl	0	0	5.6	16.2
Ag	2.6	0.5	2.9	0.9
O	19.3	13.2	27.6	33.8
Cu	6.6	2.2	6.6	11
Zn	1.7	1.4	2.3	0.3

for example at kink sites where removal of atoms is easier and (3) migration of lattice vacancies and diffusion of the more noble species (gold) to form clusters/islands. For high gold alloys, passivation is ultimately achieved, however for low gold alloys, micromorphological studies have shown ultimate disintegration of the alloy may occur through the development of tunnels or pits [14].

For heterogeneous alloys, the mechanism of dissolution becomes more complex as the tendency for corrosion for the different phases present may differ depending on the chemistry of the phases. For example the preferential corrosion of Ni-rich phases containing less gold in Ni-based white gold alloys, may occur when exposed to acid. It was further stated [3] that while this form of corrosion may occur due to the presence of heterogeneous phases in ternary, low caratage Au–Ag–Cu alloys, no studies have been reported of this effect in quaternary Au–Ag–Cu–Zn alloys (Fig. 9).

A review of the metallurgy of ternary Au–Ag–Cu alloys has shown that the phases formed can be quite complex [17–19]. For simple Au–Cu alloys, the existence of four phases are possible, namely α (Au, Cu) solid solution, Au_3Cu, $AuCu$ and $AuCu_3$ intermetallics [17]. Given that the percentage of Cu in the alloys in this study lies between 42 and 51 %, α (Au, Cu) and $AuCu$ are the most likely phases to be present and possibly Au_3Cu. For simple Au–Ag alloys, a solid solution α (Au, Ag) is expected [18, 19] and for simple Cu–Ag alloys, a two-phase alloy is expected containing immiscible Cu-rich and Ag-rich (α(Au–Cu) and α(Ag–Ag)) solutions, respectively [18, 19]. For the Au–

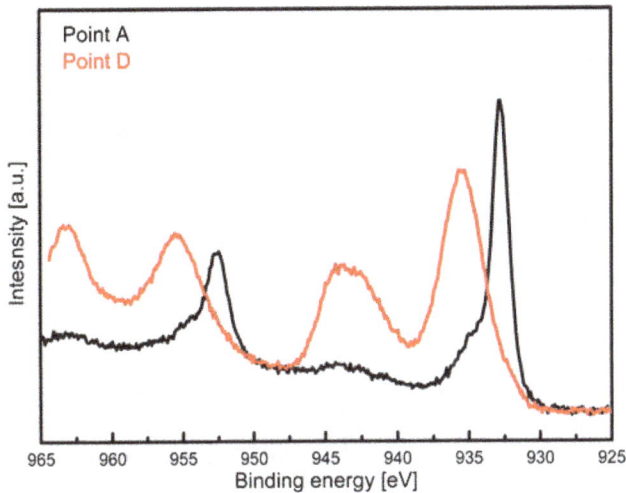

Fig. 5 Cu2p binding energy spectra of Cu on the surface of alloy #2 at position A and D of the polarisation curve

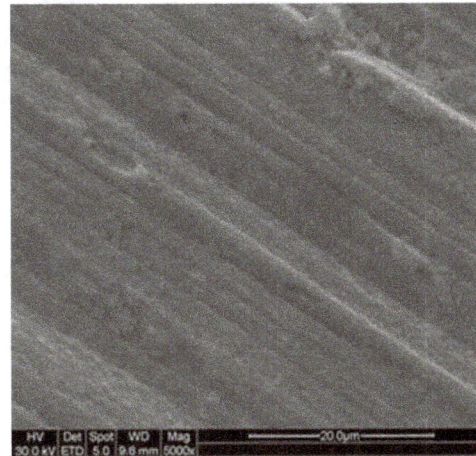

Fig. 6 SEM image of alloy #3 after 240 h in 3.5 NaCl

Ag–Cu ternary alloy, according to McDonald and Sistare [13], the parameter Ag′, defined as the ratio of Ag to Ag + Cu content, is one factor, in addition to carat value, which is important in determining the stable phases in this alloy.

In the current study, the Ag′ values were calculated as 0.20 for alloy #1 and 0.11 for alloys #2 and #3. This suggests that while alloys #2 and #3 may be homogeneous consisting of a solid solution alloy, α, of Cu, Ag and Au, alloy #1 may be heterogeneous, containing an immiscible silver-rich (Au–Ag) phase (and possibly the presence of a Cu-rich α (Cu–Au) phase cannot be neglected, although this is more prominent for alloy systems with Ag′ value 0.25–0.75) [18]. However, the presence of Zn can tend to reduce the volume of solid state immiscibility in the ternary phase diagram, driving the microstructure towards that of a homogeneous solid solution [18, 19]. This has been systematically studied and predicted by Klotz [20] who has used computer modelling techniques, particularly thermodynamic simulation to allow for calculation of phase diagrams for alloy design, heat treatment processes and understanding of segregation particularly the effects of two-phase field contraction.

Optical and scanning electron micrographs of the metallographically prepared (polished and etched) surface of alloy #1 as shown in Figs. 11 and 12 respectively, reveal the presence of a microstructure consisting of grains varying in size ranging from 25 to 100 μm diameter. EDX analysis of the light and dark regions, as shown in Fig. 12, shows no

variation in the concentration of the major alloying elements (Ag, Au, Cu and Zn). This suggests the presence of a homogeneous solid solution, the light and dark regions attributed to etch induced contrast rather than variations in elemental composition. In addition to the major alloying elements, peaks from the EDX analysis, were identified and associated with the presence of Si and ruthenium (Ru) and/or iridium (Ir), confirming the presence and use of these minor alloying additions as grain refining agents, a practice common to commercial production of gold alloys. The findings from the metallographic studies compliment the prediction of phases present from calculations of Ag′ values, in so far as a homogeneous phase structure was expected from all alloys, even from alloy #1 with an Ag′ value approaching that for alloys where binary phase structures may exist.

Identifying the possible distribution of phases and the formation of homogeneous/heterogeneous structures is of importance in determining the corrosion mechanism of these

Table 8 AAS results for Cu and Zn cation concentrations

Ion	Alloy #1	Alloy #2	Alloy #3
Cu^{2+} (mg/L)	78.8	18.6	12.2
Zn^{2+} (mg/L)	15.6	4.4	2.8

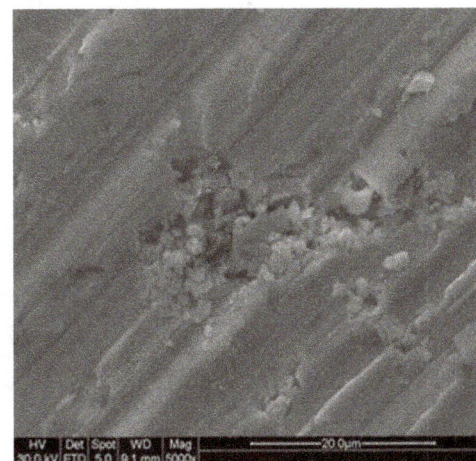

Fig. 7 SEM image of alloy #2 after 240 h in 3.5 NaCl

Table 9 EDX analysis of gold alloys after 240 h immersion in 3.5 % NaCl solution

Element	Alloy #1	Alloy #3	Alloy #4
O	10.72	1.66	12.22
Cl	4.83	1.39	5.04
Ag	7.91	6.32	9.34
Cu	23.32	26.47	21.04
Zn	3.47	5.36	3.69
Au	49.75	58.8	48.66

alloys. While this may not influence the overall corrosion behaviour of the alloys it will certainly have an influence on the mechanism by which these alloys corrode, whether through the removal of active species and diffusion/clustering of more noble species, as for alloys with a homogeneous structure, or if preferential corrosion occurs of entire phases enriched in active constituents.

Effect of composition on the corrosion behaviour of gold alloys

Corrosion of gold jewellery normally involves selective attack of the less noble alloy constituents and it is well established [3] that while the addition of silver and other base metals reduces the corrosion resistance of gold, this is not significant as long as the gold content of the alloy remains greater than 50 at% (corresponding to 15.6 carat Au–Ag alloys and 18 carat for Au–Cu alloys). However, it should be noted that this "50 at% rule of thumb" is very general and simplistic as Corti [21] reported that tarnishing of 22 carat gold can occur under relatively severe corrosion conditions in general service, as evidenced by the presence of a blackened surface on the alloy.

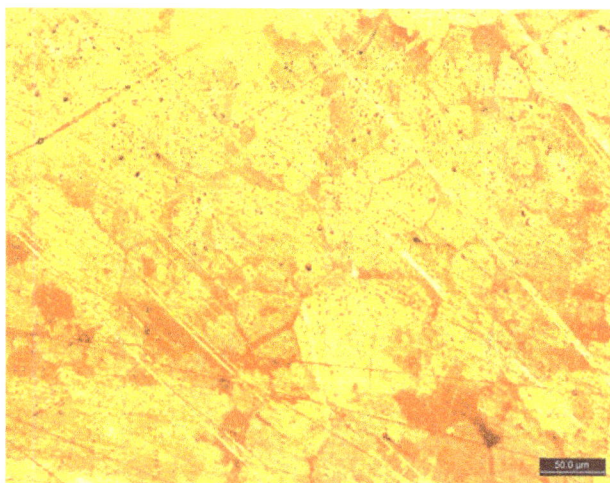

Fig. 8 Optical micrograph of alloy #1 etched in aqua regia (X340 mag)

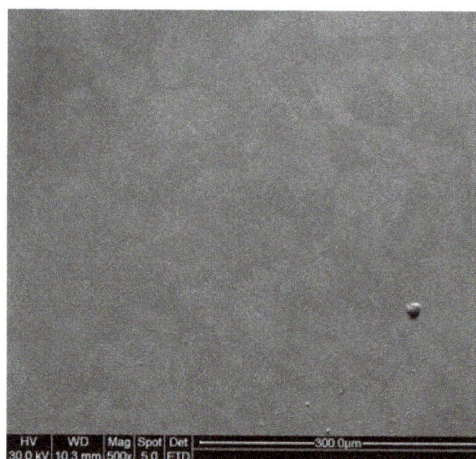

Fig. 9 SEM image of alloy #1 etched in aqua regia

The reported composition of Au in all three low carat alloys in the current study is around the 37 at%, thus making them susceptible to corrosion, particularly selective dissolution. It would appear that variations in alloying additions into the alloys not only affect the overall anodic region of the curve but also the resultant E_{corr} and I_{corr} values, as shown in Fig. 1 and Table 2.

Firstly reduced Cu (and increased Ag) content may be responsible for (1) increasing the intensity of the second transition with respect to a pronounced increase in the current density range at the commencement of this transition at −100 mV for alloy #1 compared with alloys #2 and #3 and (2) suppressing the third transition in the anodic region between +200 to +400 mV observed in alloys #2 and #3 but not alloy #1. This would suggest that reduced Cu content is responsible for intensifying the dissolution of the alloying species occurring at around −100 mV (alloy #1) and further, increased Cu content promotes the dissolution of species occurring at +200 mV.

Secondly, a reduction in the Zn content from approximately 9–10 % to 5.4 % may have been responsible for the observed more positive E_{corr} value and lower I_{corr} value observed for alloy #3 compared with the other alloys. This would suggest that the Zn content in the alloy is an important factor regarding influencing the overall corrosion properties of the alloy. Further, correlations exist between the observed I_{corr} values and results from the free corroding tests conducted. Here, the lower I_{corr} values observed for alloys #1 and #3 can be attributed to the presence of dense corrosion products on the surface as evidenced in Fig. 6 and the higher levels of O and Cl from EDX analysis compared with alloy #2 showing lower levels of O and Cl and evidence of selective dissolution (Fig. 7). This may be associated with the higher Zn content as compared with alloy #3, combined with higher Cu content as compared with alloy #1, making this alloy the most susceptible to attack and subsequent dissolution.

Dissolution mechanisms for gold alloys

The electrochemical potentials of Au, Ag, Cu and Zn are reported with values of +1,498, +799, +337 and −763 mV vs normal hydrogen electrode [22]. While these values represent ideal conditions for metals under reversible conditions in a solution of its own ions at unit activity, combined with the galvanic series for metals and alloys in salt water, they give an indication of the relative differences in the degree of activeness/inertness of each of the alloying elements [22]. In summary, it is anticipated that dissolution of Zn species would occur first, followed by Cu, then Ag and finally Au.

For sample #1, at position A (−600 mV vs SCE), approximately 250 mV more positive than the E_{corr} value, the Cu and Zn levels from EDX analysis suggest no leaching out of these species, although the formation of some oxides is evidenced. At position B (−200 mV vs SCE), the selective dissolution of both Cu and Zn, resulting in regions depleted of these species and enriched in Au and Ag, suggests that both these corrosion reactions have occurred within the 400 mV band between test positions A and B. While Zn is expected to leach out initially, the results cannot confirm this theory. While the EDX results suggest no further dissolution of species is observed upon further testing at positions C and D, the anodic potentiodynamic scans suggest further dissolution of species at point C and D. This may indicate either (1) stabilisation of the alloy with respect to alloy composition or (2) leaching out/oxidation of all species (Au and Ag in addition to Cu and Zn) thus maintaining constant alloy composition. Increase in oxide composition may suggest further oxides are formed indicating activity at these potentials.

In contrast, for alloys #2 and #3, the EDX results indicate suppression of the dissolution of Zn and Cu until more anodic potentials are achieved (between +100 mV vs SCE at point C and +600 mV vs SCE at point D). However, the anodic polarisation scans suggest that activity, possibly associated with dissolution and/or oxide formation is taking place though not to the extent as for alloy #1.

It appears that the overall corrosion characteristics of the alloys are determined by the extent of anodic reactions which induce (1) formation of oxides on the alloy surface; (2) dissolution of Zn and Cu species and (3) relative combinations of the first two. Dissolution of Zn and Cu ions appears to dominate in alloy #1, initiating in the early phase of the anodic region, as evidenced by increased concentrations of Zn and Cu from the AAS analysis of the corrosion solutions and low O and Cl concentrations from EDX analysis under free corroding conditions. This may be attributed to the possible formation of heterogeneous phases, with dissolution of Cu-rich phases and Zn regions in alloy #1. However, increased Zn levels for this alloy may also be a

contributing factor. In contrast, the formation of a homogeneous phase, likely to be evident in alloys #2 and #3, may delay dissolution of Zn and Cu species until higher anodic potentials are achieved.

The high corrosion rates observed for alloy #2 can be attributed to the lack of protective scales forming, as evidenced by the low O and Cl concentrations from the free corrosion studies. In contrast the formation of protective oxide scales (higher O and Cl levels) may account for the lower corrosion rates observed for alloy #3, which may also be attributed to higher Cu concentrations in the alloy. Further analysis from XPS suggests that at point C (approximately −200 to +100 mV vs SCE.), chlorides start to form and oxide concentrations increase suggesting the formation of stable copper oxides/chlorides and or zinc oxides/chlorides. At higher anodic potentials (+100 mV to +600 mV vs SCE) reduced Zn content could be attributed to further dissolution of Zn and/or breakdown of any stable Zn oxide layers formed. Likewise the increased Cu content could be attributed to dissolution/replating back of copper and subsequently forming copper oxides. A uniform distribution of copper and zinc oxide on the surface may be responsible for masking the Au and Ag present and thus explaining the low levels of Ag and Au detected at point D.

Correlation of the corrosion results with previous corrosion, SCC and tarnish resistance studies

The general findings from this study are that alloying additions influence the corrosion behaviour of the alloy by inducing selective dissolution of the active species, reduced corrosion resistance being observed with increased Zn levels. Laub and Stanford [15] observed a correlation between the composition of five low-carat gold dental alloys and the resultant corrosion and tarnishing results. As the gold content increased from 40 to 59.5 %, a reduction in the electrochemical corrosion rates was observed, in addition to no tarnishing or corrosion from clinical trials being observed for higher gold content alloys. While, the increased levels of gold are primarily responsible for improved corrosion and tarnish resistance, corrosion was attributed to the presence of the more active alloying additions of Cu and Ag, producing Ag-rich and Cu-rich regions which are more susceptible to corrosion. Rapson [3] reported desirable microstructural features for improved resistance to tarnish and corrosion in low-carat gold alloys are due to the presence of large grains, a minimum of phase separation and absence of precipitated and ordered phases, although it was uncertain if these conditions can apply to low-carat gold alloys containing significant percentages of Zn.

Merriman and co-workers [23] reported the time to fracture during SCC tests was much higher for the 18 carat gold alloys (several hundred hours) compared with those for lower-carat gold alloys (several minutes). In addition, the corrosion rates of the low-carat gold alloys were attributed to selective dissolution of the alloying additions, whereby dissolution of Zn occurred initially, followed by dissolution of Cu, Ni and Pd, leaving a gold-enriched layer. Hence alloys containing Zn experienced the highest corrosion rates, due to the lower oxidation potential of Zn compared with other alloying additions. Rapson [3] stated that the susceptibility of 14 carat quaternary Au–Ag–Cu–Zn alloys to SCC was influenced considerably by both the Zn content and heat treatment. Neumeyer and co-workers [10] compared the SCC behaviour of a series of 9 carat gold alloys with varying compositions of Ag, Cu and Zn. Their studies revealed that SCC behaviour appeared to be related to differences in Zn content rather than Cu content, for alloys containing the same Au and Ag content. Increased resistance to SCC was observed for gold alloys with lower Zn content. They attributed this to the low Zn content alloy promoting the formation of a heterogeneous, dual-phase structure, being more resistant to SCC, whereas at higher Zn content, the formation of a homogeneous, single-phase was promoted, thus rendering the alloy more susceptible to SCC.

In the current study the Au content was kept constant while Ag, Cu and Zn contents varied. Hence, any changes to the corrosion behaviour are predominantly dependent on the alloying additions themselves. It is clear that previous work has shown Zn content in the gold alloys to be detrimental to the corrosion rates and SCC resistance. The findings from this study correlate well with previous work, the worst corrosion resistance being observed for alloy #2 with higher Zn/lower Cu content, compared with alloy #3. It is possible that the best corrosion behaviour observed from alloy #3, being attributed to favourable Zn/Cu ratio (low Zn content and increased Cu content), may also result in these alloys showing improved SCC resistance and possibly improved tarnish resistance compared with the other alloys, although further studies would be required to confirm this theory.

Conclusions

1. Potentiodynamic scans conducted on all gold alloys in 3.5 % NaCl solution indicate the presence of similar discontinuous curves in the anodic region (a series of distinct transitions showing a rise in the current density, followed by a plateau), most likely due to selective dissolution of the alloying elements.
2. Phase 1 of the study revealed alloy #3 showed slightly more positive corrosion voltage (E_{corr}) value and lower corrosion current (I_{corr}) value (between 6X to 13X lower) than alloys #1 and #2. This may be associated with reduced Zn and increased Cu levels within the alloy

promoting the formation of dense, protective coating products on the surface.
3. Higher corrosion rates as observed for alloy #2 and to a lesser extent, alloy #1, may be associated with increased Zn and reduced Cu levels in the alloy promoting susceptibility to attack and subsequent selective dissolution.
4. Phase 2 studies revealed selective dissolution of Cu and Zn species from alloy #1 occurred early on in the anodic region of the polarisation curve. This was supported by the increased Zn and Cu levels found in solution and reduced O and Cl levels observed on the alloy surface. This may be attributed to the possible formation of a heterogeneous phase structure.
5. Dissolution of Zn and Cu species is suppressed, leaching at more positive anodic potentials, for alloys #2 and #3 compared with alloy #1. This may be attributed to the possible existence of a homogeneous phase structure that delays dissolution.
6. The corrosion behaviour of the alloys is determined by the extent of anodic reactions which induce (1) formation of oxide scales on the alloy surface and or (2) dissolution of Zn and Cu species. In general, the improved corrosion characteristics of alloy #3 can be attributed to the favourable composition of Zn/Cu in the alloy and thus favourable microstructure promoting the formation of protective oxide/chloride scales and reducing the extent of Cu and Zn dissolution.

Acknowledgement The authors would like to acknowledge Prof. Suresh Bhargava for provision of gold alloy samples. The authors further acknowledge the facilities and technical assistance of the Australian Microscopy and Microanalysis Research Facility at the RMIT Microscopy and Microanalysis Facility

References

1. Ott D (2002) Optimising gold alloys for the manufacturing process—advantages and disadvantages of small alloying additions. Gold Technol 34:37–44
2. Randin J-P (1988) Electrochemical assessment of the tarnish resistance of decorative gold alloys. Surf Coat Technol 34:253–275
3. Rapson WS (1996) Tarnish resistance, corrosion and stress corrosion cracking of gold alloys. Gold Bull 29:61–69
4. Alvarez MG, Fernández SA, Galvele JR (2000) Stress corrosion cracking in single crystals of Ag–Au alloy. Corros Sci 42:739–752
5. Chen JS, Salmeron M, Devine TM (1993) Intergranular and transgranular stress corrosion cracking of Cu-30Au. Corros Sci 34:2071–2097

6. Farina SB, Duffo GS, Galvele JR (2007) Effect of cations on stress corrosion cracking: Ag-40Cd alloy in silver ion aqueous solutions containing a variety of foreign cations. Corros Sci 49:1687–1695

7. Alvarez MG, Fernandez SA, Galvele JR (2002) Effect of temperature on transgranular and intergranular stress corrosion crack velocity of Ag–Au alloys. Corros Sci 44:2831–2840

8. Duffoó GS, Farina SB, Galvele JR (2004) Stress corrosion cracking of 18 carat gold. Corros Sci 46:1–4

9. Dugmore JMM, DesForges CD (1979) Stress corrosion in gold alloys. Gold Bull 4:140–144

10. Neumayer B, Hensler J, O'Mullane AP, Bhargava SK (2009) A facile chemical screening method for the detection of stress corrosion cracking in 9 carat gold alloys. Gold Bull 42:209–214

11. Bray W (1978) Gold working in ancient America. Gold Bull 4:136–143

12. Forty AJ (1979) Corrosion micromorphology of noble metal alloys and depletion gilding. Nature 282:597–598

13. Nakagawa M, Matsuya S, Ohta M (1992) Effect of microstructure on the corrosion behaviour of dental gold alloys. J Mater Sci Mater Med 3:114–118

14. Forty AJ (1981) Micromorphological studies of the corrosion of gold alloys. Gold Bull 14:25–35

15. Laub LW, Stanford JW (1981) Tarnish and corrosion behaviour of dental gold alloys. Gold Bull 14:13–18

16. Fujita T, Shiraishi T, Takuma Y, Hisatsune K (2011) Corrosion resistance evaluation of Pd-free Ag–Au–Pt–Cu dental alloys. Dent Mater J 30:136–142

17. Okamoto H, Chakrabarti DJ, Laughlin DE, Massalski TB (1987) The Au–Cu (gold–copper) system. Bull Alloy Phase Diagr 8:454–473

18. McDonald AS, Sistare GH (1978) The metallurgy of some carat gold jewellery alloys part I—coloured gold alloys. Gold Bull 11:66–73

19. Rapson WS (1990) The metallurgy of the coloured carat gold alloys. Gold Bull 23:125–133

20. Klotz UE (2012) Computer simulation in jewellery technology—meaningful use and limitations. Proc Santa Fe Symp 2012:297–319

21. Corti CW (2000) High carat golds do not tarnish? Proc Santa Fe Symp 2000:29–55

22. Fontana MG (1986) Corrosion engineering, 3rd edn. McGraw-Hill pp 42–43

23. Merriman CC, Bahr DF, Grant Norton M (2005) Environmentally induced failure of gold jewellery alloys. Gold Bull 38:113–119

7

Facile synthesis of AgAu alloy and core/shell nanocrystals by using Ag nanocrystals as seeds

Weiwei Xu · Jinzhong Niu · Hangying Shang · Huaibin Shen · Lan Ma · Lin Song Li

Abstract A facile seed-growth method was developed to synthesize AgAu alloy and core/shell nanocrystals (NCs) using different-sized Ag NCs (6.1, 7.4, and 9.6 nm) as seeds and octadecylamine as the reducing agent, surface ligand, and solvent. Pre-synthesized Ag NCs acted as catalysts for the reduction of Au precursors at 130 °C. Transmission electron microscopy, energy-dispersive spectroscopy, and optical absorption spectroscopy were used to characterize as-synthesized NCs. Spherical AgAu alloy NCs were obtained when pre-synthesized 6.1 and 7.4 nm Ag NCs were used as seeds. While, if 9.6 nm Ag NCs were used as seeds, cubic Ag/Au core/shell NCs were finally obtained. The shapes and structures of AgAu NCs are related to the Ag seed sizes and the growth mechanism of alloy and core/shell NCs is discussed in detail. Different reaction temperatures were tested to optimize the synthesis of AgAu alloy NCs, and it was found that the optimum reaction temperature for the growth of AgAu alloy NCs is 130 °C.

Keyword Metallic composites · Alloys · Core/shell · Nanocrystals · Seeds size

Introduction

Silver and gold nanocrystals (NCs) have attracted great attention due to their special optical [1], electronic [2], and catalytic [3] properties. In the past two decades, a myriad of chemical methods for generating faceted Ag or Au NCs with a rich variety of shapes, including sphere [4, 5], cube [6], decahedron [7, 8], icosahedrons [8], rods [9, 10], and wires [11], have been developed to tailor their properties and improve their performance in various applications. In addition, people have also developed methods for combining Au and Ag into one single system to tune the optical properties [12–17]. For example, Sun's group reported one-pot synthesis of monodisperse AuAg alloy NCs by the simultaneous reduction of gold and silver salts [13], and AuAg alloy NCs were prepared through interface diffusion of core/shell structured Ag/Au NCs [14]. Tracy and co-workers described a facile controlled synthesis of Au/Ag core/shell and AuAg alloy NCs through digestive ripening, which is a potentially general method for synthesizing alloy NCs [15]. Yang's group reported a general route to control the diameter of noble NCs (Ag, Au, and Pd), the shell thickness of the core/shell NCs (Ag@Pd, Pd@Au, Pd@Ag, Au@Pd, Au@Ag, Pt@Au, and Pt@Pd), and the composition of the alloy (Ag and Au) on the basis of the combining effect of seeding growth and digestive ripening [16]. However, there is still no report about the effect of the seed size on the growth of different shaped NCs.

In this paper, we report the seed size-dependent growth of different-shaped AgAu alloy and core/shell NCs by an organic solvothermal method. The key strategy to produce such different classes of NCs is to precisely control the sizes

W. Xu · H. Shang · H. Shen (✉) · L. S. Li
Key Laboratory for Special Functional Materials of Ministry of Education, Henan University, Kaifeng 475004, People's Republic of China
e-mail: shenhuaibin@henu.edu.cn

W. Xu · L. Ma (✉)
Life Science Division, Graduate School at Shenzhen, Tsinghua University, Shenzhen 518055, People's Republic of China
e-mail: malan@sz.tsinghua.edu.cn

J. Niu
Department of Mathematical and Physical Sciences, Henan Institute of Engineering, Zhengzhou 451191, People's Republic of China

of the Ag NCs which are synthesized by thermal reduction of $AgNO_3$ in octadecylamine (ODA). Pre-synthesized Ag NCs acted as catalysts for the reduction of Au precursors. When 6.1 and 7.4 nm Ag NCs were used as seeds, spherical AgAu alloy NCs were obtained; while, if 9.6 nm Ag NCs were used as seeds, cubic Ag/Au core/shell NCs were finally obtained.

Experimental section

Chemicals

Hydrogen tetrachloroaurate (III) ($HAuCl_4 \cdot 3H_2O$, 99.99 %), silver nitrate ($AgNO_3$, 99.8 %), and triphenylphosphine (PPh_3, 99 %) were purchased from Aldrich. Chloroform (analytical grade), ODA (95 %), hexanes (analytical grade), and methanol (analytical grade) were obtained from Beijing Chemical Reagent Ltd., China. All reagents were used as received without further experimental purification.

Au stock solution A 0.0309 g $AuPPh_3Cl$ (0.0625 mmol, which was synthesized by reacting $HAuCl_4 \cdot 3H_2O$ with PPh_3 in ethanol) was dissolved in 0.4 mL chloroform to form the Au stock solution.

Synthesis of AgAu NCs Ag NCs (6.1 and 7.4 nm) were made by adding $AgNO_3$ (0.25 g, 1.47 mmol) to ODA (10 mL, 8.6 g) at 180 °C and magnetically stirred for 20 and 60 min, respectively [18]. Ag NCs (9.6 and 11 nm) were made by dissolving $AgNO_3$ (0.02 g, 0.118 mmol) to ODA (10 mL, 8.6 g) to form a solution, which was slowly heated to 180 °C for 10 min and 1 h, respectively [13]. After cooling to 60 °C, 10 mL methanol and 10 mL chloroform were added into the solution and the NCs were separated by centrifugation (8,000 rpm, 5 min) and then dried at 45 °C. After that, 5 mL ODA was dispersed in a 25 mL flask and degassed at 130 °C for 10 min, and then 10 mg ($\sim 0.14 \times 10^{-4}$ mmol) Ag NCs (different sizes) which was dissolved in 0.2 mL chloroform was added. Au stock solution (0.1 mL, 0.0156 mmol) was added into the flask and maintained the reaction temperature at 130 °C for 20 min. During this process, the color of the reaction solution changed from yellow to purple-red. After that, Au stock solution was added again and maintained for another 20 min and the absorption peak moved to 508 nm or longer. During this process, the color of the reaction solution has no obvious change. Finally, chloroform and methanol were added to the solution to remove byproducts by centrifugation (8000 rpm, 5 min). UV–vis absorption spectra were recorded using an Ocean Optics spectrophotometer (mode PC2000-ISA). All optical measurements were performed at room temperature under ambient conditions. Transmission electron microscopy (TEM) images were obtained with a JEOL 2010 F microscope operated at accelerating voltage of 200 kV. For each NC sample, the diameter and standard deviation were determined by averaging measurements of 100 NCs.

Results and discussion

In the control experiments, large Au NCs with diameter of 80 nm were obtained after the injection of Au precursors ($AuPPh_3Cl$) to ODA solution at 180 °C (see Electronic supplementary material (ESM) Fig. S1). If the Au precursor was injected into ODA at 150 °C, the color of the reaction solution did not change after the injection, which implies that the reduction of Au precursors did not occur at this temperature. If the pre-synthesized Ag NCs were used as seeds, Au was deposited by reduction of $AuPPh_3Cl$ with ODA at 130 °C. At this reaction temperature, the reduction of Au precursors by ODA in the organic phase was not feasible in the absence of the Ag seeds as mentioned above, indicating that Ag NCs acted as catalysts for the reduction of Au precursors. A similar phenomenon was also reported in previous report [19].

As shown in Fig. 1a–c, the AgAu NCs have typical size distributions of approximately ±50 % (Fig. 1b) and ±10 % (Fig. 1c) after the first and second round addition of Au precursors when 6.1 nm Ag NCs were used as seed (Fig. 1a). During the reaction process, the color of the reaction solution changed gradually from yellow to purple-red, indicating the composition of AgAu alloy NCs changed from Ag rich to Au rich. The Au/Ag ratio was changed from 0.52/0.48 to 0.76/0.24, which was measured by energy-dispersive X-ray spectroscopy (EDS) attached to a scanning electron microscope (see ESM Fig. S3). When 7.4 nm Ag NCs were used as seed (Fig. 1d), the growth process of AgAu NCs with the addition of Au precursors was similar to that on 6.1 nm Ag seed. The final average size of AuAg NCs was 16 nm (Fig. 1f), which was bigger than that shown in Fig 1c with an average size of 13 nm. If the average size of Ag seed was increased to 9.6 nm, faceted AgAu NCs (Fig. 1h) were formed after the injection of Au precursors for the first time, and then most of the faceted NCs turned into cubic NCs (Fig. 1i) with the continuous addition of Au precursors for the second time during the growth process. The molar ratio of Ag/Au changed from 63:37 (Fig. 1h) to 43:57 (Fig. 1i) with the addition of Au precursors. After that, the shape of AgAu NCs became uncontrollable if we continued to add excess Au precursors (see ESM Fig. S4). When 11 nm Ag NCs (ESM Fig. S5a) were used as seed, the color of the reaction solution changed gradually from yellow to purple-red to black and formed a precipitate after adding the Au precursors. The AgAu NCs were collected from the bottom of the beaker (see ESM Fig. S5).

Fig. 1 TEM images of as-prepared NCs. (**a**) 6.1 nm Ag NCs; (**b**) and (**c**) AgAu alloy NCs synthesized in use of 6.1 nm Ag NCs; (**d**) 7.4 nm Ag NCs; (**e**) and (**f**) AgAu alloy NCs synthesized in use of 7.4 nm Ag NCs; (**g**) 9.6 nm Ag seed; (**h**) and (**i**) Ag/Au core/shell NCs synthesized in use of 9.6 nm Ag NCs. The compositions of each sample measured by EDS are indicated

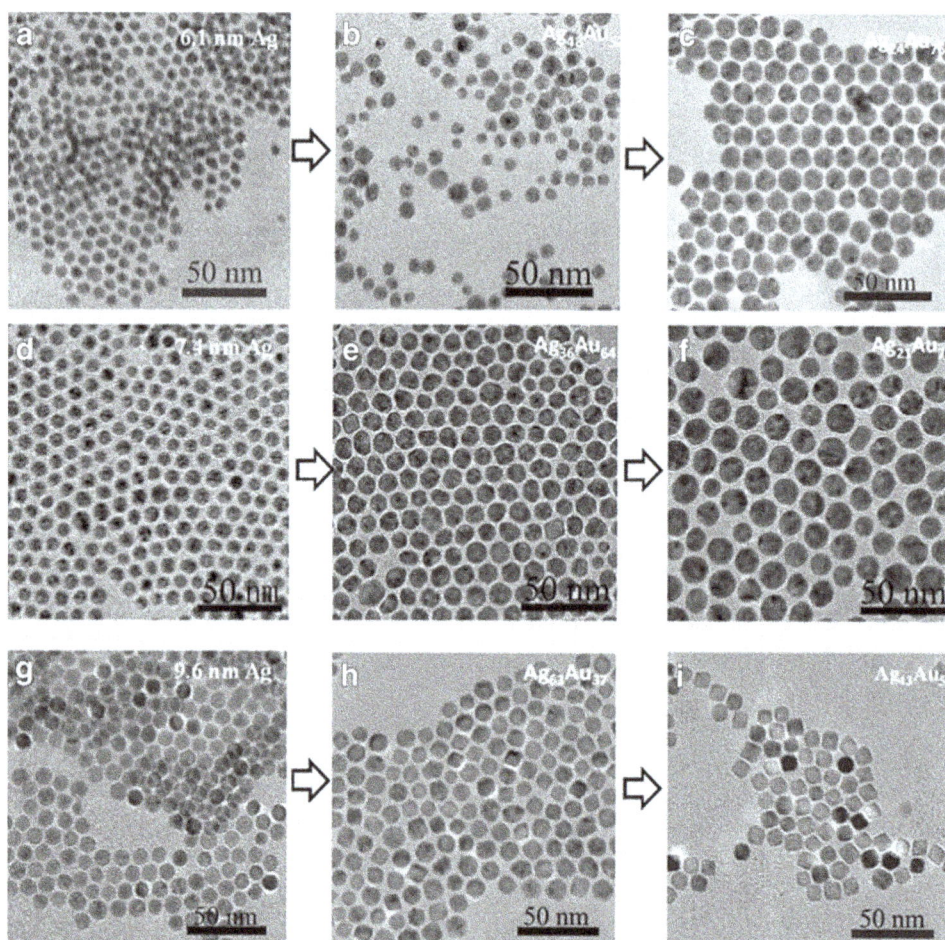

The formation of the AgAu alloy and core/shell structures can be demonstrated by the absorption peaks in the UV–Vis spectra. Figure 2a and b show the absorption peaks of 6.1 nm and 7.4 nm sized spherical Ag NCs, which are both located at 405 nm. After the addition of Au precursors, the absorption peaks red shifted to 494 nm, and then to 508 nm, respectively.

The red shift of absorption peaks indicates that the composition of spherical AgAu NCs changed from Ag rich to Au rich, which is in accordance with the EDS results of AgAu NCs, and confirms the AgAu NCs are the alloy structure. When metal NCs are enlarged, their optical properties change only slightly as observed. However, when anisotropy is added to the

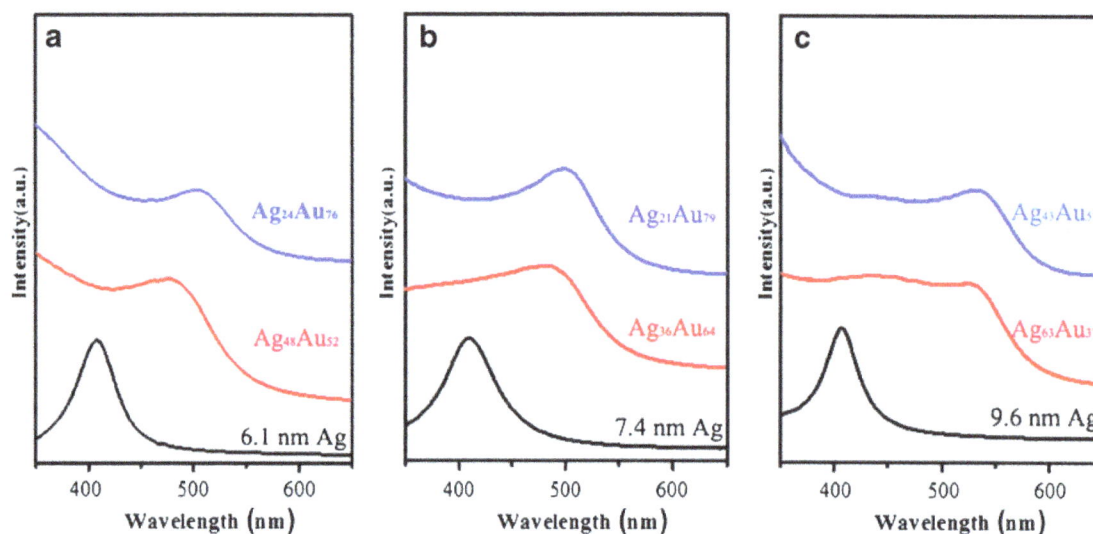

Fig. 2 UV–vis absorption spectra of as-prepared AgAu NCs by using (**a**) 6.1 nm, (**b**) 7.4 nm, and (**c**) 9.6 nm Ag NCs

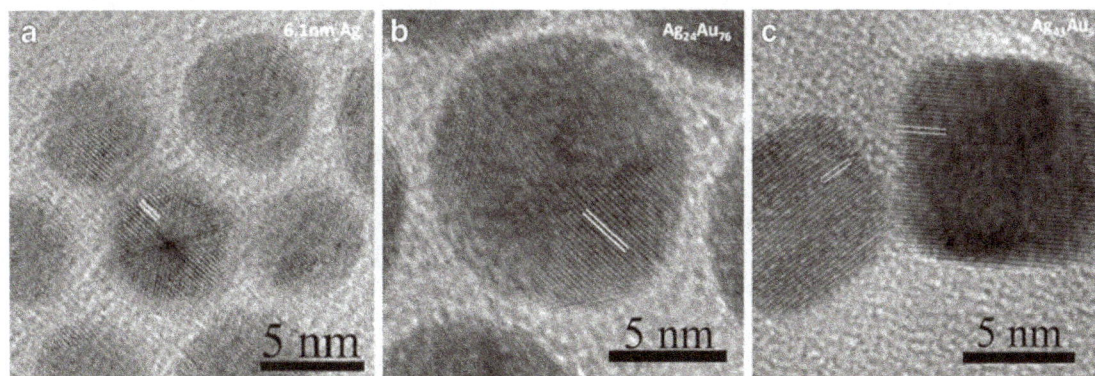

Fig. 3 High resolution TEM images of as-prepared (**a**) 6.1 nm Ag NCs, (**b**) $Ag_{24}Au_{76}$ alloy NCs, and (**c**) Ag_{43}/Au_{57} core/shell NCs

NCs, the optical properties of the NCs change dramatically. For example, spherical Ag/Au core/shell nanocrystals (shell thickness around 2 nm and size around 15 nm) have an absorption peak at 524 nm [14], and hollow Au octahedra NCs (shell thickness around 3 nm and size around 15 nm), which were synthesized by a replacement reaction performed on spherical Ag NCs, have absorption peaks at 430 and 720 nm [20]. Cubic Au/Ag core/shell NCs (thickness around 1 nm and size around 15 nm) have absorption peak at 410 and 510 nm [17]. In our result, Fig. 2c shows the absorption peak of 9.6 nm Ag NCs located at 408 nm. After the first round addition of Au precursors, the absorption peak red shifted and broadened. There are two absorption peaks (Fig. 1h): the one at 433 nm is the absorption peak for AgAu alloy NCs, and the other one is at 529 nm which is close to the absorption peak for Au NCs (520 nm) at this size. With the second round addition of Au precursors (Fig. 1i), the absorption signal at 433 nm became weaker and the absorption signal at 529 nm became stronger than before.

Due to the two absorption peaks, we think that the core within the Au shells is AgAu alloy rather than pure Ag cores in the cubic structure, and this is confirmed by a simple calculation of the atomic ratio for Au to Ag. We assume that the cubic AgAu core/shell NCs (Fig. 1i) are formed by a 9.6 nm Ag core and Au shells which surrounding the Ag core. Based on this model, the Au/Ag atomic ratio should be 0.911. This value is much smaller than the result of 1.325 by EDS measurement (see ESM Fig. S3f). It indicates that Au is actually richer than in the simple model with spherical Ag core/Au shell structure. So the initial pure spherical Ag cores must have been partly replaced by Au atoms to form the AgAu alloy cores, which is in accordance with the absorption peak of AgAu alloy NCs at 433 nm (Fig. 2c). Similar calculation of the atomic ratio for Ag/Au core/shell structure has been reported in previous literature [14].

We speculate the growth mechanism of different structured NCs is as follows. When 6.1 nm Ag NCs were used as seed, the HRTEM images show that the finally formed AgAu NCs (13 nm) have high crystallinity with continuous lattice fringes (*d* spacing of 0.203 nm) throughout the whole particles, and the shape does not change between Fig. 3a and b. During this process, the reduction of Au precursors, the atom diffusion and NCs growth happened at the same time. When Au precursor is catalyzed by Ag seed, the motion of Ag

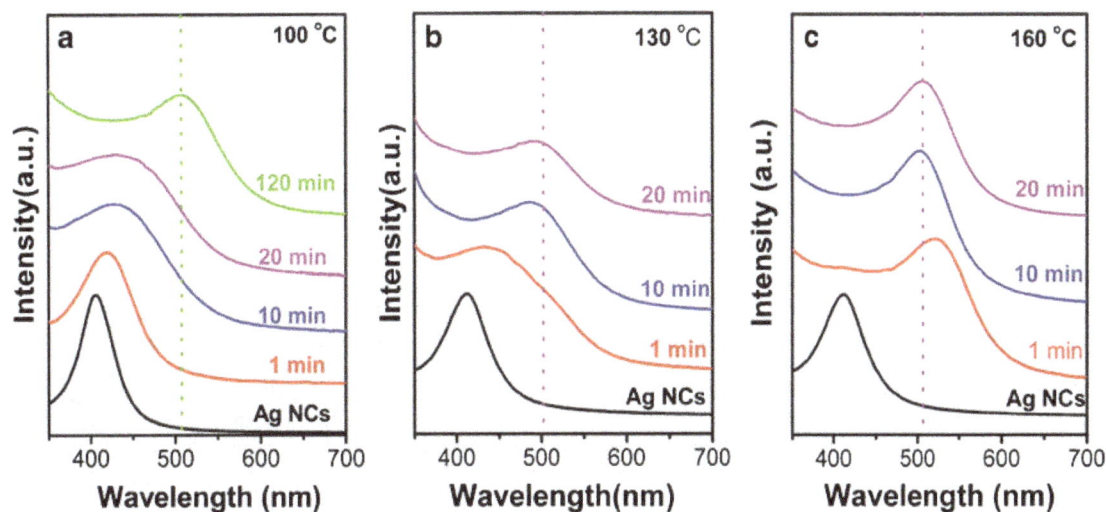

Fig. 4 UV–vis absorption spectra of spherical AgAu alloy NCs prepared at (**a**) 100 °C, (**b**) 130 °C, and (**c**) 160 °C

atoms facilitates the diffusion of Au atoms into the Ag seed in the growth process. A fast diffusion competes with the Au deposition process which leads to the growth of alloy AgAu NCs (as shown in Fig. 1b), and then AgAu NCs reached 13 nm with the continuous increase of Au precursors. If 9.6 nm Ag NCs were used as seed, the spherical Ag NCs were turned to cubic Ag/Au core/shell NCs (Figs. 1i and 3c). During this process, the diffusion rate of Au atoms into Ag seeds is slower than the Au deposition process because of the decrease of surface energy with the increase of Ag size, and thus a layer of Au shell was formed on the AgAu alloy cores.

Meanwhile, different reaction temperatures were tested to investigate their influence on the formation of AgAu alloy NCs (Fig. 4). When the injection temperature was set at 100 °C, the color of the reaction solution changed slowly and the absorption peak red shifted to 501 nm after the reaction proceeded 2 h (Fig. 4a). If the injection temperature was set at 130 °C, the color of the reaction solution quickly changed and the absorption peak red shifted with the increase of reaction time, and was steady at 501 nm after the reaction proceeded 20 min (Fig. 4b). When the injection temperature was set at 160 °C, the AgAu NCs had broad size distributions after the addition of Au precursors at 1 min (Fig. S7a) and the absorbance band was located at 522 nm (Fig. 4c). The reason for this phenomenon is that reduction of the Au precursor occurred rapidly, and formed Ag_{core}/Au_{shell} structure. After reaction for 10 min, the absorption peak blue shifted slightly to 509 nm, indicating the end of the diffusion process. The size distribution of AgAu alloy NCs synthesized at 130 °C is better than the results at 160 °C (Fig. 1c). So, 130 °C is the proper reaction temperature to synthesize AuAg alloy NCs.

Conclusion

In conclusion, a facile seed-growth method was developed to synthesize AgAu alloy and core/shell NCs by using ODA as the reducing agent, surface ligand, and solvent. Pre-synthesized Ag NCs acted as catalysts for the reduction of Au precursors. Through controlling the size of the Ag seed, size and shape control for AgAu NCs with alloy or core/shell structures was obtained. Different reaction temperatures were tested to optimize the synthesis of AgAu alloy NCs, and it was found that a proper reaction temperature for the growth of AgAu alloy NCs is 130 °C.

Acknowledgments This work was financially supported by the research project of the National Natural Science Foundation of China (21071041) and Program for New Century Excellent Talents in University of Chinese Ministry of Education.

References

1. Sanders AW, Routenberg DA, Wiley BJ, Xia Y, Dufresne ER, Reed MA (2006) Observation of plasmon propagation, redirection, and fan-out in silver nanowires. Nano Lett 6:1822–1826
2. Bishop PT, Ashfield LJ, Berzins A, Boardman A, Buche V, Cookson J, Gordon RJ, Salcianu C, Sutton PA (2010) Printed gold for electronic applications. Gold Bull 43:181–188
3. Juliusa M, Robertsa S, Fletchera JCQ (2010) A review of the use of gold catalysts in selective hydrogenation reactions Lynsey McEwana. Gold Bull 43:298–306
4. Zheng N, Fan J, Stucky GD (2006) One-step one-phase synthesis of monodisperse noble-metallic nanoparticles and their colloidal crystals. J Am Chem Soc 128:6550–6551
5. Chen M, Feng YG, Wang X, Li TC, Zhang JY, Qian DJ (2007) Silver nanoparticles capped by oleylamine: formation, growth, and self-organization. Langmuir 23:5296–5304
6. Skrabalak SE, Chen J, Sun Y, Lu X, Au L, Cobley CM, Xia Y (2008) Gold nanocages: synthesis, properties, and applications. Acc Chem Res 41:1587–1595
7. Zhang Q, Xie J, Yu Y, Yang J, Lee JY (2010) Tuning the crystallinity of Au nanoparticle. Small 6:523–527
8. Seo D, Yoo CI, Chung IS, Park SM, Ryu S, Song H (2008) Shape adjustment between multiply twinned and single-crystalline polyhedral gold nanocrystals: decahedra, icosahedra, and truncated tetrahedra. J Phys Chem C 112:2469–2475
9. Ming T, Feng W, Tang Q, Wang F, Sun L, Wang J, Yan C (2009) Growth of tetrahexahedral gold nanocrystals with high-index facets. J Am Chem Soc 131:16350–16351
10. Khanal BP, Zubarev ER (2008) Purification of high aspect ratio gold nanorods: complete removal of platelets. J Am Chem Soc 130:12634–12635
11. Wang C, Hu Y, Lieber CM, Sun S (2008) Ultrathin Au nanowires and their transport properties. J Am Chem Soc 130:8902–8903
12. Xing S, Feng Y, Tay YY, Chen T, Xu J, Pan M, He J, Hng HH, Yan Q, Chen H (2010) Reducing the symmetry of bimetallic Au@Ag nanoparticles by exploiting eccentric polymer shells. J Am Chem Soc 132:9537–9539
13. Wang C, Yin H, Chan R, Peng S, Dai S, Sun S (2009) One-pot synthesis of oleylamine coated AuAg Alloy NPs and their catalysis for CO oxidation. Chem Mater 21:433–435
14. Wang C, Peng S, Chan R, Sun S (2009) Synthesis of AuAg alloy nanoparticles from core/shell-structured Ag/Au. Small 5:567–570
15. Shore MS, Wang J, Johnston-Peck AC, Oldenburg AL, Tracy JB (2011) Synthesis of Au(core)/Ag(shell) nanoparticles and their conversion to AuAg alloy nanoparticles. Small 7:230–234
16. Yang Y, Gong X, Zeng H, Zhang L, Zhang X, Zou C, Huang S (2010) Combination of digestive ripening and seeding growth as a generalized route for precisely controlling size of monodispersed noble monometallic, shell thickness of core-shell and composition of alloy nanoparticles. J Phys Chem C 114:256–264
17. Ma Y, Li W, Cho EC, Li Z, Yu T, Zeng J, Xie Z, Xia Y (2010) Au@Ag core-shell nanocubes with finely tuned and well-controlled sizes, shell thicknesses, and optical properties. ACS NANO 4:6725–6734
18. Wang D, Xie T, Peng Q, Li Y (2008) Ag, Ag2S, and Ag2Se nanocrystals: synthesis, assembly, and construction of mesoporous structures. J Am Chem Soc 130:4016–4022
19. Yang J, Ying JY (2009) Room-temperature synthesis of nanocrystalline Ag$_2$S and its nanocomposites with gold. Chem Commun 22:3187–3189
20. Yin Y, Erdonmez C, Aloni S, Alivisatos AP (2006) Faceting of nanocrystals during chemical transformation: from solid silver spheres to hollow gold octahedra. J Am Chem Soc 128:12671–12673

8

Sites for the selective hydrogenation of ethyne to ethene on supported NiO/Au catalysts

S. A. Nikolaev · D. A. Pichugina · D. F. Mukhamedzyanova

Abstract Au, NiO, and NiO/Au clusters of 2.5–16 nm, supported on Al_2O_3, ZrO_2, TiO_2, and ZnO, were studied in the purification of ethene feedstock from ethyne by hydrogenation at 357 K. The Au, NiO, and NiO/Au catalysts possessed 100 % selectivity to ethene. As the size of NiO clusters decreased from 7 to 3 nm, the turnover frequency (TOF) decreased from 812–1,023 to 276 h^{-1}. In contrast with NiO, Au activity increased with decreasing particle size. NiO/Au catalysts possessed higher stability and activity in comparison with Au and NiO catalysts. The synergistic gain on NiO/Au clusters (SG) calculated as $TOF_{NiO/Au}-TOF_{Au}-TOF_{NiO}$ was 1,466; 1,147; 563; and 569 h^{-1} for $NiO/Au/Al_2O_3$, $NiO/Au/TiO_2$, $NiO/Au/ZnO$, and $NiO/Au/ZrO_2$, respectively. The reasons of the observed catalytic trends and the origin of the most active and selective sites are discussed.

Keywords Synergism · Ethyne hydrogenation · DFT · NiO/Au · Au · Heat of adsorption

Introduction

Ethene feedstock for polymerization coming from the stage of hydrocarbons cracking contains 0.5–3 % of ethyne compounds [1, 2]. Even such a small content of ethyne impurities quickly poisons Ziegler–Natta polymerization catalysts [2, 3]. Thus, preliminary removal of alkynes via hydrogenation is necessary [1–4]. Traditionally, the Ag/Pd catalysts

S. A. Nikolaev (✉) · D. A. Pichugina · D. F. Mukhamedzyanova
Department of Chemistry, M.V. Lomonosov Moscow State University,
1 Leninskie Gory,
119991, Moscow, Russia
e-mail: serge2000@rambler.ru

HO-21 (BASF), G-58, and G-83 (Süd-Chemie) are employed for the hydrogenation of C≡C bond to C=C bond. However, turnover frequency (TOF) and stability of Ag/Pd catalysts are often low. Moreover, undesirable "green-oil" and ethane by-products are usually formed on palladium surface [3, 4]. Nowadays, the studies directed to improving of catalytic properties of palladium-based catalysts are becoming less popular, giving way to the intensive searching for alternative and high-performance palladium-free catalysts for the ethyne hydrogenation [1–5].

Nano-sized gold is an effective catalyst for oxidation [6], isomerization [7], cyclization [8], hydrodechlorination [9], and other reactions [10, 11]. Composites based on bare gold clusters demonstrate high activity and/or selectivity in hydrogenation of 1,3-butadiene [2], 1-propyne [12], and ethynylbenzene [13] under the mild conditions. Bimetallic gold-nickel catalysts also may possess the enhanced catalytic properties in conversion of different hydrocarbons: n-butane [14], allylbenzene [5, 7], and 2,4-dichloropfenol [9]. Thus, the application of gold and gold-nickel clusters in selective hydrogenation is very promising.

This work is an extension of studies of the catalytic properties of Au and NiO/Au clusters in the hydrogenation of ethyne derivatives [13, 15]. It was reported that Au/Al_2O_3 catalysts possess high activity (up to 0.142 s^{-1}) in the hydrogenation of ethynylbenzene into styrene at 423 K. A synergistic effect was revealed for $NiO/Au/Al_2O_3$ in the hydrogenation of ethyne into ethene [15]: The conversion of ethyne on the NiO/Au catalysts was higher than the sum of conversions on Au and Ni catalysts. It was found that synergistic gain on NiO/Au clusters was maximized at 357 K, and Au/Ni ratio equaled 1:1 [15]. According to the analysis of oxidation state of metals in gold-contained catalysts [7], it was proposed that synergistic effect could be caused by the formation of new $Au^{\delta+}$ catalytic sites. The goals of the present research were: (1) to study the regularities of catalytic action of Au and NiO/Au clusters (Au/Ni=1:1), supported on Al_2O_3, TiO_2, ZrO_2, and ZnO, in the hydrogenation of ethyne–ethene mixture at 357 K; (2) to study the impact of clusters shape and size to the catalysis by gold, and (3) to determine the origin of the active gold sites by density functional theory (DFT).

Experimental

Al_2O_3 ("Katalizator," AOK-63-11) with a 160 m^2/g surface area and an isoelectric point (IEP) equal to 7.0, TiO_2 ("Degussa," Aerolyst®7710, 50 m^2/g, 6.0 IEP), ZrO_2 (Degussa, Aerolyst®6100, 30 m^2/g, 4.5 IEP), and ZnO (Degussa, Aerolyst®9000, 36 m^2/g, 10.0 IEP) were used as the supports for metal clusters. Au/supports and Pd/Al_2O_3 were produced by deposition–precipitation [15, 16]. In typical synthesis, 50 ml of the aqueous solution of $HAuCl_4$ (or $PdCl_2$) with 5.8×10^{-6} g/ml metal content was adjusted to pH equal to the IEP of the support by adding the aqueous solution of NaOH (0.1 M). Then 1 g of the support was dispersed in the solution with stirring for 1 h at 307 K. The precursor obtained was washed to remove NaCl. The degree of NaCl removing was controlled via $AgNO_3$ test. Then the sample was dried for 24 h at 298 K and calcined for 3 h in air at 623 K.

Monometallic NiO samples were prepared by impregnation of the calcined support (3 h, air, 623 K) with the aqueous solution of $Ni(NO_3)_2$, corresponding to the pore volume of the support, followed by drying for 24 h at 298 K and calcining for 3 h in air at 623 K [15]. Bimetallic NiO/Au/supports and Ag/Pd/Al_2O_3 were produced by pore volume impregnation of fresh Au/supports and Pd/Al_2O_3 with the aqueous solution of $Ni(NO_3)_2$ and $AgNO_3$, respectively. The precursors obtained were dried for 24 h at 298 K, and calcined for 3 h in air at 623 K.

The metal contents in catalysts were determined by atomic absorption on a Thermo iCE 3000 AA spectrometer. The metals were preliminarily removed from the support by washing with aqua regia (HCl/HNO_3=4:1). The relative error of metal content determination was less than 1 %. X-ray diffraction (XRD) patterns were accumulated with a Rigaku D/MAX 2500 instrument using Cu K(a) radiation with a step size of 0.02° two-theta (2θ) over the range 35–70°. Transmission electron microscopy (TEM) and energy-dispersive X-ray (EDX) analysis of catalysts were carried out on a JEOL JEM 2100F/UHR microscope with 0.1 nm resolution and a JED-2300 X-ray spectrometer, respectively. The size of spherical (SPH) and distorted (DIS) particles was calculated as diameter and maximum linear size, respectively. For each catalyst, 300–380 particles were processed to determine the particle size distribution. The

average particle size was determined as the average size of the most frequent particles. The concentration of SPH particles in the sample was calculated as $C(SPH)=n(SPH) \times N \times 100$ %, where $n(SPH)$ is the number of SPH particles, N (300–380) is the number of processed particles. The concentration of DIST particles was calculated in the same manner.

The hydrogenation of ethyne–ethene mixture was carried out using a fixed-bed flow reactor. A catalyst (1 g) was placed in a quartz reactor and heated to 357 K for 1 h in a stream of H_2. The gas mixture containing C_2H_2, H_2, and C_2H_4 in a 1:2:20 ratio was passed through the catalyst at a flow rate of 720 h^{-1}. The reactor effluents were analyzed with Tcvet-800 gas chromatograph equipped with a flame ionization detector and a 30-m PoropakT capillary column. The ethyne conversion was calculated as $\Delta(C_2H_2)=[\chi_0(C_2H_2)-\chi_i(C_2H_2)] \times [\chi_0(C_2H_2)]^{-1} \times 100$ %, where $\chi_0(C_2H_2)$ is the molar concentration of C_2H_2 in the initial gas mixture, and $\chi_i(C_2H_2)$ is the molar concentration of C_2H_2 in the products after time on stream (i). The selectivity of the ethene formation was determined as $S(C_2H_4)=[(\chi_i(C_2H_4)-\chi_0(C_2H_4)] \times \Delta(C_2H_2)^{-1}$, where $\chi_0(C_2H_4)$ and $\chi_i(C_2H_4)$ are the molar concentrations of C_2H_4 in the initial gas mixture and in the products after reaction time (i), respectively; $\Delta(C_2H_2)$ is the ethyne conversion. The reproducibility of the catalytic measurements was ±2 % with respect to the absolute value of conversion and selectivity.

TOFs were calculated by equation $TOF=A \times B^{-1} \times t^{-1}$ as described in Mohr et al. and Okumura et al. [17, 18], where $[A]=[mole]$ is the amount of ethyne converted per the reaction time $[t]=[h]$, $[B]=[mole]$ is the amount of the surface atoms in deposited clusters. B was calculated as $(B_{total}(Au)+B_{total}(Ni)) \times D$, where $B_{total}(Au)$ and $B_{total}(Ni)$ is the total moles of Au and Ni in the sample, respectively, $[D]$ is the surface-to-volume ratio for clusters with different average sizes given in Mohr et al. and Okumura et al. [17–19].

The calculations of C_2H_2, C_2H_4, and H_2 adsorption on Au_{12} were performed within DFT framework using Perdow–Burke–Ernzerhof functional [20]. Relativistic effects of gold were taken into account in scalar–relativistic approach using Dirac–Coloumb–Breit Hamiltonian modified by Dyall [21]. The basis sets as implemented in the Priroda 06 program [22] were used in the calculations (Au {30s29p20d14f}/[8s7p5d2f]; H {6s2p}/[2s1p]; C {10s7p3d}/[3s2p1d]). Au_{12} cluster was considered as a model. It has a dynamic structure [23] and co-exists in flat (two-dimensional (2D)) and three-dimensional (3D) structures that allowed us to study the dependence of adsorption properties on the structural features of gold particles.

The adsorption energy of RH molecule ($E_{ad}(RH)$, RH= C_2H_2, C_2H_4, H_2) on Au_{12} was calculated as $E_{ad}(RH)=E(Au_{12})+E(RH)-E(Au_{12}RH)$, where $E(Au_{12})$ is the total energy of an isolate gold cluster, $E(RH)$ is the total energy of isolate hydrocarbon, $E(Au_{12}RH)$ is the total energy of the hydrocarbon absorbed on Au_{12}. All the total energies included zero-point vibration energies. The heats of adsorption (Q) were calculated at 298 K on the basis of the formulas of statistical thermodynamics in the approximation of rigid rotato-harmonic oscillator. The calculation of all the structures was performed in singlet ground state. The types of stationary points on the potential energy surface were determined from the analysis of Hess matrix; the second derivatives were calculated analytically [23].

Results and discussion

In the present work, some M/X catalysts (M=Au, NiO, NiO/Au; X=Al₂O₃, TiO₂, ZnO, ZrO₂) numbered from №1 to №12 and reference Ag/Pd/Al₂O₃ catalyst ([Pd]= 0.021 wt.%; [Ag]=0.006 wt.%) were synthesized. The actual metal loadings in the samples №s 1–12 are presented in Table 1. The actual gold content in the samples prepared by deposition–precipitation was 0.018–0.024 wt.% that is lower than theoretical loading. The loss of gold is attributed to washing treatment during the preparation [10]. The actual concentration of Ni in the impregnated samples was 0.006–0.007 wt.% that is in agreement with theoretical weight percent. The molar ratio of Ni/Au in bimetallic samples was close to the desired 1:1 ratio.

The size and shape of supported NiO, Au, and NiO/Au particles

The structural features of supported particles in the samples №s 1–12 are presented in Figs. 1, 2, 3, and 4 and Table 1.

The particle size distributions in Ni samples were monomodal (see Fig. 1 as representative). The sizes of detectable nickel particles varied from 2 to 17 nm (Table 1). The average particle size in NiO/Al₂O₃, NiO/TiO₂, NiO/ZnO, and NiO/ZrO₂ was 3, 4.5, 7, and 7 nm, respectively (Table 1). The increase in particle size of NiO during transition from Al₂O₃ to ZrO₂ could be caused by the decrease of the surface area of a support, which favors segregation of clusters during the calcination [3–5]. The shape of the NiO particles, deposited on Al₂O₃, was nearly spherical (Fig. 1). The similar shape was observed for NiO particles immobilized on TiO₂, ZnO, and ZrO₂ (Table 1). The obtained results are in agreement with the shape of nickel particles immobilized by impregnation from the aqueous solution of Ni(NO₃)₂ [9].

The particle size distributions in Au/Al₂O₃ (Fig. 2, №1) and monometallic gold samples №s 4, 7, and 10 were monomodal and varied from 2 to 23 nm (Table 1). The average particle size of Au in Au/Al₂O₃, Au/TiO₂, Au/ZnO, and Au/ZrO₂ was 2.5, 8, 10, and 9 nm, respectively. The shapes of the supported Au particles were spherical (Table 1) that is in agreement with the shape of Au particles on the same supports immobilized by deposition–precipitation from HAuCl₄×aq [10, 24].

The deposition of NiO particles on the Au catalysts resulted in a broadening of particle size distribution (see Figs. 1, 2, and 3 and Table 1). The shift of the average particle size to higher values was detected in bimetallic samples in comparison with corresponding monometallic ones (Table 1). The observed trend could be caused by the formation of decorated M_1/M_2 clusters, whose sizes are larger than M_1 and M_2 [3–5, 14]. The presence of NiO/Au particles in the bimetallic samples was confirmed by the TEM-EDX analysis. The TEM micrograph of the gold–nickel catalyst and the EDX elemental maps of Au and Ni

Table 1 The structural properties of supported particles M (M=Au, NiO, NiO/Au) and synergistic gain (SG) on NiO/Au particles	№	Support	[Au]	[Ni]	Sizes	L	M shape, %		TOF	SG
			wt.%	wt.%	nm	nm	SPH	DIS	h⁻¹	h⁻¹
	1	Al₂O₃	0.018	0	2–6	2.5	100	0	36	1,466
	2		0	0.006	3–10	3	100	0	276	
	3		0.018	0.006	2–13	5	60	40	1,778	
	4	TiO₂	0.02	0	4–12	8	100	0	28	1,147
	5		0	0.006	4–17	4.5	100	0	402	
The synergistic gain on NiO/Au clusters (SG) was calculated as differences between TOF$_{NiO/Au}$ −TOF$_{Au}$−TOF$_{NiO}$ after 735 min of reaction	6		0.02	0.006	3–25	13	68	32	1,577	
	7	ZnO	0.024	0	3–18	10	100	0	0	563
	8		0	0.006	2–16	7	100	0	1,023	
	9		0.024	0.007	3–23	14	74	26	1,586	
L average particle size, SPH spherical shape, DIS distorted (nonspherical) shape, TOF turnover frequency	10	ZrO₂	0.021	0	2–21	9	100	0	0	569
	11		0	0.006	3–12	7	100	0	812	
	12		0.021	0.006	2–28	16	81	19	1,381	

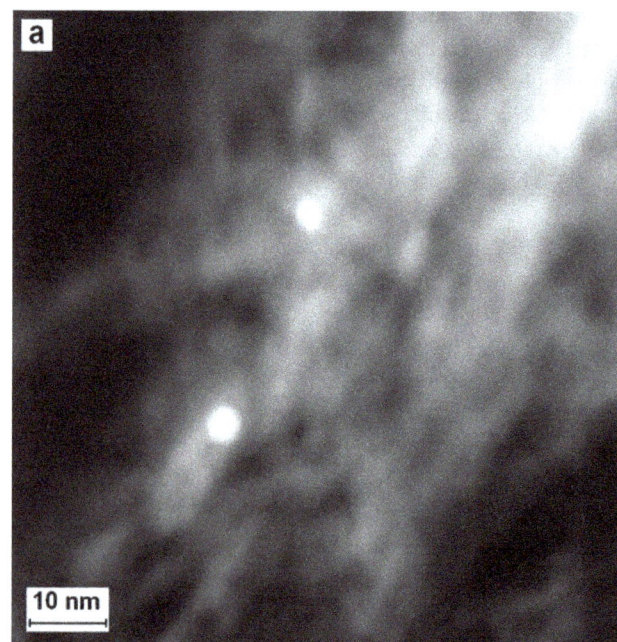

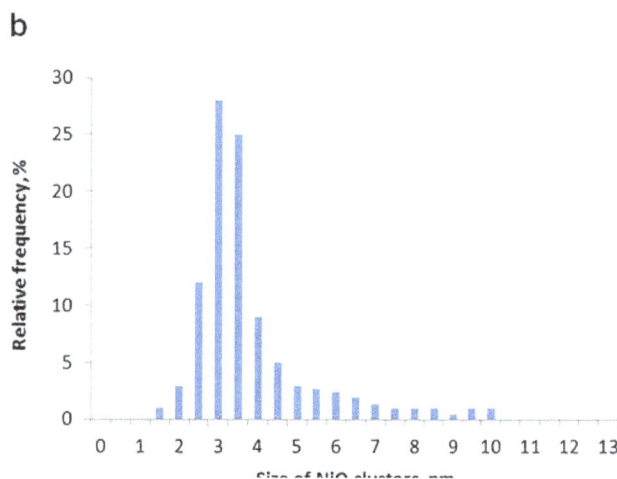

Fig. 1 TEM image of the NiO/Al$_2$O$_3$ (**a**) and particle size distribution (**b**)

The catalytic properties of NiO, Au, and NiO/Au nanocomposites

The conversion, selectivity, and durability of catalysts №s 1–12 tested in hydrogenation of ethyne–ethene mixture at 357 K are presented in Figs. 5, 6, and 7, the calculated TOFs are summarized in Table 1. After hydrogen treatment, NiO particles with size of 3–4.5 nm supported on Al$_2$O$_3$ and TiO$_2$ showed 18–22 % ethyne conversion (Fig. 5). At these conversions, the selectivity to ethene on NiO catalysts was high—95–99 % that is in agreement with the results obtained in Zhang et al. [25]. The initial TOF of NiO/Al$_2$O$_3$ and NiO/TiO$_2$ was 497 and 608 h^{-1}, respectively (Table 1). The TOFs of 3 nm NiO particles are lower by

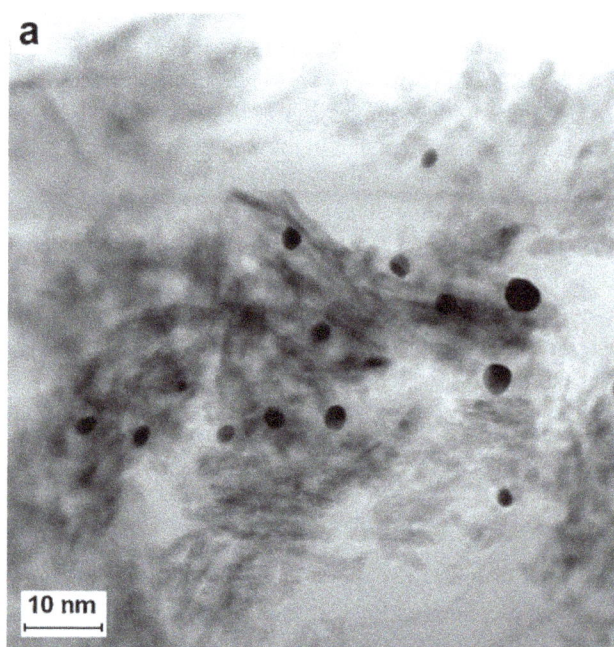

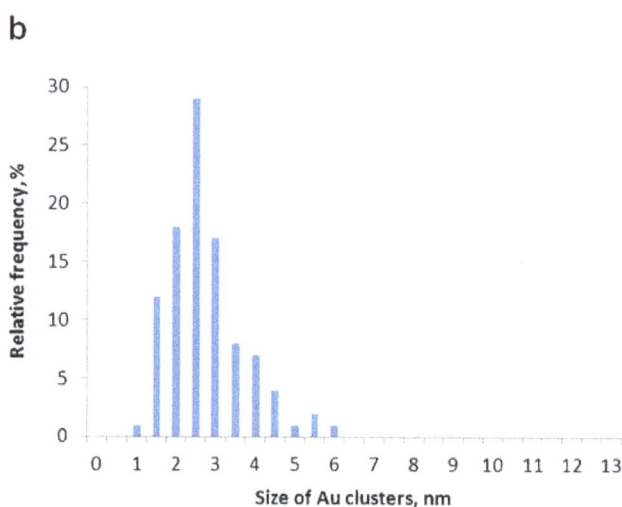

Fig. 2 TEM image of the Au/Al$_2$O$_3$ (**a**) and particle size distribution (**b**)

are presented in Fig. 4. The elemental maps of bimetallic sample contained individual particles of Au and NiO (Au Mα and Ni Kα signals from these clusters do not overlap) along with mixed particles of NiO/Au (overlay Au Mα and Ni Kα signals from the same particle).

The fraction of NiO/Au particles in the bimetallic catalysts №s 3, 6, 9, and 12 ranged from 40 % to 19 % from the total number of supported particles. The size of NiO/Au particles varied from 4 to 20 nm, and the average size was always larger in comparison with Au and NiO. An important feature of NiO/Au clusters was their distortion (deviation from spherical shape) (see Fig. 4), which may be caused by the inclusion of NiO clusters into the surface of Au nanoparticles during the stage of catalyst preparation.

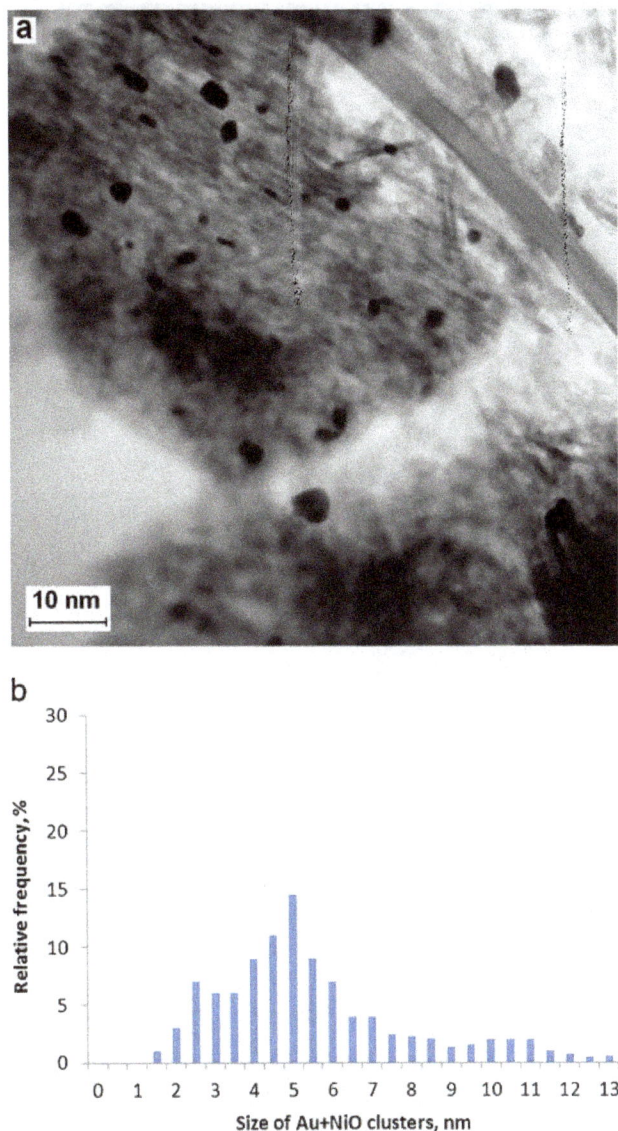

Fig. 3 TEM images of the NiO/Au/Al$_2$O$_3$ (**a**) and particle size distribution (**b**)

one to two orders of magnitude in comparison with C≡C bond hydrogenation activity of 3–10 nm clusters of Pt [26] and Ag/Pd [25] supported on Al$_2$O$_3$.

The 7 nm NiO particles supported on ZnO and ZrO$_2$ showed better activity in comparison with 3–4.5 nm NiO particles supported on Al$_2$O$_3$ and TiO$_2$ (Table 1). The negative-size effect of activity is the known feature of Pd, Pt, and Ni catalysts attributed to irreversible adsorption of unsaturated compounds on small clusters [3, 4, 10]. According to data presented in del Angel et al. [26], the TOF of Pt and Pd clusters decreases from 780 to 480 and from 9,910 to 8,640 h^{-1} as the cluster size decreases from 12 to 3 and from 10 to 3 nm, respectively. The reason of the better activity of larger NiO particles can also be influenced by their reducibility. The rate-determining step in hydrogenation is usually

a dissociative adsorption of H$_2$ that takes place at Ni^0 sites not at Ni^{2+} [4]. Keane and Medina showed that large NiO clusters are reduced much easier to Ni^0 in comparison with small NiO clusters that strongly interact with surface groups of the support [9, 27]. Therefore, the 7 nm NiO of the sample №s 8 and 11 possess higher activity in hydrogenation in comparison with the 3 and 4.5 nm NiO of the sample № 3 and № 5 (Table 1).

The stability of supported NiO clusters was low (Fig. 5). After 735 min of the reaction, the color of NiO/Al$_2$O$_3$ and NiO/TiO$_2$ became darker, and TOFs decreased by approximately 55 % and 49 %, respectively. The same trends were observed for NiO/ZnO and NiO/ZrO$_2$ (Fig. 5). The most probable mechanism of nickel deactivation is due to green-oil formation as described in Borodziński et al. and Nikolaev et al. [3, 4].

A 1.5 % conversion of ethyne with 100 % selectivity to ethene was detected on 2.5 nm Au deposited on Al$_2$O$_3$ (Fig. 6). The high selectivity of Au/Al$_2$O$_3$ is in agreement with results of Jia [28] and Gluhoi [29] reported that ethyne was hydrogenated to ethene on Au/Al$_2$O$_3$ with 100 % selectivity at 313–523 K. The initial TOF for 2.5 nm Au was 36 h^{-1} that is in agreement with gold hydrogenation activity [1, 13, 15, 18] and lower in comparison with the activity of either NiO (Table 1) or Pt and Pd clusters of the same size deposited on alumina [26, 30].

The positive-size effect of activity was found for gold catalysts. As the size of gold decreases from 8 to 2.5 nm, the initial TOF increases from 28 to 36 h^{-1} (Table 1). The better hydrogenation activity of 2.5 nm gold particles can be explained if the chemisorption of H$_2$ is dependent on gold particle size. Our recent research on the TOF of Au particles of different sizes (2–30 nm) in hydrogenation of ethynylbenzene [13] and studies by Bus [30, 31], Serna [32], Boronat [33], and Jia [28] have proved the positive influence of small gold clusters on hydrogenation rate. It was concluded that H$_2$ is dissociatively adsorbed only at the corners and edges of the supported gold particles [34, 35]. Gold atoms at the corner and edge sites have a low coordination number compared with face atoms and thus have a more reactive d-band, resulting in these atoms to be able to interact more easily with H$_2$ [30] and facilitate the H$_2$ dissociation.

It is interesting to note that almost identical 8–10 nm Au clusters supported on ZrO$_2$, ZnO, and TiO$_2$ possess different hydrogenation properties. Whereas Au/ZrO$_2$ and Au/ZnO are inactive, Au/TiO$_2$ converts 0.5 % of ethyne into ethene. It seems that the higher hydrogenation activity of Au/TiO$_2$ is due to specific metal-support interaction. Fujitani [36] find no HD formation between H$_2$ and D$_2$ other than on Au/TiO$_2$. No HD formation was observed at any single-crystal surface (Au(111), Au(311), and TiO$_2$(110)). This suggests that the bulk gold and TiO$_2$(110) surface did not make H$_2$ dissociate, whereas the Au/TiO$_2$ surface did. Recent DFT

Fig. 4 TEM image of the NiO/
Au/Al$_2$O$_3$ (**a**); EDX elemental
maps of the Au Mα (**b**) and Ni
Kα (**c**) and typical TEM images
of the NiO/Au particles

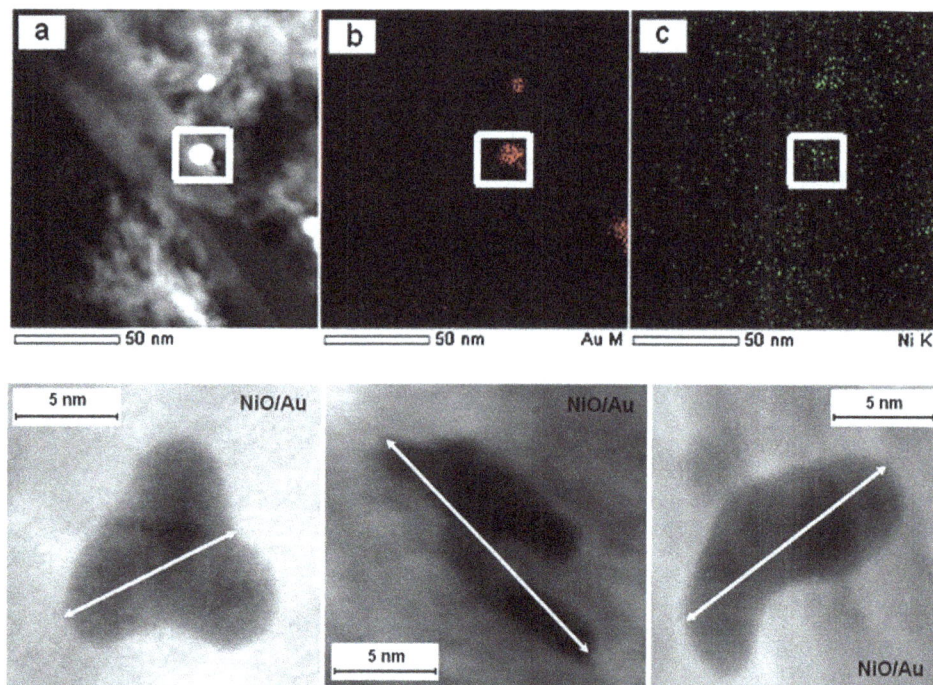

study [37] proved this phenomenon. Yong revealed the favored heterolytic dissociation of H$_2$ with E_a=0.37 eV at the perimeter sites of Au/TiO$_2$, where Au atom and a nearby surface O$_{(s)}^{2-}$ were involved. As a result of this process, oxygen atoms near the perimeter were passivated. Further H$_2$ dissociation occurred on pure gold atoms via a homolytic mode with E_a=0.64 eV [37].

Some contribution to the observed decreasing of TOFs in the row Au/Al$_2$O$_3$>Au/TiO$_2$>Au/ZnO>Au/ZrO$_2$ (Table 1) could be provided by OH groups and metal cations of the supports. Thus, Zhang reported that dehydroxylated Au/ZrO$_2$ was inactive in semi-hydrogenation of 1,3-butadiene but became active after regeneration of the surface –M–O–

M– groups by water treatment [38]. Hydroxyl groups on ZrO$_2$ were proposed to supply protons for hydrogenation and according to H/D exchange reactions between D_2 and the –OH groups occurred at above 323 K [38]. The specific surface of supports decreases in the row Al$_2$O$_3$>TiO$_2$>ZnO >ZrO$_2$, thus the concentration of hydroxyl groups in Au catalysts decreases during the transition from Al$_2$O$_3$ to ZrO$_2$. As a result, the positive contribution of hydrogen from hydroxyl groups to the hydrogenation decreases.

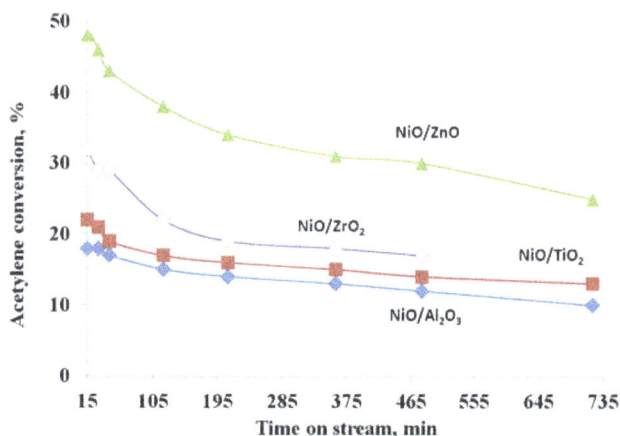

Fig. 5 Hydrogenation of ethyne-ethene mixture at 357 K over NiO catalysts № 2, 5, 8, 11. The metal loading of Ni in samples is 0.006 wt.%

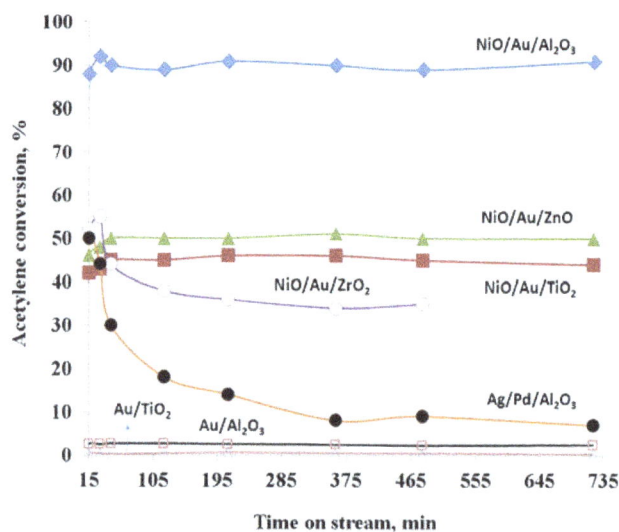

Fig. 6 Hydrogenation of ethyne-ethene mixture at 357 K over Au catalysts № 1, 3, 4, 6, 9, 12 and Ag/Pd catalyst. The metal loadings of Pd and Ag are 0.021 and 0.006 wt.%, respectively. The metal loadings of Au and Ni are 0.018–0.02 wt.% and 0.006–0.007 wt.%, respectively

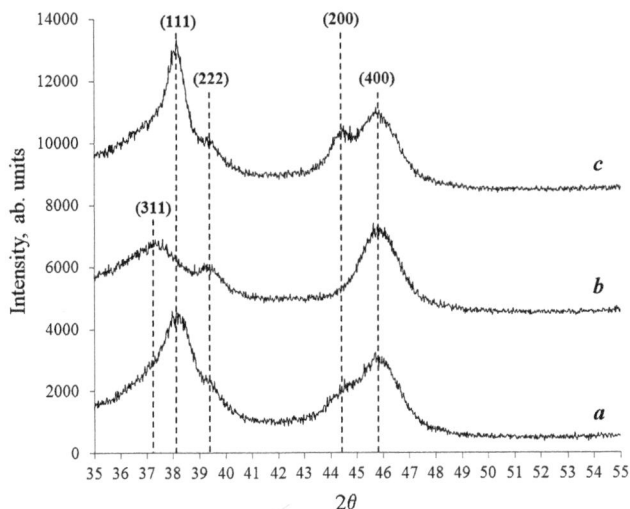

Fig. 7 Powder X-ray diffraction patterns for **a** 0.21 % Au/Al$_2$O$_3$, **b** 0.15%Ni/Al$_2$O$_3$, and **c** 0.21%Au+0.21%Ni/Al$_2$O$_3$. Planes (*111*), (*200*) are attributed to gold clusters. Planes (*311*), (*222*), (*400*) are attributed to alumina support

The metal cations of the supports are of importance in the adsorption and activation of reactants [39, 40]. It was suggested that cooperation of coordinatively unsaturated Au atoms and the acid–base pair site (Al$^{\delta+}$–O$^{\delta-}$) play an important role as in H$_2$ dissociation step and preferential adsorption of hydrogenated group [39]. Alumina is the oxide with strong Lewis acid sites that able to activate the C–H bond of ethyne directly on the M$^{\delta+}$–O$^{\delta-}$ pairs [40]. A decrease in the Lewis coordinative activity was found for the metal cation of basic oxides (CaO, ZnO, and MgO). The weakening of Lewis acid sites decreases the ability to form the acetylide species (ZrO$_2$) up to the full inactivity (SiO$_2$).

The durability of active Au/TiO$_2$ and Au/Al$_2$O$_3$ catalysts was higher in comparison with nickel samples: No decrease in conversion occurred during 735 min (Figs. 5 and 6). The high durability of gold catalysts in hydrogenation of ethyne at 357 K is in good agreement with results of Jia [28], Zhang [38], and Azizi [41]. For example, Azizi and co-workers investigated the hydrogenation of ethyne and the formation of carbonaceous deposits using Au/CeO$_2$. The catalysts were prepared by direct anionic exchange, which produced catalysts with an average gold particle size of 2 nm. H$_2$/ C$_2$H$_2$ ratios between 3 and 60 were tested in the temperature range 300–673 K. An ethene selectivity of 100 % was observed for all H$_2$/C$_2$H$_2$ ratios up to 573 K. Above 573 K, the deactivation occurred swiftly and was accompanied by the decreased selectivity.

The decoration of Au by NiO led to a sharp increase in ethyne conversion up to 90 % while the selectivity still remains 100 %, and the time of stable operation remains at least 735 min (Fig. 6). Moreover, the activity and stability of NiO/Au catalysts at 357 K were found to be higher than the activity and stability of reference Ag/Pd catalyst for

selective hydrogenation prepared by the same method as NiO/Au catalyst (Fig. 6). The conversion of ethyne on NiO/Au catalysts was higher than the sum of conversions on NiO and Au catalysts (Fig. 6) that points out the synergistic activity of NiO/Au catalysts. The synergistic gain on NiO/Au clusters (SG) after 735 min of time on stream was 1,466; 1,147; 563; and 569 h^{-1} for NiO/Au/Al$_2$O$_3$, NiO/Au/ TiO$_2$, NiO/Au/ZnO, and NiO/Au/ZrO$_2$, respectively (Table 1).

The synergistic effects of activity resulted from the interaction between M$_1$ and M$_2$ metals can be attributed to the formation of alloy [3, 5, 8, 14, 42]. Besenbacher observed the formation of Au–Ni alloy in Au–Ni/MgAl$_2$O$_4$ catalyst with extremely high metal loading (Au+Ni=17 wt.%) after reduction at 823 K [14]. The obtained catalyst possessed the enhanced catalytic properties in *n*-butane steam-reforming. Zhang prepared Au$_7$Ni$_3$ intermetallic alloy by melting the mixture of gold and nickel at 1,333 K [42]. This alloy exhibited the synergistic activity in electro-oxidation of formic acid. It was reported that, compared with the standard spectrum of pure Au, the diffraction angles of Au–Ni sample were shifted to the higher positions, reflecting Au–Ni alloy with lattice contraction due to the partial substitution of Au by Ni [42]. The attempts to obtain the XRD reflexes from supported metals of the sample №s 1–12 were failed due to the low concentration of gold (0.018–0.024 wt.%) and/or nickel (0.006–0.007 wt.%). The XRD patterns of specially prepared samples with relatively high metal contents (0.15–0.42 wt.%) are presented in Fig. 7. The XRD pattern of NiO/Au sample (Au=Ni=0.21 wt.%) was described by the sum of XRD patterns of the gold and nickel catalysts (Fig. 7). No new reflexes or shifts of Au reflexes toward larger values pointed at alloy formation, as described in Lijuan et al. [42], were detected. The obtained XRD results allowed us to exclude the formation of new Au–Ni compounds in our samples with a high degree of certainty. Moreover, the synthesis of Au–Ni alloy [14, 42] was carried out at 823–1333 K and high metal contents (17–100 wt.%). Thus, alloy formation is not possible at the temperatures and metal loadings employed in our study.

The synergism in M$_1$–M$_2$ activity also can be attributed to the formation of the new structural sites. For example, Keane linked the increase of the rate of 2,4-dichlorophenol hydrodeclorination on Au/Ni/Al$_2$O$_3$ with decoration of nickel clusters with gold ones [9]. It was proposed that the Au component served to adsorb and activate the C–Cl bond whereas the H$_2$ dissociated at the Ni^0. The addition of tin to the Pt/SiO$_2$ catalyst produced a remarkable positive shift in both the product distribution and TOF in the liquid-phase hydrogenation of unsaturated aldehyde to corresponding alcohols [43, 44]. The simultaneous increase in both activity and selectivity indicates the formation of a new type of mixed (M^{x+})/(Pt)^0 clusters, different in nature when

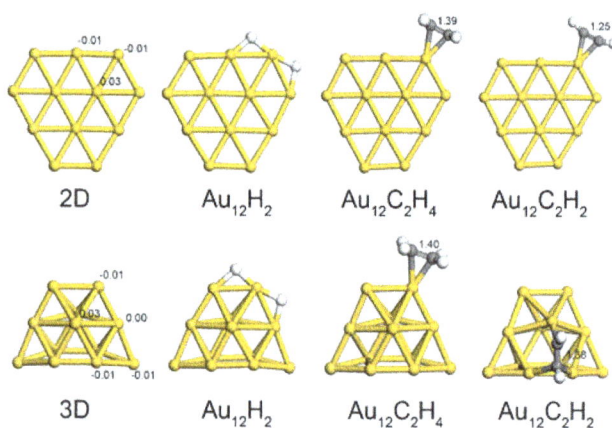

Fig. 8 The optimized structures of flat (2D) and three-dimensional (3D) Au$_{12}$ clusters and the most stable complexes of Au$_{12}$ with hydrogen, ethene, and ethyne molecules

compared with the original Pt and M ones. The promotion effect was proposed due to a (M^{x+}) activating the C=O bond, which becomes easily hydrogenated. The building of mixed (M$_1$)/(M$_2$) clusters is very similar to what is occurring in the well-known strong metal support interaction between reducible oxide support and deposited metal clusters [1–5, 10].

Our investigations of model Au, Ni/Au, and Ni catalysts by XPS, DRIFT, and XAS techniques showed that nickel in both Ni/Au and Ni catalysts exists as NiO, gold in the Au catalyst exists as Au^0 nanoclusters, whereas the Au^0 nanoclusters co-exist with Au^{3+} cations in the NiO/Au samples [7]. Then, the first probable explanation (I) of synergistic activity between NiO and Au in ethyne hydrogenation (Table 1) could be due to the electron transfer from the electron-rich Au^0 particles to the electron-deficient NiO that give new Au$^{\delta+}$ catalytic sites.

According to the TEM data of the present research, there is a new significant fraction of distorted Au structures in NiO/Au catalysts (Table 1). Moreover, the concentration of distorted particles is correlated with SG values. It is known that the deviation from spherical shape of the particle led to increase in surface corners and edges [5]. Thus, the second explanation (II) of synergistic activity can be linked with the formation of new Au structural sites with a low coordination number in NiO/Au clusters.

The nature of Au activation in NiO/Au clusters

The DFT calculation of H$_2$, C$_2$H$_2$ and C$_2$H$_4$ on flat Au$_{12}$ (2D) and Au$_{12}$ (3D) containing flat, corner and edge atoms, were performed to test the reasonability of the explanations (I) and (II). The optimized structures of the most stable (2D) and (3D) Au$_{12}$ isomers are shown in Fig. 8. The calculated heats of adsorption of substrates are listed in Table 2.

The calculated heat (Q) of dissociative adsorption of H$_2$ on the flat Au$_{12}^0$ (2D) was equal to 40 kJ/mol (Table 2). During transition to uncharged Au$_{12}^0$ (3D) structures that contain flat elements (planes) and structural defects (corners and edges), the Q(H$_2$) increases up to 78 kJ/mol. The obtained results confirmed the existing hypotheses on the positive impact of cluster defects to the activation of H$_2$ [13, 30, 32–35] and are in agreement with explanation of NiO/Au synergy (II).

The presence of a positive charge on the Au$_{12}$ (2D) and Au$_{12}$ (3D) clusters led to an increase in the Q(H$_2$) by 23 and 2 kJ/mol, respectively (Table 2). These results are in good agreement with the results of Zhang et al. [45, 46]. This group used Au/ZrO$_2$ to investigate the effects of Au^{3+}/Au^0 ratio on 1,3-butadiene hydrogenation. It was reported that metallic Au^0 atoms at the surface of the gold particles were not the only active catalytic sites and that isolated Au^{3+} ions were the most active [45, 46]. The enhanced activation of H$_2$ on charged gold could be explained by mechanism of Au$_n$–H$_2$ formation. Recently, Gao and Lyalin in their DFT calculation [47] showed that H$_2$ dissociates at the low coordinated corner Au atom with formation of the slightly bent H–Au–H bond. The main feature of the formation of this hydride structure is the catalytic oxidation–reduction cycle M$^n \rightarrow$ M$^{n+2} \rightarrow$ M^n. In the case of non-charged zero-valence gold nanoparticles, this cycle should look like (Au$_{n+1}$)$^0 \rightarrow$ (Au$_n$)Au$^{+2} \rightarrow$ (Au$_{n+1}$)^0. But these valence changes seem to be improbable for gold, as the typical gold oxidation states are 0, +1, and +3 rather than +2. However, if supported gold nanoparticles may carry a positive charge +3 or +1, then a highly probable catalytic cycle with common gold valences will be obtained: (Au$_n$)Au$^{+1} \rightarrow$ (Au$_n$)Au$^{+3} \rightarrow$ (Au$_n$)Au^{+1}.

Table 2 The calculated heats of adsorption (Q) of C$_2$H$_2$, C$_2$H$_4$, H$_2$ (kilojoules per mole) on the flat zero-valence Au$_{12}^0$ (2D), positively charged Au$_{12}^+$ (2D), three-dimensional zero-valence Au$_{12}^0$ (3D), and three-dimensional positively charged Au$_{12}^+$ (3D) clusters

RH	Type of adsorption	Q (RH) on Au$_{12}^Z$ (2D)		Q (RH) on Au$_{12}^Z$ (3D)	
		$Z=0$	$Z=+1$	$Z=0$	$Z=+1$
H$_2$	Dissociative	40	63	78	80
C$_2$H$_4$	π Complex	88	132	90	147
C$_2$H$_2$	π Complex	75	113	73	129
	di-σ Complex	59	102	98	134
	Bridge complexes (⊥ and ∥)	–	–	139 ⊥	172 ∥

The adsorption of ethene on Au_{12}^0 (2D) was accompanied by the formation of π-complex with one of the atoms in the cluster, while the adsorption of ethyne can be realized through π- and di-σ bonding (Table 2). The adsorption of C_2H_4 on Au_{12}^0 (3D) occurred through the formation of π-complex, and $Q(C_2H_4)$ was slightly higher than $Q(C_2H_4)$ on Au_{12}^0 (2D) (Table 2). The increase in the heat of adsorption was also detected for ethyne π-complex during transition from Au_{12}^0 (2D) to Au_{12}^0 (3D). It was also revealed that ethyne on Au_{12}^0 (3D) can form a new type of coordination—perpendicular ($\perp$) and parallel ($\parallel$) bridges (Fig. 8), with the high heat of adsorption (Table 2). Our calculation trends are in good agreement with results obtained in Jia et al. and Segura [28, 48]. Jia et al. [28] have shown that, at 273 K, the amount of ethyne adsorbed on 3.8-nm gold particles immobilized on Al_2O_3 was 18 times greater than that of ethene. Moreover, in contrast to ethene, adsorption of ethyne was irreversible. Segura [48] has demonstrated that 4-nm gold nanoparticles supported on CeO_2 are extremely selective in the hydrogenation of C≡C bond in ethyne–ethene mixtures. His DFT calculation shows that differences in binding energy $E_b(C_2H_2)$–$E_b(C_2H_4)$ on flat Au_{19}^0 (2D) surface and on Au_{19}^0 (3D) cluster are 0.08 and 0.66 eV, respectively. Segura attributed the high selectivity of gold to the preferential adsorption of C≡C bond compared with C=C bond at the edges of gold clusters. Thus, Au defects favor the selective ethyne adsorption and its further conversion.

Ethyne and ethene are Lewis bases sensitive to electron acceptors [4, 5, 40]. An increase in the surface positive charge on Au due to electron transfer [7] improves the electron acceptor properties of gold and thereby should lead to stronger adsorption of ethyne or ethene on $Au^{\delta+}$ than on Au^0 and, therefore, increase the probability of the chemical reaction. Table 2 shows that positive charge on the flat as well as on the Au_{12}^+ (2D) and Au_{12}^+ (3D) cluster leads to an increase in the heat of adsorption of ethene in π-complex up to 132 and 147 kJ/mol, respectively. Surprisingly, the calculated heats of adsorption of ethyne (113 and 102 kJ/mol for π- and di-σ form, respectively) were smaller in comparison to the heat of adsorption of ethene on flat gold surface, but the positive impact to growth of heat of ethyne adsorption remained unchanged. Moreover, the heat of ethyne adsorption in the most stable ($\parallel$) bridge form on Au_{12}^+ (3D) increased up to 172 kJ/mol, which is larger by 25 kJ/mol in comparison with the heat of ethene adsorption on Au_{12}^+ (3D). Thus, positive charge on 3D clusters of Au favors preferential ethyne adsorption and increases its further conversion.

Conclusion

Monometallic Au and NiO particles supported on Al_2O_3, TiO_2, ZnO, and ZrO_2 possess 95–100 % selectivity to ethene in hydrogenation of ethyne–ethene mixture at 357 K.

The negative-size effect was revealed for NiO particles: As the size of NiO clusters decreases from 7 to 3 nm, the TOF decreases from 812–1,023 to 276 h^{-1}. In contrast to NiO, the size effect of activity was found to be positive for Au. As the size of gold decreases from 8 to 2.5 nm, the TOF increases from 0–28 to 36 h^{-1}. The observed size trends of TOF result from different abilities of gold and nickel to adsorb and activate the ethyne, ethene, and hydrogen.

The bimetallic NiO/Au catalysts possess 100 % selectivity to ethene and higher activity and stability in comparison with mono- and bimetallic catalysts based on Pd, Ag, and Ni. The decoration of Au by NiO led to increase in TOFs of NiO/Au catalysts up to 1,778 h^{-1}. The synergistic gain on NiO/Au clusters calculated as $TOF_{NiO/Au}-TOF_{Au}-TOF_{NiO}$ was 1,466; 1,147; 563; and 569 h^{-1} for NiO/Au/Al_2O_3, NiO/Au/TiO_2, NiO/Au/ZnO, and NiO/Au/ZrO_2, respectively. The formation of $Au^{\delta+}$ in NiO/Au particles and deviation of NiO/Au clusters shape from the spherical were detected. Using the methods of quantum chemistry, it was shown that distorted 3D and/or positively charged gold structures favor a dissociative adsorption of H_2 and preferential adsorption of ethyne from ethyne–ethene mixture.

Acknowledgments This work was supported by the Russian Foundation for Basic Research (Grant № 11-01-00280, № 10-03-00999, № 11-03-01011, and № 11-03-00403), the Russian Federation Ministry of Education and Science (State Contracts № 16.513.11.3137) and by the grants from the Russian Federation President (MK-107.2011.3, MK-2917.2012.3, and MK-1621.2012.3).

References

1. McEwan L, Julius M, Roberts S, Fletcher JCQ (2010) A review of the use of gold catalysts in selective hydrogenation reactions. Gold Bull 43:298–306.
2. Hugon A, Delannoy L, Louis C (2008) Supported gold catalysts for selective hydrogenation of 1,3-butadiene in the presence of an excess of alkenes. Gold Bull 41:127–138.
3. Borodziński A, Bond GC (2008) Selective hydrogenation of ethyne in ethene-rich streams on palladium catalysts, part 2: steady-state kinetics and effects of palladium particle size, carbon monoxide, and promoters. Catal Rev Sci Eng 50:379–469.
4. Nikolaev SA, Zanaveskin LN, Smirnov VV, Averyanov VA, Zanaveskin KL (2009) Catalytic hydrogenation of alkyne and alkadiene impurities in alkenes. Practical and theoretical aspects. Russ Chem Rev 78:231–247.
5. Rostovshchikova TN, Lokteva ES, Nikolaev SA, Golubina EV, Gurevich SA, Kozhevin VM, Yavsin DA, Lunin VV (2011) New approaches to design of nanostructured catalysts. In: Song M (ed) Catalysis: principles, types and applications. Nova Science Publishers, New York, pp 245–306

6. Grisel R, Weststrate K-J, Gluhoi A, Nieuwenhuys BE (2002) Catalysis by gold nanoparticles. Gold Bull 35:39–45.

7. Tkachenko OP, Kustov LM, Nikolaev SA, Smirnov VV, Klementiev KV et al (2009) DRIFT, XPS and XAS investigation of Au-Ni/Al_2O_3 synergetic catalyst for allylbenzene isomerization. Topics in Catalysis 52:344–350.

8. Hashmi A, Stephen K, Hutchings GJ (2006) Gold catalysis. Angew Chem Int Ed 45:7896–7936.

9. Keane MA, Gómez-Quero S, Cárdenas-Lizana F (2009) Alumina-supported Ni–Au: surface synergistic effects in catalytic hydrodechlorination. Chem Cat Chem 1:270–278.

10. Bond GC, Louis C, Thompson DT (2006) Catalysis by gold. Imperial College Press, London, 366 P

11. Simakova OA, Campo B, Murzin DY (2010) Gold on carbon catalysts. In: Chow PE (ed) Gold nanoparticles: preparation, characterization and fabrication. Nova Science Publishers, New York, pp 147–171

12. Lopez-Sanchez JA, Lennon D (2005) The use of titania- and iron oxide-supported gold catalysts for the hydrogenation of propyne. Appl Catal A 291:230–237.

13. Nikolaev SA, Smirnov VV (2009) Selective hydrogenation of phenylacetylene on gold nanoparticles. Gold Bull 42:182–189.

14. Besenbacher F, Chorkendorff I, Clausen BS, Hammer B, Molenbroek AM, Nørskov JK, Stensgaard I (1998) Design of a surface alloy catalyst for steam reforming. Science 279:1913–1915.

15. Nikolaev SA, Smirnov VV (2009) Synergistic and size effects in selective hydrogenation of alkynes on gold nanocomposites. Catal Today 147S:S336–S341.

16. Haruta M (2004) Nano particulate gold catalysts for low-temperature CO oxidation. J New Mat Elect Syst 7:163–172

17. Mohr C, Hofmeister H, Claus P (2003) The influence of real structure of gold catalysts in the partial hydrogenation of acrolein. J Catal 213:86–94.

18. Okumura M, Akita T, Haruta M (2002) Hydrogenation of 1,3-butadiene and of crotonaldehyde over highly dispersed Au catalysts. Catal Today 74:265–269.

19. Mohr C, Hofmeister H, Radnik J, Claus P (2003) Identification of active sites in gold-catalyzed hydrogenation of acrolein. J Am Chem Soc 125:1905–1911.

20. Perdew JP, Burke K, Ernzerhof M (1996) Generalized gradient approximation made simple. Phys Rev Lett 77:3865–3868.

21. Visscher L (2002) Chapter 6. Post Dirac-Hartree-Fock methods—electron correlation. Theor and Comput Chem 11:291–331.

22. Laikov DN (1997) Fast evaluation of density functional exchange-correlation terms using the expansion of the electron density in auxiliary basis sets. Chem Phys Lett 281:151–156.

23. Mukhamedzyanova DF, Ratmanova NK, Pichugina DA, Kuz'menko NE (2012) A structural and stability evaluation of Au_{12} from an isolated cluster to the deposited material. J Phys Chem C 116:11507–11518.

24. Claus P (2005) Heterogeneously catalysed hydrogenation using gold catalysts. Appl Catal A 291:222–229.

25. Zhang Q, Li J, Liu X, Zhu Q (2000) Synergetic effect of Pd and Ag dispersed on Al_2O_3 in the selective hydrogenation of acetylene. Appl Catal A 197:221–228.

26. Del Angel G, Benitez JL (1993) Selective hydrogenation of phenylacetylene on Pd/Al_2O_3: effect of the addition of Pt and particle size. React Kinet Catal Lett 51:547–553.

27. Medina F, Salagre P, Sueiras J-E, Fierro J-LG (1994) Characterization of several γ-alumina-supported nickel catalysts and activity for selective hydrogenation of hexanedinitrile. J Chem Soc Faraday Trans 90:1455–1459.

28. Jia J, Haraki K, Kondo JN, Domen K, Tamaru K (2000) Selective hydrogenation of acetylene over Au/Al_2O_3 catalyst. J Phys Chem B 104:11153–11156.

29. Gluhoi AC, Bakker JW, Nieuwenhuys BE (2010) Gold, still a surprising catalyst: selective hydrogenation of acetylene to ethylene over Au nanoparticles. Catal Today 154:13–20.

30. Bus E, Prins R, van Bokhoven JA (2007) Origin of the cluster-size effect in the hydrogenation of cinnamaldehyde over supported Au catalysts. Catal Comm 8:1397–1402.

31. Bus E, Miller JT, van Bokhoven JA (2005) Hydrogen chemisorption on Al_2O_3-supported gold catalysts. J Phys Chem B 109:14581–14587.

32. Boronat M, Concepción P, Corma A, González S, Illas F, Serna P (2007) A molecular mechanism for the chemoselective hydrogenation of substituted nitroaromatics with nanoparticles of gold on TiO2 Catalysts: a cooperative effect between gold and the support. J Am Chem Soc 129:16230–16237.

33. Boronat M, Lllas F, Corma A (2009) Active sites for H_2 adsorption and activation in Au/TiO_2 and the role of the support. J Phys Chem A 113:3750–3757.

34. Boronat M, Concepción P, Corma A (2009) Unraveling the nature of gold surface sites by combining IR spectroscopy and DFT calculations. Implications in catalysis. J Phys Chem C 113:16772–16784.

35. Kartusch C, van Bokhoven JA (2009) Hydrogenation over gold catalysts: the interaction of gold with hydrogen. Gold Bull 42:343–348.

36. Fujitani T, Nakamura I, Akita T, Okumura M, Haruta M (2009) Hydrogen dissociation by gold clusters. Angew Chem Int Ed 48:9515–9518.

37. Yong B, Cao X-M, Gong X-Q, Hu P (2012) A density functional theory study of hydrogen dissociation and diffusion at the perimeter sites of Au/TiO_2. Phys Chem Chem Phys 114:3741–3745.

38. Zhang X, Shi H, Xu B-Q (2011) Vital roles of hydroxyl groups and gold oxidation states in Au/ZrO_2 catalysts for 1,3-butadiene hydrogenation. J Catal 279:75–87.

39. Shimizu K-i, Yamamoto T, Tai Y, Satsuma A (2011) Selective hydrogenation of nitrocyclohexane to cyclohexanone oxime by alumina-supported gold cluster catalysts. J Mol Catal A 345:54–59.

40. Ivanov AV, Koklin AE, Uvarova EB, Kustov LM (2003) A DRIFT spectroscopic study of acetylene adsorbed on metal oxides. Phys Chem Chem Phys 5:4718–4723.

41. Azizi Y, Petit C, Pitchon V (2008) Formation of polymer-grade ethylene by selective hydrogenation of acetylene over Au/CeO_2 catalyst. J Catal 256:338–344.

42. Lijuan Z, Tian Ruili HP, Yuru M, Dingguo X (2010) A gold-nickel alloy as anodic catalyst in a direct formic acid fuel cell. Rare Metal Mat Eng 39:945–948.

43. Merlo AB, Machado BF, Vetere V, Faria JL, Casella ML (2010) PtSn/SiO_2 catalysts prepared by surface controlled reactions for the selective hydrogenation of cinnamaldehyde. Appl Catal A 383:43–49.

44. Coq B, Figueras F (2001) Bimetallic palladium catalysts: influence of the co-metal on the catalyst performance. J Mol Catal A 173:117–134.

45. Zhang X, Shi H, Xu B-Q (2005) Catalysis by gold: isolated surface Au^{3+} ions are active sites for selective hydrogenation of 1,3-butadiene over Au/ZrO_2 catalysts. Angew Chem Int Ed 44:7132–7135.

46. Zhang X, Shi H, Xu B-Q (2007) Comparative study of Au/ZrO_2 catalysts in CO oxidation and 1,3-butadiene hydrogenation. Catal Today 122:330–337.

47. Gao M, Lyalin A, Taketsugu T (2011) Role of the support effects on the catalytic activity of gold clusters: a density functional theory study. Catalysts 1:18–39.

48. Segura Y, López N, Pérez-Ramírez J (2007) Origin of the superior hydrogenation selectivity of gold nanoparticles in alkyne+alkene mixtures: triple-versus double-bond activation. J Catal 247:383–386.

9

On the hot pressing of coloured high-gold alloys powder compacts applied to the manufacturing of innovative jewellery items

B. Henriques · P. Pinto · J. Souza · J. C. Teixeira · D. Soares · F. S. Silva

Abstract Innovative design through colour has always been a very important feature in decorative items, jewellery inclusive. The use of coloured gold alloys (e.g. red, white, green) in the manufacture of jewellery artifacts is nowadays generalized, and one can often see jewellery items that combine more than a single colour. However, multi-coloured jewellery artifacts are generally made of parts produced separately and bonded together in a later step. The ability to produce jewellery sub-components or final pieces using a powder metallurgy process aimed at producing innovative multi-coloured gold jewellery items with colour gradients is explored in this paper. Four 19.2 ct coloured gold powders (yellow, green, red and white) were hot pressed at different times and temperatures and the resulting samples were analyzed in terms of their microstructure, hardness, porosity, microstructure and optical properties. A manufacturing route to obtain multi-coloured jewellery pieces with colour gradients is also presented.

Keywords Gold alloys · Powder metallurgy · Hot pressing · Optical properties · Microstructure · Hardness

Introduction

Gold and gold alloys have been for ages the metal of choice for jewellery manufacturing. The jewellery items are classified in terms of gold content, described in terms of caratage, and a range of caratages are used depending on the country in which the jewellery items are traded. High caratages (24, 22 and 21 ct) are mostly traded in Asian countries whereas lower caratages (18, 14, 9 ct) are mostly found in western countries [1]. Nevertheless, there are some exceptions to this rule and Portugal is one of them. The standard caratage in Portugal is 19.2 ct (800 fineness), although sale of other caratages is now allowed.

Coloured gold is an aesthetic feature that has been used by jewellery makers and designers in the creative process in order to make jewellery more appealing to customers. Gold and copper are the only two pure metals with intrinsic colour, all other pure metals being white or grey. Moreover, gold alloys can be produced in several colours depending on the alloying elements that are added. The conventional carat gold colours often used in jewellery making are yellow, green and red, and they are all based in the Au–Ag–Cu ternary system. White carat golds have also been used since 1920s when there was a need for platinum substitution. There are two main types of white golds: the nickel whites and palladium whites [2]. Both nickel and palladium have a strong bleaching effect in gold. Gold can assume almost any colour in the visible spectrum but most of the unusual coloured alloys lack ductility and toughness, and cannot undergo considerable plastic deformation, which make them inappropriate for the production of jewellery. Blue and purple golds are two good examples of fragile coloured golds, which are formed by intermetallic compounds. Its application has been shown possible after being faceted and used as gemstones and inlays [3]. More recently, Fischer-Bühner et al. [4] and Klotz [5] have shown some improvements in the production of plated and cast blue and purple golds.

Innovative design plays a determinant role in a fast-changing industry that the jewellery industry is [6]. The demand for up-to-date jewellery items in shorter time frames has put the production technology in the centre of the jewellery business, as a key element of competitiveness.

B. Henriques (✉) · P. Pinto · J. Souza · J. C. Teixeira · D. Soares · F. S. Silva
Center for Mechanical and Materials Technologies,
Universidade do Minho, Campus de Azurém,
4800-058 Guimarães, Portugal
e-mail: brunohenriques@dem.uminho.pt

The introduction of non-conventional technologies such as lasers, CAD/CAD systems, electroforming and powder metallurgy (P/M) in the production of jewellery are just a few examples of the technological tools available today in the jewellery industry [7]. This work intends to demonstrate P/M as a technological tool to create innovative jewellery items. P/M has been presented on several occasions and discussed at jewellery forums, such as the Santa Fe Symposium [8–12] and World Gold Council Technical Conferences [13, 14]. The P/M technique has also been shown to be cost effective in the production of carat gold wedding rings, in which their mechanical properties compared favorably with those produced by lost wax casting technique [9]. The production of products which combine several colours can be accomplished by this technique, either in a random way as demonstrated by Taylor [15], or in a continuous transition in colour, as proposed by Böhm [7]. The multi-coloured products thus produced have the advantage over any coloured coatings of being permanent, i.e. the aesthetic feature results in the bulk material and it cannot be removed by any erosive agent, thus lasting for the piece's lifetime.

The present work is devoted to the study of the hot pressing conditions of several coloured 19.2 ct gold alloys and the influence of processing conditions on the microstructure, hardness and optical properties of the hot pressed samples. The aim was to determine the appropriate processing conditions for the production of innovative jewellery pieces using a P/M process. Several examples of products thus obtained are disclosed in the last section of the paper.

Material and methods

Powders

In this study, four different 19.2 ct gold alloys were used in the form of metal powders. Table 1 shows the chemical composition of the alloys obtained by energy-dispersive X-ray spectroscopy (Nova 200, FEI, Oregon, USA). Powders were obtained by a process of solid-state reduction by the means of in-house equipment that uses a file to produce powders on a small scale. At this point, it is important to state that no significant iron traces were found in the EDS analysis of powders. The morphology of the irregular-shaped powders thus produced is presented in Fig. 1. Powders can be spherical or irregular in geometry. Air atomized powders are spherical while water atomized powders or those obtained from a filing process are irregular. The morphology of powders can have a strong influence on the mechanical strength of pre compacted powder compacts, also called green compacts. Fine and irregular powders are preferred to coarse and spherical powders due to higher contact points between particles and higher ability for mechanical interlocking to occur. In this study, despite powders having an irregular shape, which might have resulted in higher green strength, the powders were not subjected to pre-compaction for shaping before the hot pressing process. As will be presented ahead in this paper, they rather underwent a hot compaction inside the graphite die during the initial stage of the hot pressing process. High temperatures make powders softer and easy to densify. Therefore, in this case as well as in other similar situations, the shape of the powders did not assume a critical role. Less geometric and size specifications of powders processed by hot pressing may result in relevant cost savings in the manufacturing process.

The size distribution of powders was determined by the means of a laser particle size analyzer (Malvern HSD2600, Malvern Instruments Ltd, England). The particle size distribution of powders is shown in Table 2.

Processing conditions

Several gold coloured specimens were obtained by the hot pressing P/M technique. Powders were hot pressed under a range of selected conditions of pressure, time and temperature, which are shown in Table 3. Hot pressing was performed

Table 1 Chemical composition of 19.2 ct coloured gold alloys

Alloy	Colour	Au	Cu	Ag	Ni	Zn
A	Yellow	80 (79)	10(11)	10		
B	Green	80(81)		20 (19)		
C	Red	80 (81)	(20) 19			
D	White	80 (76)	10 (12)		5 (7)	5(5)

Numbers in brackets indicate the composition, in wt.%, of the alloys measured by EDS analysis

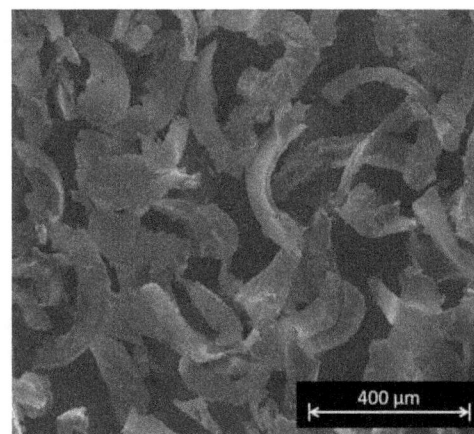

Fig. 1 SEM micrograph showing the morphology of the 19.2 ct gold alloy powders (×250)

Table 2 Powders size distribution of several 19.2 ct coloured gold alloys [in micrometer]

Alloy	Colour	d_{10}	d_{50}	d_{90}
A	Yellow	64	158	223
B	Green	66	161	224
C	Red	53	129	219
D	White	50	114	209

d_{10}, d_{50}, d_{90} particle size values indicating that, respectively, 10, 50 and 90 % of the distribution is below this value, i.e. a d_{10} of 64 μm means 10 % of the sample is below 64 μm in size (using a volume-based calculation)

under vacuum (10^{-2} mbar) using in-house equipment (Fig. 2) based on a vacuum chamber and an induction heating furnace (Ameritherm Easyheat 5060). The die was made of graphite and had a cavity with 4 mm diameter. Specimens had therefore the diameter (4 mm) and also 4 mm height.

Microstructure and chemical analysis

The hot pressed compacts were analyzed by optical microscopy (Axiotech, Carl Zeiss, USA) and SEM/EDS (Nova 200, FEI, Oregon, USA). For that, all samples were embedded in cold-curing-type epoxy resin and subjected to the wet grinding process. The samples were first ground on SiC paper down to a 1200-grit finish and then successively polished using a polishing cloth filled with diamond suspension with 6 and 1 μm particles. Then, all specimens were ultrasonically cleaned in an alcohol bath for 10 min and rinsed in distilled water for another 10 min to remove contaminants. Afterwards, they were dried with adsorbent paper towels.

The microstructures of hot pressed specimens were examined after etching the surface with a solution based on 100 g hydrochloric acid+3 g chromic acid.

Porosity measurement

The porosity existing in the hot pressed powder compacts was measured by the means of image analysis (ImageJ). The porosity of each specimen was calculated as an average of six measurements at random locations on the specimen's surface. Prior to image analysis, all specimens had undergone a further polishing cycle down to a mirror finish state, followed by another cleaning step.

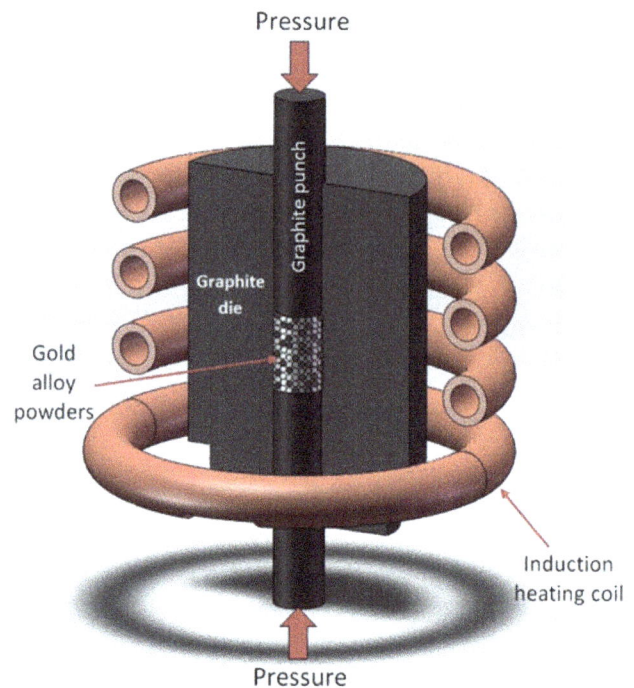

Fig. 2 The hot pressing process: schematic of the apparatus

Hardness measurement

The Vickers hardness evaluation (microhardness tester, type M, Shimadzu, Japan) of the hot pressed compacts was performed for all specimens. Five indentations were made in each specimen and the mean value and standard deviation were calculated. The load used was 1,000 g during 15 s.

Spectrophotometric colourimetry

Colour measurement of the samples was performed in a computer-controlled spectrophotometer (Shimadzu UV-2450, Shimadzu Corporation, Kyoto, Japan). Spectral reflectance data from the mirror-polished flat surface were recorded at 1 nm wavelength intervals from 380 to 780 nm with the geometry of diffuse illumination and 10° viewing. The three-dimensional colour coordinates using the CIELab system, L^* (lightness), a^* (red–green chromaticity index) and b^* (yellow–blue chromaticity index), were obtained. The exposed area of the hot pressed specimens for colour measurement was 7 mm^2.

Results and discussion

Microstructure

The microstructures of the hot pressed gold coloured samples are shown in Fig. 3. They all exhibited similar microstructures, regardless of the hot pressing time and temperature. Similar grain sizes were found among the different alloys

Table 3 Hot pressing conditions of the gold alloy powders

Pressure (MPa)	Time (min)	Temperature (°C)
35	10	700
	30	800
	60	

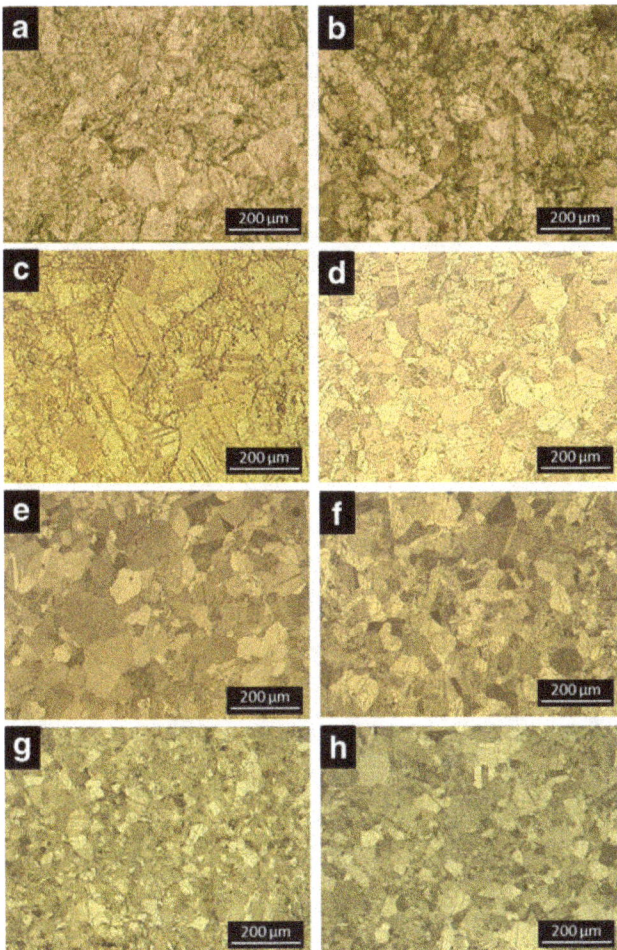

Fig. 3 Microstructures of 19.2 ct coloured gold alloys hot pressed for 30 min at the temperatures of 700 and 800 °C: **a** green gold, 700 °C; **b** green gold, 800 °C; **c** yellow gold, 700 °C; **d** yellow gold, 800 °C; **e** red gold, 700 °C; **f** red gold, 800 °C; **g** white gold, 700 °C; **h** white gold, 800 °C (×200)

and no significant grain growth could be observed for higher temperatures (800 °C) and longer hot pressing times (60 min). The similarity to the grain size of the hot pressed coloured samples contrasted with the different grain structures of the as-cast bulk alloys, where the green gold (alloy A) exhibited the finest grains and the red gold (alloy C) exhibited the coarser grains. It has been demonstrated that a fine-grain structure is preferred to a coarse grain one, as it positively impacts the mechanical properties, such as tensile strength and elongation [16]. The surfaces of fine-grained gold alloys have also been reported to be more easily polished [17]. However, fine-grain structures in gold alloys (as also in copper alloy castings) do not seem to improve other properties like the hardness and yield strength [16, 18], at least in the practical range of grain refinements commonly used in the jewellery manufacturing. The grains size of the gold hot pressed samples shown in Fig. 3 is below 100 μm, which is regarded as fine grains in precious metals technology standards [16].

No dendritic segregations were found among the hot pressed samples, contrary to what is often found in cast samples [19]. The microstructures of all samples also revealed the presence of twins within the grains, which are often found in gold–silver–copper alloys. The appearance of twins as well as the lack of as-cast dendritic structure might be synonym of recrystallization having occurred, hence lack of segregation seen. This might also indicate that recrystallization is dynamic during hot pressing (hot working).

Porosity

Figure 4 shows the porosity measured in the samples for the different alloys and processing conditions. The porosity is lower than 0.5 %, except for the case of alloy A (green gold) that showed higher porosity (~0.9 %) for the hot pressing condition of 700 °C, 10 min. Results showed that no significant difference could be observed in the porosity level of the hot pressed samples, regardless the processing conditions used in this study. This means that at these temperature levels, the sintering time did not substantially impact the porosity of the compacts. On the other hand, it can significantly influence other mechanical properties of the samples, such as tensile and rupture strength, but that issue was not assessed in this study. The hot pressing technique is known for producing fully dense parts [20–26] and the results reflect well this feature of the process.

Hardness

Hardness is perhaps the most important property among the mechanical properties that can be assessed in an alloy for jewellery making: first, because it is easy to measure and, second, because it greatly impacts the fabrication and the performance of the jewellery item when worn by the customer [27].

The hardness data of hot pressed samples obtained in different conditions are shown in Fig. 5. The data shows that the hardness values are not significantly changed with the different processing conditions within each type of gold alloy. These results may be related with the results exhibited in Fig. 4 that show that all specimens are fully dense ($d > 99$ %), regardless the processing condition. Also, at such high hot pressing temperatures (700 and 800 °C) these alloys, all containing Cu, do not develop any age hardening mechanism. The main hardening mechanisms are therefore the solid solution and the ordering of the Cu-rich phase, which accounts for the different hardness of the alloys.

Age hardening occurs either in high or low carat alloys during the annealing of the alloy at moderate temperatures, ranging from 300 to 500 °C, which causes the formation of two phases, silver- and copper-rich solid solutions [28]. The hardening of the alloy is due to the formation of these two

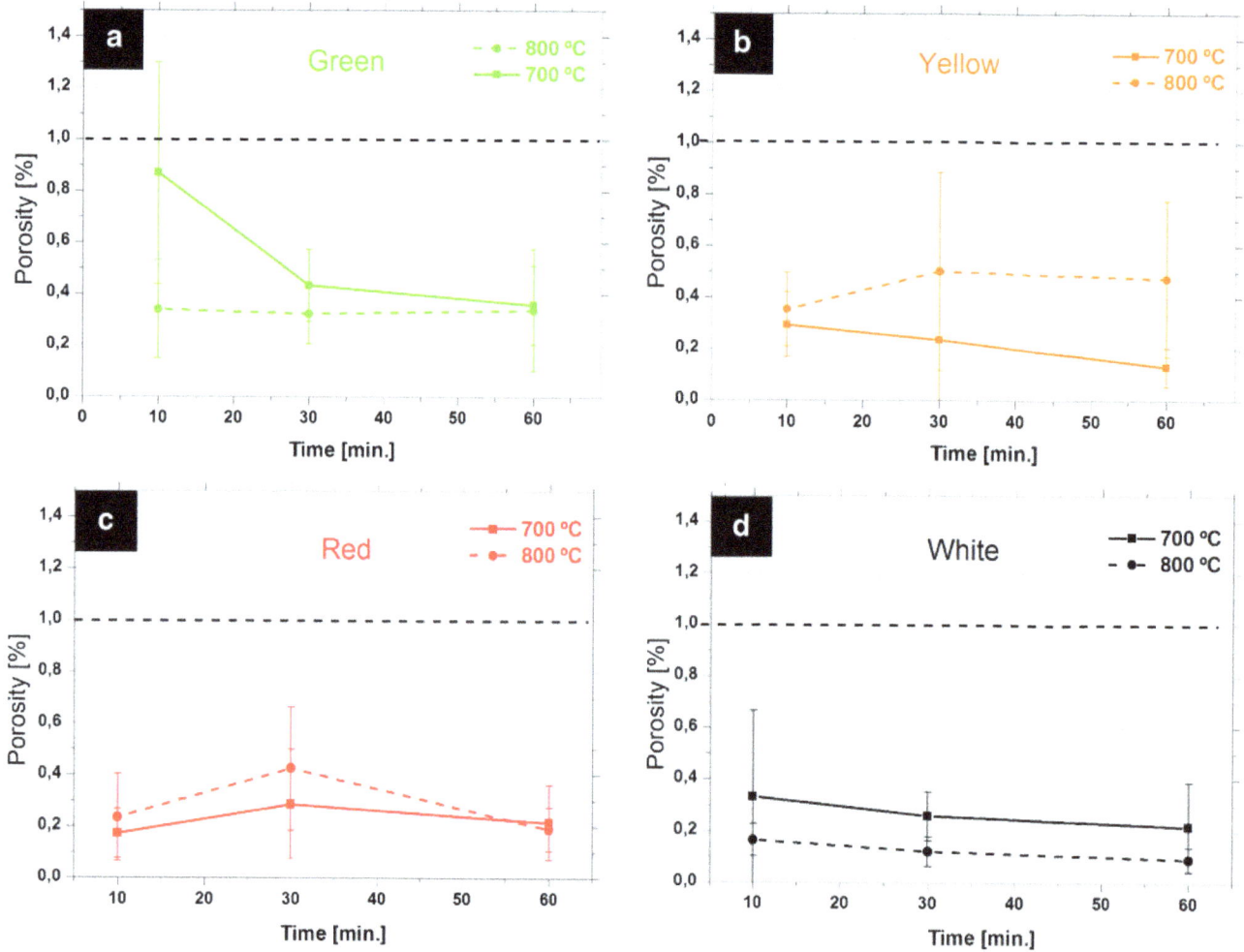

Fig. 4 Influence of the processing parameters (time and temperature) on the porosity of the hot pressed 19.2 ct gold samples of different chemical compositions

phases, the silver- and the copper-rich solid solutions, where the precipitation of the latter phase accounts for the hardening

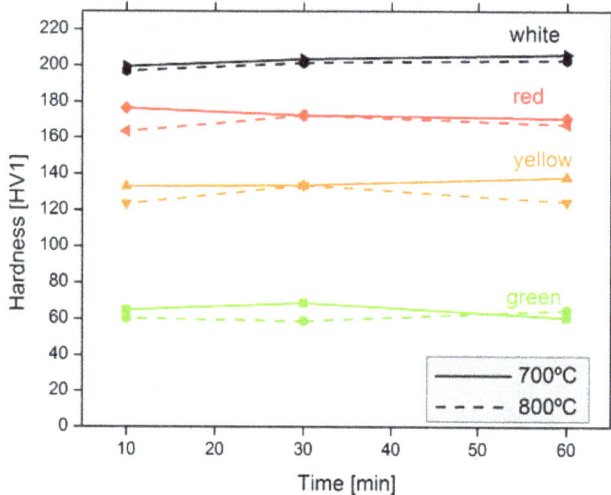

Fig. 5 Influence of the processing parameters (time and temperature) on the hardness of the hot pressed 19.2 ct coloured gold alloy powder compacts

of the alloy. Age hardening in copper-containing alloys is also due to ordering reaction (order/disorder). It is important to point out that in the present study the samples were annealed at 700 and 800 °C (during the hot pressing process with duration of 10, 30 and 60 min) and not age hardened.

The alloys A (green gold), B (yellow gold) and C (red gold) are based on the ternary Au–Ag–Cu systems and they consequently follow the typical properties of these alloys. In this system, the strength and hardness of the alloy increases with increasing content of copper [28]. The strengthening and hardening mechanisms caused by the silver and copper addition to gold is called solid solution strengthening. However, due to copper's smaller atoms size and higher shear modulus relative to silver ones, its substitution in the gold crystal lattice results in a greater hardening effect. Hence, alloy A (green gold), which contain no copper, exhibited the least hardness, 61–69 HV. The alloy B (yellow gold), containing 10 wt.% copper, presented higher hardness (124–138 HV). Finally, alloy C, which contained 20 wt.% copper displayed the highest hardness among the alloys based in the Au–Ag–Cu system, between 165 and 176HV. The alloy D (white gold) is a nickel-white gold and presented the highest hardness among the alloys tested (197–206 HV). Nickel is added to gold to provide it with a white colour along with increased strength and hardness [2].

Spectrophotometric colourimetry

Figure 6 shows a 3D plot comprising the three colour co-ordinates of the CIELab system (L^*, a^* and b^*), measured for the hot pressed coloured gold samples processed in different conditions (Table 3). The L^* coordinate expresses the luminance (brightness) and it can vary between 0, where no light is reflected by the sample, and 100, meaning that all incident light is reflected. The L^* value registered for the different samples did not significantly differ within each group, ranging from ~55 to ~59. These luminance values are relatively lower than those reported by Shirishi et al. [29], but such differences are related to colour measuring experimental set up, particularly with the reduced area of the samples that was used for measuring the colour.

The a^* coordinate measures the intensity of the green (negative) or red (positive) component of the spectrum, whereas b^* measures the blue (negative) or yellow (positive) component part. Figure 6 shows a significant variation in a^* and b^* coordinates between the four type of alloys A–D but no significant changes within the same group of specimens. Thus, the alloy A exhibited a low values for a^* coordinate, from 0.46 to 2.39, and higher b^* coordinates (20.9 to 23.1), which reveal a strong intensity of the green colour. The alloy

B exhibited higher values for a^* coordinates (4.4 to 5.5) and lower b^* coordinates (17.2 to 19.2) relative to alloy A, which resulted in the measurement of an intense yellow component coupled with a moderate component of red, thus providing to this alloy a reddish yellow colour. The alloy C displayed the highest value for a^* coordinate (6.5 to 7.2) and relatively low values for b^* coordinate, which makes this alloy looks red. Finally, the alloy D exhibited low a^* (2.3 to 2.7) and b^* (10.1 to 11.8) coordinate values, which is typical of metals with white colour such as pure silver and pure platinum [29].

Figure 7 shows the reflectance curves for the hot pressed coloured gold samples, processed at 700 °C for 30 min. The reflectance data obtained in this particular processing condition was considered to be representative of the global reflectance behaviour exhibited by the other specimens when processed under the other conditions. The spectral reflectance of the gold alloys used in this study revealed a general trend of higher reflectance for long wavelengths and lower reflectance for short wavelengths within the visible spectrum. Thus, the reflectance exhibited a decreasing trend from long wavelengths towards short wavelengths, with a higher decreasing slope being visible in the wavelengths interval comprised between 500 and 60 nm for alloys A, B and C, but less pronounced in alloy D.

The higher reflectance for high wavelengths exhibited by alloys A, B and C are in agreement with Shiraishi et al. [29] findings for gold alloys with high gold content. The same authors showed that the three pure elements composing these alloys, i.e. Au–Ag–Cu, all exhibited the same reflectance pattern for high wavelengths. In contrast, at low wavelengths, the alloy A displayed lower reflectance than

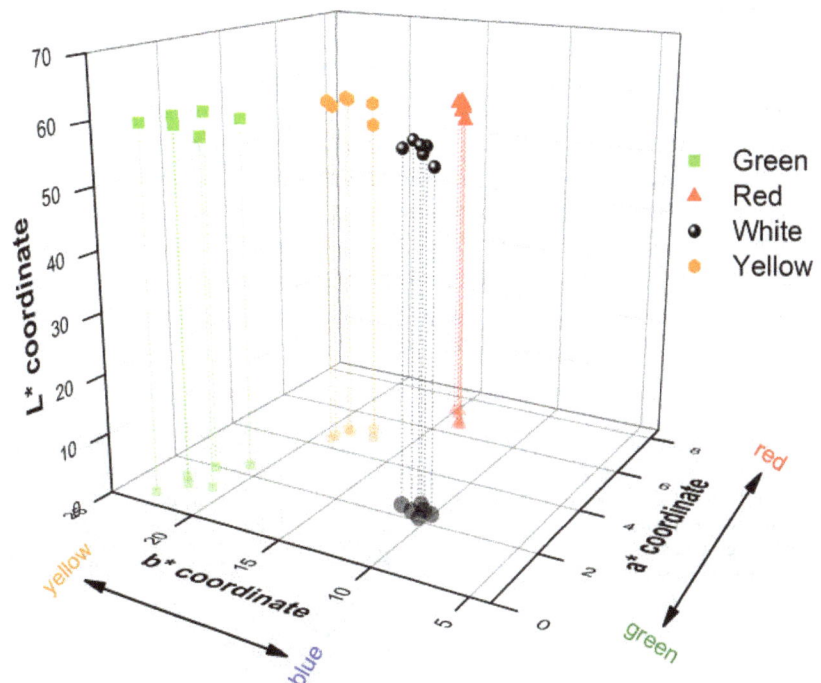

Fig. 6 Colour comparison of the 19.2 ct coloured gold hot pressed samples plotted in the CIELab system

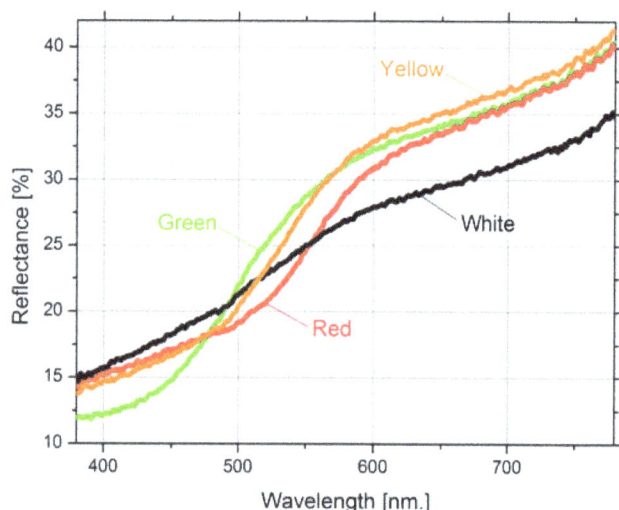

Fig. 7 Reflectance of hot pressed coloured gold alloys plotted against the wavelength in the region of the visible spectrum (380–780 nm)

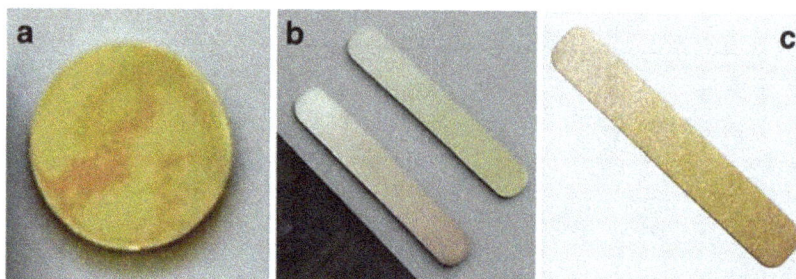

Fig. 9 Multi-coloured gold ring exhibiting parts with colour gradients fabricated by powder metallurgy. Courtesy of Grad'Or Jewels, Portugal (www.gradorjewels.com)

those of alloys B and C. The behaviour of alloy A is explained by the presence in the alloy of 20 wt.% of Ag, as single alloying element. It was demonstrated by Shiraishi et al. [29] that, in Au–Ag alloys, the addition of Ag to pure Au up to a maximum content of 70 wt.%, resulted in a reflectance drop for low wavelengths in comparison to that observed for pure gold.

The alloy D showed similar behaviour to alloys A, B and C, with low reflectance values for low wavelengths. Regarding high wavelengths, the reflectance was lower than that exhibited by the other alloys and the steep reflectance transition between low and high wavelengths could not be observed. The spectral reflectance curve of this alloy showed to be highly influenced by the presence of Ni as alloying element, once its behaviour is very similar to that exhibited by pure nickel [30].

The reflectance data displayed in Fig. 7 may seem lower than expected, particularly when compared to those presented elsewhere [29], where significantly higher reflectance values are reported. This fact is explained by the reduced surface area of the specimens used in this study. As the L value regards to the amount of radiation of the light source that is reflected from the samples surface, the reduced area of analysis used in this study (7 mm^2) showed to have influence on the amount of reflected light that was measured by the spectrophotometer sensor. This fact was confirmed when a higher surface area was tested for a couple of samples and higher L values were obtained. As a consequence of this fact, the analysis of results displayed in Fig. 7 should therefore assume a relative dimension in the comparison of the different gold samples and no definite conclusions about the absolute reflectance values of the specimens should be taken.

Application to the production of jewellery items

P/M can be used as a route for the production of innovative jewellery products, taking advantage of the mixture of powders with different colours to produce patterns with an aesthetic end. Figure 8 shows gold coloured samples produced by the hot pressing powder metallurgy process, using the same gold powders that were studied above. Figure 8a shows a round part with a bicolour random pattern, yellow and red. This component was obtained by randomly arranging the yellow and red gold powders in a graphite die, followed by hot pressing treatment at 700 °C during 30 min.

Figure 8b shows two parts exhibiting a gradient in colour from one end of the component to the other: from green to white, and from red to white. The transition between the two

Fig. 8 Carat gold components obtained by powder metallurgy: **a** random pattern in yellow and red colours; **b** colour gradient between the red–white and green–white, ground surface; **c** colour gradient between yellow and white, grit-blasted surface

colours in the parts' tips was made in four steps, i.e. four mixtures of the two coloured powders with different volume fractions were used between the two main colours. Powders were hot pressed in a graphite die to final shape at 700 °C for 30 min. To achieve the appropriate colour contrast, the surface of the part should be in matte finish, which can be also used for masking any surface defect. Parts in Fig. 8b were ground using a 600-grit SiC paper. A sandblasting surface treatment can also be applied to the surface of component as shown in Fig. 8c. In this case, the component with colour gradient transition between yellow and white was grit-blasted with 250 µm alumina particles.

These are a few examples of how powder metallurgy involving coloured carat gold can be applied to the production of innovative designs for jewellery. Figure 9 shows a jewellery ring where the concept of colours and colour gradients had been applied. It is inspired in a flower where the top petals display a radial continuous colour transition between white and yellow.

Conclusions

Within the limitations of this study, the following conclusions can be drawn:

1. The different hot pressing conditions did not produce significant changes in microstructure among the tested alloys;
2. The mean porosity measured in the hot pressed samples was lower than 1 %;
3. The hardness of the hot pressed samples was not significantly affected by the different processing conditions;
4. The coloured gold alloys showed different coordinates in the CIELab system and the colour of each alloy was not significantly affected by the different hot pressing conditions;
5. The alloys based on the ternary systems Au–Ag–Cu (A, B and C) exhibited similar spectral reflectance curves, whereas that of the alloy D was different, especially for long wavelengths. This behaviour was explained by the presence of nickel in the alloy.
6. The innovation possible in the design of jewellery items through the use of powder metallurgy was demonstrated in this study.

Acknowledgments This work has been supported by post-doc grant of FCT (Portuguese Foundation for Science and Technology) with the reference SFRH/BPD/87435/2012. The authors would like to address special thanks to Doctor Christopher W. Corti and to Engineer Paolo Battaini for their kind advice in this paper.

References

1. Cretu C, van der Lingen E, Glaner L (2000) Hard 22 carat gold alloy. Gold Technol (No. 29, Summer):p25–28
2. Cretu C, van der Lingen E (1999) Coloured gold alloys. Gold Bull 32(4):115–126
3. Poliero M (2001) White golds for investment casting. Gold Technol (No. 31, Spring):p10–20
4. Fischer-Bühner J, Basso A, Poliero M (2010) Metallurgy and processing of coloured gold intermetallics—part II: investment casting and related alloy design. Gold Bull 43(1):11–20
5. Klotz UE (2010) Metallurgy and processing of coloured gold intermetallics—part I: properties and surface processing. Gold Bull 43(1):4–10
6. Corti CW (2003) Technology is irrelevant to jewellery design—or is it? In: Bell E (ed) Proceedings of the Santa Fe Symposium, pub. Met-Chem Research Inc, Albuquerque, NM, USA. p15–28
7. Böhm W (1998) Design opportunities through production technology. Gold Technol (No. 23, April):p8–11
8. Strauss JT (2003) P/M (powder metallurgy) in jewelry manufacturing; current status, new developments, and future projections. In: Bell E (ed) Proceedings of the Santa Fe Symposium, pub. Met-Chem Research Inc, Albuquerque, NM, USA, p387–412
9. Raw P (2000) Mass production of gold and platinum wedding rings using powder metallurgy. In: Bell E (ed) Proceedings of the Santa Fe Symposium, pub. Met-Chem Research Inc, Albuquerque, NM, USA, p251–270
10. Wiesner K (2003) Metal injection molding (MIM) Technology with 18-ct Gold-A Feasible Study. In: Bell E (ed) Proceedings of the Santa Fe Symposium, pub. Met-Chem Research Inc, Albuquerque, NM, USA, p443–462
11. Strauss JT (1996) Metal injection molding for gold jewelry production. Gold Technol (No. 20, November):p17–29
12. Raw P (1999) Gold wedding rings from powder—tomorrow's technology today. Gold Technol (No. 27, November):p2–8
13. Strauss JT (2004) The potential of MIM for the manufacture of precious metal components. Adv Powder Metall Part Mater
14. Strauss JT (2007) Application of MIM for jewelry manufacturing, presented at PIM2007, Orlando, Florida
15. Raw P (2000) Development of a powder metallurgical technique for the mass production of carat gold wedding rings. Gold Bull 33:79–88
16. Taylor SS (1997) Decorative precious metals composites give a new dimension to jewellery design. In: Bell E (ed) Proceedings of the Santa Fe Symposium, pub. Met-Chem Research Inc, Albuquerque, NM, USA, p443–465
17. Zito D (2001) Coloured carat golds for investment casting. Gold Technol (No. 31, Spring):p35–42
18. Ott D, Raub CJ (1981) Grain size of gold and gold alloys. Gold Bull 14:69–74
19. Ott D, Schindler U (2001) Metallography of gold and gold alloys. Gold Technol (No. 33 Winter):p6–11
20. Henriques B, Soares D, Silva F (2011) Optimization of bond strength between gold alloy and porcelain through a composite interlayer obtained by powder metallurgy. Mater Sci Eng, A 528:1415–1420
21. Henriques B, Soares D, Silva F (2011) Shear bond strength of a hot pressed Au–Pd–Pt alloy–porcelain dental composite. J Mech Behav Biomed Mater 4(8):1718–1726
22. Henriques B, Soares D, Silva F (2012) Microstructure, hardness, corrosion resistance and porcelain shear bond strength comparison between cast and hot pressed CoCrMo alloy for metal-ceramic dental restorations. J Mech Behav Biomed Mater 12:83–92
23. Henriques B, Soares D, Silva F (2012) Experimental evaluation of the bond strength between a CoCrMo dental alloy and porcelain

through a composite metal-ceramic graded transition interlayer. J Mech Behav Biomed Mater 13:206–214

24. Henriques B, Soares D, Silva F (2012) Shear bond strength comparison between conventional porcelain fused to metal and new functionally graded dental restorations after thermal-mechanical cycling. J Mech Behav Biomed Mater 13:194–205

25. Henriques B, Soares D, Silva F (2012) Influence of preoxidation cycle on the bond strength of CoCrMo–porcelain dental composites. Mater Sci Eng C 32(8):2374–2380

26. Henriques B, Soares D, Silva F (2013) Hot Pressing effect on the bond strength of a CoCrMoSi alloy to a dental porcelain. Mater Sci Eng C 33(1):557–563

27. Corti CW (2008) The role of hardness in jewellery alloys. In: Bell E (ed) Proceedings of the Santa Fe Symposium, pub. Met-Chem Research Inc, Albuquerque, NM, USA, p103–120

28. Grimwade M (2009) Introduction to precious metals. Metallurgy for jewelers and silversmiths, first printing. Brynmorgen

29. Shiraishi T, Takuma Y, Fujita T, Miura E, Hisatsune K (2009) Optical properties and microstructures of Pd-free Ag–Au–Pt–Cu dental alloys. J Mater Sci 44:2796–2804

30. Ahmad N, Stokes J, Fox NA, Teng M, Cryan MJ (2012) Ultra-thin metal films for enhanced solar absorption. Nano Energ 1:777–782

10

A comparative study of classical approaches to surface plasmon resonance of colloidal gold nanorods

Ngac An Bang · Phung Thi Thom · Hoang Nam Nhat

Abstract We report the errors in the evaluation of the surface plasmon resonance of gold nanorods by three classical approaches: the Gans model, the Discrete Dipole Approximation and the Surface Integral method. Using these methods, which are based on the propagation of an electromagnetic wave through a composite medium with different refractive indices, might result in an inaccurate prediction of absorption maxima. For test samples of nanorods prepared by a seed-mediated method, whose homogeneity and quality were also fully demonstrated in this study, the mismatches in the wavelengths of absorption maxima $|\Delta\lambda_{max}|$ between experimental and theoretical data were observed to be greater than 50 nm. In general, the observed surface plasmon resonances exhibit two distinctive bands corresponding to the transverse and longitudinal modes. The weak transverse mode was located in the region from 510 to 518 nm and varied slightly with the aspect ratio of the rods. In contrast, the longitudinal mode showed a strong dependence on aspect ratio and ranged from 658 to 768 nm. We demonstrated that the mismatches may be sufficiently reduced if the interdependence between these two modes is taken into account.

Keywords Gold · Nanorods · Plasmon · Gans · Discrete Dipole Approximation · Surface Integral

PACS 36.40.Gk · 42.25.Bs · 61.46.Km · 61.46.Bc

N. A. Bang · P. T. Thom
VNU-University of Science, 334 Nguyen Trai,
Thanh Xuan, Hanoi, Vietnam

H. N. Nhat (✉)
VNU-University of Engineering and Technology, 144 Xuan Thuy,
Cau Giay, Hanoi, Vietnam
e-mail: namnhat@gmail.com

Introduction

Colloidal gold nanostructures have great potential for application in nanomedicine, photonics, and optoelectronics owing to their surface plasmon resonance (SPR) properties [1, 2]. SPR is an optical phenomenon that arises from the interaction between an incident electromagnetic field and conduction electrons in metals. Under light irradiation, the conduction electrons are driven by the electric field to collectively oscillate at a resonant frequency relative to the energy of the incident light. At this frequency, the incident radiation is strongly absorbed, inducing a particular resonance behavior in a given nanostructure. This effect is of great interest for both analytical sensing and nanostructure functionalization in modern nanomedicine. The SPR cross-section generally depends strongly on the shape and size of a given gold nanostructure, as well as on the dielectric constant of the surrounding medium. There are two ways to understand SPR. At a low level theory, SPR is caused by the absorption of conduction electrons circulating in the delocalized molecular orbitals (MOs) at the surface of a metallic nanostructure. Because a number of valence band MOs are always available, there are also plenty of possible circulation routes around the nanoparticle surface. This appears to be a natural explanation for multipole resonance in metallic nanoclusters. The specific geometry of gold nanorods usually induces two main circulation paths, one corresponding to the transverse and the other to the longitudinal SPR mode. The modes are interdependent, and both vary with the aspect ratio (which is defined as the ratio of the length to the diameter of a rod). Qualitative evaluation may be performed using a technique such as time-dependent density functional theory (TD-DFT). Recent theoretical predictions of SPR based on the aspect ratio (AR) rely on Maxwell's theory of wave propagation through a composite medium with two different refractive indices. Absorption occurs at the interface where the refractive index of metallic nanoclusters abruptly changes to the value of vacuum (or that of the surrounding medium). The loss of momentum is one reason why absorption occurs. In this study, the SPR phenomenon of anisotropic gold nanostructures, gold nanorods prepared by a modified seed-mediated method, was investigated. We first discuss the careful characterization of sample quality and size homogeneity and show that both the transverse and longitudinal modes depend on the AR. Whereas a minor change in the transverse mode induced no observable shift in color, the large variation in the longitudinal mode (approximately 10 times larger) shifted the color of the colloids from green to bright red. The experimental results were then compared to the predictions made by classical approaches to show that within the framework of Maxwell's theory (where no correspondence between the two modes is taken into account), the deviation between theory and experiment might be large. The failure of classical theory lies in its fundamental assumption of a unique refractive index for an entire nanostructure, whereas it is clear that the density of the electronic cloud surrounding a metal nanostructure varies with position; thus, if SPR is being interpreted in terms of the refractive index, then one should take into account the variation of refractive index across the surface of a particular nanostructure.

Material preparation and characterization

Gold nanorods are commonly prepared by the seed-mediated method, which was first introduced by Jana et al. [3] and later modified and improved by Nikoobakht et al. [4]. A recent modification of the method to improve the reproducibility of gold nanorods in terms of their shape and size was reported in ref. [5]. Briefly, seed gold nanoparticles were prepared by adding an ice-cold aqueous solution of sodium borohydride ($NaBH_4$, Kanto) to a mixed aqueous solution of cetyltrimethylammonium bromide (CTAB, Aldrich) and hydrogen tetrachloroaurate hydrate ($HAuCl_4$, Sigma-Aldrich), resulting in the formation of a brownish-yellow solution containing seed gold nanoparticles. All chemicals used were of a high grade of purity. The age of the seed solution is an important factor in reproducing gold nanorod samples. In our experiments, the seed solution was used within 5 to 10 min of preparation, whereas in the original procedure, the solution was used within 2 to 5 h. A small amount of the gold seed solution was then added to a growth solution containing CTAB, $HAuCl_4$, silver nitrate ($AgNO_3$, Merck, 99.9 %), and ascorbic acid ($C_6H_8O_6$, Kanto, 99.5 %; silver nitrate was used as a growth catalyst and an agent for controlling the aspect ratio of the nanorods). The color of the solution changed over time depending on the final size of the gold nanorods. The nanorods were allowed to grow overnight

without stirring at 24 °C. The resulting solution was centrifuged to remove the surfactants and possible by-products such as large particles and cubes. For comparison, spherical gold nanoparticle samples were synthesized using the citrate reduction method [6]. This method, pioneered by Turkevich et al. [6], is one of the most popular and environmentally friendly ways to synthesize gold nanospheres. We modified the original procedure to improve the control over the size and size distribution of the resultant gold nanoparticles. Briefly, an aqueous mixture composed of hydrogen tetrachloroaurate hydrate ($HAuCl_4$, Sigma-Aldrich, 99.9 %) and sodium citrate ($Na_3C_6H_5O_7$, Scharlau, 99.5 %) in a certain molar ratio was heated to boiling under vigorous stirring. At 72 ± 1 °C, within a few minutes, the solution first became colorless, then turned dark blue, and finally became brilliant red or purple, depending on the molar ratio between the precursors. The gold nanoparticle sample was then left to cool naturally to room temperature. The average particle diameter can be tuned over a wide range (from approximately 10 to 100 nm) by varying the concentration ratio between the Au salt and sodium citrate.

The crystal structure and morphology of the synthesized samples were characterized using a Siemens D5005 XRD diffractometer and JEOL JEM-1010 transmission electron microscope (TEM), respectively. The absorption spectra of the samples were measured at room temperature using a Shimadzu UV–vis-2450PC spectrometer.

Figure 1a, b shows the typical XRD patterns of the synthesized gold nanorods and nanospheres, respectively. The pattern exhibits well-resolved (111) and (200) diffraction peaks, which could be well indexed to the faced-centered cubic (fcc) phase of metallic gold (PDF 04–0784, ICDD). It is worth pointing out that for the nanorods, the ratio of the intensity of the (200) diffraction peaks to the (111) diffraction peaks is substantially higher than both the standard value (0.53) given in PDF 04–0784 (ICDD) and that of the gold nanospheres shown in Fig. 1b. A sum of two Gaussians representing the XRD peaks and a decaying exponential function simulating the background was fitted to the XRD data. As a result of the fit, for the gold nanorods, the relative intensity ratio of the (200) and (111) diffraction peaks was found to be approximately 2.29. The (200) diffraction peak became the most prominent peak, which may indicate anisotropic growth of the crystal along the [100] direction, resulting in the formation of rod-like particles. In case of the spherical gold nanoparticles, the ratio of intensity between the (200) and (111) diffraction peaks is considerably lower than the standard value (0.39 versus 0.53). This observation supports the theoretical prediction [7] and the results of HRTEM analysis described elsewhere [8] that suggest that noble-metal nanocrystals are exclusively composed of the lowest-index crystal planes and, in general, the thermodynamic equilibrium shape is a truncated octahedron bound by {111} and {100} planes.

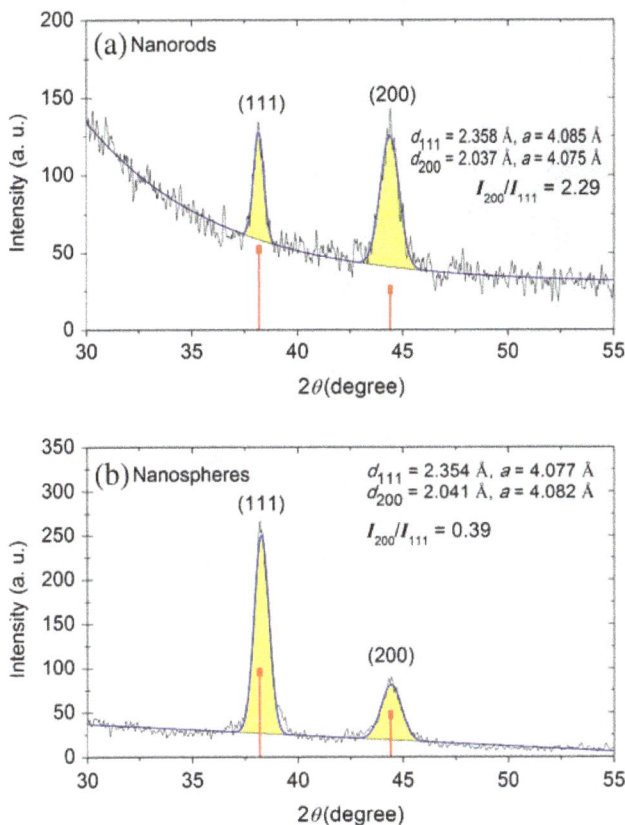

Fig. 1 The typical XRD patterns of the prepared gold nanorods (**a**) and gold nanospheres (**b**)

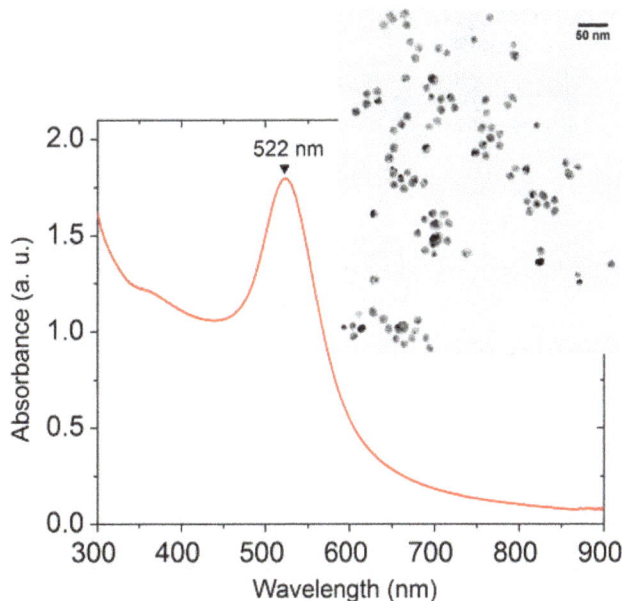

Fig. 2 The absorption spectrum of the gold nanospheres; the *inset* shows the corresponding TEM image

The morphology of the synthesized gold nanorods and particles was examined using TEM images with the help of the image processing program ImageJ [9]. Figure 2 shows the absorption spectrum of the gold nanospheres with a representative TEM image (the inset). As shown, the absorption maximum occurs only at 522 nm, which signifies a single resonance mode. The TEM image shown in the set demonstrates that the particles were well dispersed, with a narrow size distribution. The nanoparticles appeared to be spherical in shape, and their average size was determined to be 15.4 nm with a standard deviation of 2.3 nm. A total of six nanorod samples were produced: CR658, CR687, CR696, CR726, CR745, and CR768; TEM images of the samples are shown in Fig. 3a. The samples mostly consist of relatively well-dispersed gold nanorods, although by-products such as large particles and cubes were not negligible, especially in the case of samples CR695 and CR726. As was revealed by TEM, the synthesized gold nanorods are somewhat spherically or ellipsoidally capped cylinders. The average ARs of samples CR658, CR687, CR696, CR726, CR745, and CR768 were estimated from their corresponding TEM images to be 2.3 ± 0.1, 2.8 ± 0.1, 2.9 ± 0.1, 3.1 ± 0.1, 3.40 ± 0.15, and 3.8 ± 0.2, respectively. Sample CR768, shown in the lower-right corner of Fig. 3a, which exhibited the highest AR (3.8 ± 0.2), showed a broad AR distribution,

leading to a large standard error. However, compared with the results reported by other authors, the errors obtained in this study are quite smaller, which indicates that good size and shape homogeneity were achieved.

The SPR phenomenon of gold nanorods was monitored using standard UV–vis–NIR spectroscopy. The normalized absorption spectra of the gold nanorods produced in this study are shown in Fig. 3b. While symmetric gold nanospheres exhibit a single SPR peak at approximately 522 nm, anisotropic nanoparticles such as rods, cubes, triangular plates, and pyramids exhibit multiple SPR peaks due to highly localized charge polarizations at corners and edges [1, 2]. In all six spectra, there are two distinctive SPR bands corresponding to the oscillations of conduction electrons along and perpendicular to the long axis of the rods. The weak transverse surface plasmon resonance (TSPR) peak occurs in the region from 510 to 518 nm and does not depend strongly on the aspect ratio of the rods. However, the peak is clearly blue-shifted from the value of 522 nm observed for the nanospheres. The longitudinal surface plasmon resonance (LSPR) mode shows a strong resonance peak at longer wavelengths and depends strongly on the aspect ratio of the rods. As the average AR increases from 2.3 to 3.8, the LSPR peak is red-shifted from 658 to 768 nm, leading to an increase in the brightness of the colloid. By varying the AR, the LSPR peak can be effectively tuned over a wide wavelength range from the visible to near-infrared region, making gold nanorods very attractive candidates for applications in photonics, optoelectronics, and biotechnology [1, 2]. It is evident from Fig. 3b that the measured shifts, blue for the transverse mode and red for the longitudinal mode, appear to increase in magnitude with the aspect ratio.

Fig. 3 The TEM images of the nanorod samples CR658, CR687, CR696, CR726, CR745 and CR768 (**a**) and the corresponding normalized absorption spectra (**b**)

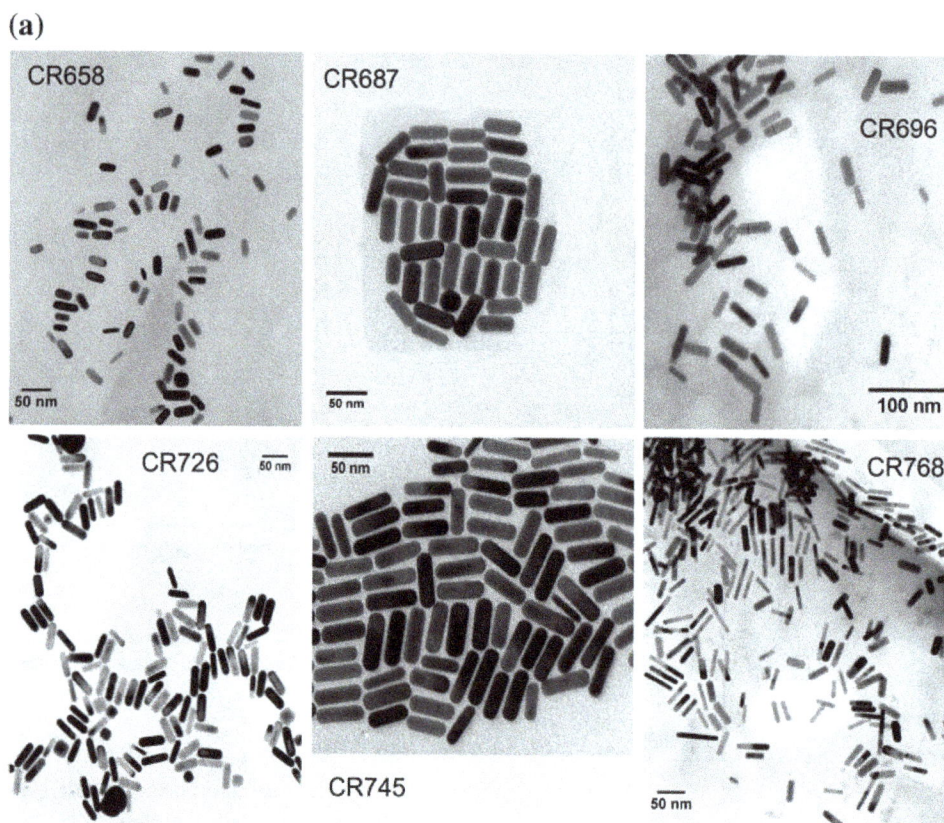

Discussion

The experimental results regarding the dependence of the position of the LSPR peak on the aspect ratio were compared to the theoretical predictions made by the Gans model [10], Discrete Dipole Approximation (DDA) [11], and the Surface Integral (SI) approach [12]. As an extension of Mie's theory, the Gans model predicts how the LSPR wavelength λ_{LSPR} varies with AR for small ellipsoids suspended in water (high dielectric constant of approximately 80.1 at

20 °C). The experimentally measured SPR peaks appear to be red-shifted from the values predicted by the Gans model, as shown in Fig. 4. There are a number of factors that account for this disagreement, but first and foremost, the Gans model treats particles as ellipsoids suspended in water, which is not the case for the studied samples.

The discrepancy between the experimental data and theoretical predictions is well-known, and several groups have tried to use computational methods such as the DDA [11] and SI [12] approaches to solve the problem. Taking into

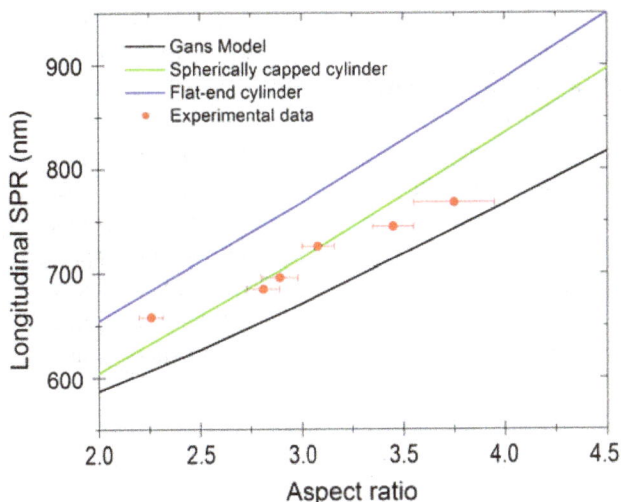

Fig. 4 The comparison of experimental longitudinal SPR peak positions with the results of Gans and DDA simulations for various shapes of gold nanorods

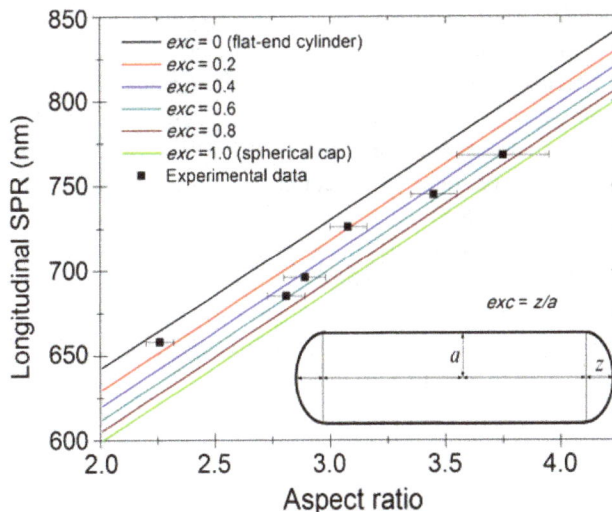

Fig. 5 The comparison of experimental longitudinal SPR peak positions with the results of SI calculations for various end-cap geometries of smooth gold nanorods

account the actual shapes and sizes of rods, Prescott and Mulvaney [11] demonstrated that the AR alone cannot be used to accurately predict λ_{LSPR}. In addition, the position of the LSPR peak is very sensitive to the end-cap geometry of rods. Detailed analysis and discussions can be found in [11]. Figure 4 shows two theoretical simulations for two different morphologies of gold nanorods, a flat-end cylinder and a spherically capped cylinder, suspended in a medium with a refractive index of 1.33 [11]. The latter appears to be a more appropriate model, as shown in Fig. 4. It is also reasonable to treat the synthesized rods as spherically capped cylinders, as suggested by TEM analysis. However, overall, the DDA approach still does not accurately predict the experimental data. Moreover, the predicted rate of change of λ_{LSPR} with the AR is too high compared with the experimental data.

Using the SI approach, Pecharroman et al. [12] recently concluded that not only the AR but both the end-cap geometry and surface roughness of rods play an important role in determining the extinction spectrum. Detailed analysis and discussions are reported in [12]. Fig. 5 compares the experimental data from this analysis with the results of the SI calculations reported by Pecharroman et al. [12] for smooth rods with an end-cap geometry featuring different degrees of flatness. The flatness can be modeled by using the parameter exc=z/a (z and a denoted in the figure). As shown, the best fit to the experimental data is obtained when exc is between 0.4 and 0.6: that is, when the rod is treated as neither a flat-end nor spherically capped cylinder but something in between. In Fig. 6, we compare the experimental data with the SI predictions for spherically capped rods with different degrees of surface roughness $\rho=r/a$ [12]. The roughness ρ was considered to be smaller than 10 % of the nanorod diameter. It is clear that the experimental data, spread over

the range of ρ from 0.02 to 0.08, and the rate of change of λ_{LSPR} with AR were not well modeled.

Thus, neither DDA simulations nor the SI approach accurately predicted the experimental results. It is crucial to determine whether the discrepancy between the experimental results and theoretical predictions is due to physical effects or some inadequacy in the experimental measurements. The fact that gold nanorods synthesized via the seed-mediated method are believed to be coated with a thin layer of surfactant CTAB should be taken into account when modeling the refractive index of the rods' surrounding medium. Moreover, the problems of polydispersity and plasmon coupling between rods may also play a non-negligible role. These problems, however, should introduce

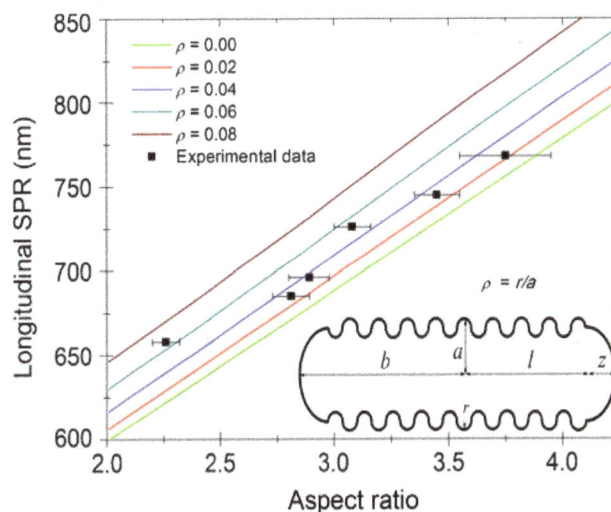

Fig. 6 The comparison of experimental longitudinal SPR peak positions with the results of SI calculations for spherically capped gold nanorods with different values of surface roughness ρ

a systematic error that affects the overall estimation accuracy, but this is not the case. Therefore, the only remaining reason seems to lie in the inadequacy of the formulation of SPR by classical theory, but because a discussion of this subject extends beyond the scope of this article, we only show that, if we take into account the interdependence between the two SPR modes, we are able to sufficiently reduce the discrepancies listed above.

Recall that the shifts in both λ_{TSPR} and λ_{LSPR} scale with the AR; thus, by extension, λ_{TSPR} may be used to linearly scale λ_{LSPR} to an expected value. Assuming the simplest case, the relation may be written as follows.

$$\lambda_{LSPR}^{Expected} = \left(\frac{\lambda_{TSPR}^{Measured}}{\lambda_0} \right) \lambda_{LSPR}^{Measured}, \qquad (1)$$

where λ_0 is a constant. Within the narrow range of experimental ARs (Figs. 4–6), we may safely consider the expected values of λ_{LSPR} to scale linearly with the AR, that is, $\lambda_{LSPR}^{Expected} = k a_r$ (where k is a scaling constant and a_r is an aspect ratio). Substituting this expression into (1) yields

$$\lambda_0 = \lambda_{TSPR}^{Measured} \left(\frac{\lambda_{LSPR}^{Measured}}{k a_r} \right). \qquad (2)$$

Therefore, each measured value $a_r(i)$ ($i=1,\ldots,6$ in our case) provides one value $\lambda_0(i)$. If the physical correspondence between the two SPR modes really exists, then λ_0 will be determined with high accuracy. The temporal, incomplete numerical evaluation indeed showed that the relative error of the estimation of λ_0 was less than 5 %, and λ_0 had a value of 514.13 nm, which is very close to the mean value of λ_{TSPR} (514.33 nm). The predicted values $\lambda_{LSPR}^{Expected}$ obtained from (1) using the λ_0 calculated for our samples showed that the relative error $|\Delta\lambda_{LSPR} / \lambda_{LSPR}| \times 100 \% < 0.65 \%$. For the accuracy obtained, we believe that some physical correspondence between the two resonance modes truly exists. A full discussion of this conclusion in the light of density functional theory will be presented elsewhere; to date, TD-DFT has been used only for small clusters with a few atoms (less than 20, e.g., Ref. [13] for Ag_3Au_{10} cluster, Ref. [14] for Ag_mNi_p, $m+p \leq 8$, and Ref. [15] for Au_{20} single chain clusters). Because TD-DFT calculations are still difficult for clusters large enough to provide meaningful results (such as the Ag clusters with diameters of up to 9 nm in Ref. [16] and [17]), the interpretation of the interplay between the SPR modes according to relation (1) can provide a rough, phenomenological estimate of this coupling effect.

Conclusion

The optical properties of gold nanorods are dominated by the SPR phenomenon. As anisotropic particles, gold nanorods exhibit two characteristic SPR peaks corresponding to the transverse and longitudinal SPR modes. There is clearly a discrepancy between the experimental results and theoretical predictions made by Gans model, DDA and SI calculations. In our opinion, the Gans model is too simple, whereas the Surface Integral approach appears somewhat arbitrary when modeling surface topology and roughness; thus, the Discrete Dipole Approximation is the most appropriate method for SPR simulation at the classical theory level. Furthermore, the refractive index of the surrounding medium, as well as the problems of polydispersity and plasmon coupling between nanorods, might sufficiently account for the discussed discrepancies. We suggest that the interdependence between the two resonance modes might play an important role and should be taken into account.

Acknowledgments The authors wish to thank Brian Korgel and Danielle Smith at University of Texas at Austin for fruitful discussions. Thanks must also be given to Paul Mulvaney (University of Melbourne), Stuart Prescott (Bristol University), and Carlos Pecharroman (Instituto de Ciencia de Materiales de Madrid) for providing their theoretical simulations and detailed discussions. One of the authors (HNN) would like to acknowledge the financial support from the National Foundation for Science and Technology Development (NAFOSTED) of Vietnam, the project "Nanofluids and Application" (code 103.02.19.09). The financial support from Vietnam National University (Project QG, 2013) is gratefully acknowledged by the other authors (NAB and PTT).

References

1. Daniel M, Astruc D (2004) Chem Rev 104:293
2. Perez-Juste J, Pastoriza-Santos I, Liz-Marzan LM, Mulvaney P (2005) Coord Chem Rev 249:1870
3. Jana NR, Gearheart L, Murphy CJ (2001) J Phys Chem B 105:4065
4. Nikoobakht B, El-Sayed MA (2003) Chem Mater 15:1957
5. Smith DK, Korgel BA (2008) Smith and Brian A. Korgel, Langmuir 24(3):644–649
6. Turkevich J, Stevenson PC, Hillier J (1951) Discuss Faraday Soc 11:55
7. Tao A, Habas S, Yang P (2008) Small 4:310
8. Wang ZL, Mohamed MB, Link S, El-Sayed MA (1999) Surf Sci 440:L809
9. Abramoff MD, Magelhaes PJ, Ram SJ (2004) Biophoton Int 11:36
10. Gans R (1912) Ann Phys 37:881
11. Prescott SW, Mulvaney P (2006) J App Phys 99:123504
12. Pecharroman C, Perez-Juste J, Mata-Osoro G, Liz-Marzan LM, Mulvaney P (2008) Phys Rev B 77:035418
13. Chena FY, Johnston RL (2007) Appl Phys Lett 90:153123
14. Harb M, Rabilloud F, Simon D (2007) Chem Phys Lett 449:38–43
15. Lian K-Y, Sałek P, Jin M, Ding D (2009) J Chem Phys 130:174701
16. Zuloaga J, Prodan E, Nordlander P (2010) ACS Nano 4(9):5269–5276
17. Zuloaga J, Prodan E, Nordlander P (2009) Nano Letetrs 9(2):887–891

11

Colorimetric assay of lead using unmodified gold nanorods

Guozhen Chen · Yan Jin · Wenhong Wang · Yina Zhao

Abstract Lead poisoning in adults can affect the peripheral and central nervous systems, the kidneys, and blood pressure. Thus, the development of environment-friendly and simple methods for Pb^{2+} detection is of great importance. Herein, a label-free colorimetric method has been developed for the detection of Pb^{2+} based on the conformational switch from single-stranded DNA to G-quadruplex. The electrostatic interactions between DNA probe and gold nanorods (GNRs) induce GNRs to space closely. However, the electrostatic interaction is not strong enough to change the suspension state of GNRs. In the presence of Pb^{2+}, the formation of G-quadruplexes increases the surface charge density around DNA, which is expected to strengthen the electrostatic interaction between the GNRs and the DNA. Therefore, the longitudinal absorption of GNRs decreased because the stronger interaction induced aggregation of GNRs. Importantly, the decrease in longitudinal absorption is proportional to concentration of Pb^{2+}. By monitoring the change of absorbance, Pb^{2+} can be detected at a level of 3 nM with a linear range from 5 nM to 1 μM. The overall test only takes a few minutes and very little interference is observed from other metal ions. The major advantages of this method are its low cost, convenience, simplicity, sensitivity, and specificity.

Keywords Pb^{2+} · Gold nanorods (GNRs) · G-quadruplex · DNA

Introduction

Heavy metal pollution in the environment attracts increasing attention because it has severely adverse effect on human health. The contamination by heavy metal ions, particularly Pb^{2+}, poses a serious threat to human health and environment. As lead is nondegradable, its persistence in the environment can produce toxic effects to plants and animals. Once introduced into the body, Pb^{2+} is a potential neurotoxin that can cause chronic inflammation of the kidney and heart, inhibit brain development, and decrease nerve conduction velocity [1–3]. The maximum level of lead in drinking water permitted by the US Environmental Protection Agency is 15 μg/L (~72 nM) [4, 5]. These environmental and health problems of Pb^{2+} have prompted researchers to develop efficient methods for selective and sensitive assay of the heavy metal ion to understand its distribution and pollution potential. Therefore, development of a simple, sensitive, selective, economical, and practical method is highly demanded for environmental monitoring, food industry, and clinical diagnostics.

To monitor Pb^{2+} level, several methods have been developed, including inductively coupled plasma mass spectrometry [6, 7], atomic fluorescence spectrometry [8, 9], atomic absorption spectroscopy [10], reversed-phased high-performance liquid chromatography [11], and so on. Even with sensitivity and accuracy, there also share some disadvantages, such as time-consuming, expensive, and/or require sophisticated equipment, etc. To overcome these limitation and drawbacks, a variety of sensors have been developed to rapidly detect lead

G. Chen · Y. Jin (✉) · W. Wang · Y. Zhao
Key Laboratory of Applied Surface and Colloid Chemistry,
Ministry of Education, Key Laboratory of Analytical Chemistry
for Life Science of Shaanxi Province, School of Chemistry
and Chemical Engineering, Shaanxi Normal University,
Xi'an 710062, China
e-mail: jinyan@snnu.edu.cn

G. Chen · Y. Jin · W. Wang · Y. Zhao
China State Key Laboratory of Chemo/Biosensing
and Chemometrics, Hunan University,
Changsha 410082, People's Republic of China

with high selectivity and sensitivity [12–26]. Among them, colorimetric sensors offer a promising approach for facile tracking of metal ions in biological, toxicological, and environmental samples. Lu and co-workers have developed a series of functional DNAzyme-based Pb^{2+} sensors by using GNPs. The detection range of the sensor could be tuned from 3 nM to 1 μM [19–25]. Dong and co-workers reported a DNAzyme-based colorimetric sensor for Pb^{2+}; the detection limit was 32 nM [26]. In this paper, we aim to develop an ultrasensitive, environmentally friendly, yet simple method for colorimetric detection of Pb^{2+} ion.

Among the various nanostructures of gold, nanorods have attracted wide attention due to their versatile structures and special localized surface plasmon resonance [27, 28]. Gold nanorods (GNRs) have two directional electron oscillations in response to the polarization of the incident light. The absorption along the longer axis referred to as the longitudinal band is stronger (usually >600 nm) and the shorter axis one (around 520 nm) is called the transverse band. GNRs are normally passivated by positively charged surfactants, which give not only high stability but also positively charged surface. Gold nanorods as a typical anisotropic metal nanostructure possess many unique physical properties that have been widely applied in the field of medical imaging [29, 30] and biological sensors [31–39]. Mann et al. reported the specific organization of short GNRs into anisotropic three-dimensional aggregates by DNA hybridization [36]. He et al. utilized the optical and chemical properties of the GNRs designed a GNRs-quantum dots (QDs) quenching system for sensitive DNA detection [37]. Zhu et al. demonstrate a near infrared sensing system for the detection of human IgG based on the FRET between QDs and GNRs [38]. Ma et al. took advantage of the localized surface plasmon resonance properties of unmodified GNR detection DNA [39].

Herein, a label-free colorimetric method has been developed for the detection of Pb^{2+} in aqueous solution using gold nanorods as signal probe. By monitoring the change of longitudinal absorption of GNRs, the detection limit of Pb^{2+} was 3 nM. Circular dichroism (CD) and transmission electron microscope (TEM) measurements were adopted to further confirm these occurred interactions. Experimental results revealed that the developed method could be applied to monitor the existence of traces of Pb^{2+} ions in aqueous solution with high selectivity.

Experimental section

Chemicals Unless otherwise indicated, all reagents and solvents were purchased in their highest available purity and used without further purification or treatment. $HAuCl_4 \cdot 4H_2O$ was purchased from Sigma (San Diego, USA). DNA (5'-GGGTGGGTGGGTGGGT-3', DNA) was synthesized by Shanghai Sangon Biotechnology Co. (Shanghai, China) and used without further purification. The oligonucleotide stock solutions were prepared with a Tris–HAc buffer (pH 7.4) and kept frozen. Metal ion solutions were prepared from nitrate salts. Millipore Milli-Q (18.2 MΩ cm) water was used in all experiments.

Instrumentation UV–Vis absorption spectra were recorded by using a Hitachi U-3900H UV–Vis Spectrophotometer (Tokyo, Japan). CD spectra were measured on a Chirascan Circular Dichroism Spectrometer (Applied Photophysics Ltd, London, England). The TEM images of GNRs were taken by using a JEM-2100 transmission electron microscope (Jeol Co. Ltd, Tokyo, Japan).

Synthesis of gold nanorods Gold nanorods were synthesized by using a seed-mediated, surfactant-assisted growth method in a two-step procedure [31, 32]. Briefly, colloidal gold seeds were first prepared by mixing aqueous solutions of cetyltrimethylammonium bromide (CTAB, 0.1 M, 7.5 mL) and hydrogen tetrachloroaurate (III) hydrate (1 %, 0.098 mL). Freshly prepared aqueous solution of sodium borohydride (0.01 M, 0.6 mL) was then added. The colloidal gold seed solutions (0.215 mL) were then injected into an aqueous growth solution of CTAB (0.1 M, 47.6 mL), hydrogen tetrachloroaurate (III) hydrate (1 %, 0.788 mL), silver nitrate (0.01 M, 0.3 mL), and freshly prepared ascorbic acid (0.1 M, 0.32 mL). The nanorods were purified by several cycles of suspension in ultrapure water, followed by centrifugation. They were isolated in the precipitate, and excess CTAB was removed in the supernatant. Then, it was stored in a refrigerator at 4°C before being used. The nanorods were characterized by absorption spectroscopy.

UV–Vis measurements Typically, the UV–Vis spectra of GNRs were recorded on UV–Vis spectrophotometer (Tokyo, Japan). Then, a certain quantity of GDNA was added into GNR suspension solution. After thorough mixing, a suitable Pb^{2+} was added to the mixture. The solution is vortexed thoroughly and used for the UV–Vis absorption spectra measurement experiment. The UV–Vis spectra of GNRs were recorded on a UV–Vis spectrophotometer with the wavelength range from 400 to 900 nm.

CD spectroscopy The CD spectra of DNA oligonucleotides were measured for 2 μM DNA total strand concentration using a Chirascan Circular Dichroism Spectrometer (Applied Photophysics Ltd, London, England). CD spectra were recorded using a quartz cell of 1-mm optical path length and an instrument scanning speed of 100 nm/min with a response time of 2 s at room temperature. CD spectra were obtained by taking the average of three scans made from 200 to 350 nm. All DNA samples at a final concentration of 2 μM were dissolved in Tris–HCl buffer and heated to 90 °C for 5 min, gradually cooled to room temperature, and incubated at 4 °C overnight.

TEM TEM images were obtained using a JEM-2100 transmission electron microscope (Jeol Co. Ltd, Tokyo, Japan). Samples were prepared on 400 mesh Cu grids coated with a thin layer of carbon (EM Sciences). The solution (5.00 μL) was pipetted onto the surface of the grid and allowed to dry in air. GNRs can grow out to 10 nm wide and up to 30 nm long with an aspect ratio of ~3.

Results and discussion

Sensing mechanism

The design rationale is illustrated in Fig. 1. The assay is based on positively charged GNRs having a higher affinity to G-quadruplex DNA than ssDNA because the surface charge density of G-quadruplex DNA is much larger than that of ssDNA [40, 41]. In the absence of Pb^{2+}, the electronic attraction between DNA and GNRs brought the DNA and GNRs into close proximity, which induced slight change of absorption spectrum of GNRs. However, upon the addition of Pb^{2+}, electrostatic interactions obviously strengthened due to the formation of G-quadruplex, resulting in a decrease of longitudinal absorption of GNRs. Therefore, Pb^{2+} can be simply and directly detected by monitoring the change of absorbance or color.

It is known that the UV absorption spectra of GNRs display two bands assigned to the transversal and longitudinal modes of electronic oscillations, and the locations depend on the aspect ratio. As Fig. 2 shows, the two absorption bands for our prepared GNRs are located around 520 and 786 nm. The broadening of the transverse band of GNRs between 520 and 600 nm is ascribed to the presence of GNRs with different aspect ratios and gold nanoparticles with different sizes and shapes. When DNA was added to the GNR suspension, they attracted GNRs near because of the electrostatic interactions between the anionic backbone phosphates of oligonucleotides and the cationic surfactant bilayer around the nanorods. However, the electrostatic interaction is not strong enough to change the suspension state of GNRs. So, there is little change in the UV–Vis absorption spectra. Curve C in Fig. 2a shows the absorption spectrum of the GNRs in the presence of ssDNA upon addition of 5 nM Pb^{2+}. Upon the addition of the Pb^{2+}, the transverse absorption peak of the GNRs shows

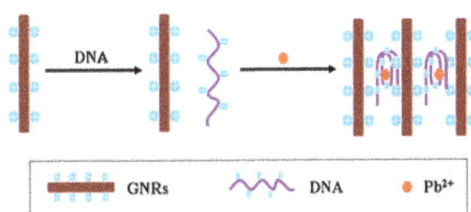

Fig. 1 Schematic illustration of GNR-based colorimetric strategy for Pb^{2+} detection

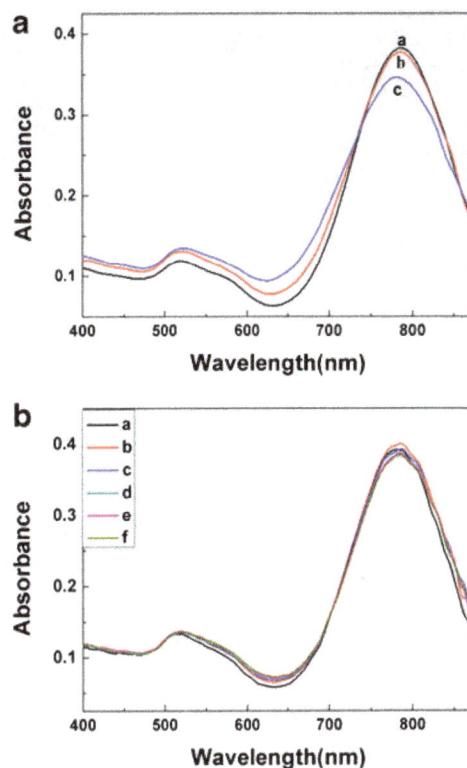

Fig. 2 UV–Vis absorption spectra of GNRs under different conditions. **a** (*a*) GNRs, (*b*) GNRs + ssDNA, and (*c*) GNRs + ssDNA + 5 nM Pb^{2+}. The concentration of ssDNA was 20 nM. **b** GNRs with the addition of Pb^{2+}. From *a* to *f*, the concentration of Pb^{2+} is 0, 5, 50, 100, and 500 nM and 1 μM, respectively

slight red shift, while the intensity of the longitudinal absorption shows a distinct decrease and a blueshift. This result suggests that GNRs assembled in the side-by-side mode due to the electrostatic interaction between the positive charge of GNRs and the negative charge of the phosphate backbone of the G-quadruplex DNA induced by Pb^{2+}. It is consistent with the previous reports that the longitudinal band of GNRs is more sensitive to the local environment around GNRs than the transverse band [42–45]. This result suggested that the aggregation of GNRs takes place in a side-by-side manner [46, 47].

To further study the rationality and reliability of this assay, more control experiments have been performed. First, the influence of Pb^{2+} on the absorption spectrum of GNRs was studied. We investigated the absorption spectrum of GNRs in the presence of Pb^{2+} without ssDNA. As shown in Fig. 2b, the absorption spectrum of GNRs had no fundamental change without ssDNA as the concentration of Pb^{2+} increased. It is therefore obvious that the Pb^{2+} itself did not influence the longitudinal absorption of GNRs, but rather the decrease of absorbance in the presence of Pb^{2+} (Fig. 2a) is mainly caused by G-quadruplex formation which strengthened electrostatic interaction between GNRs and DNA. Then, the sensing mechanism was further studied by TEM images. From images a and b of Fig. 3, we found

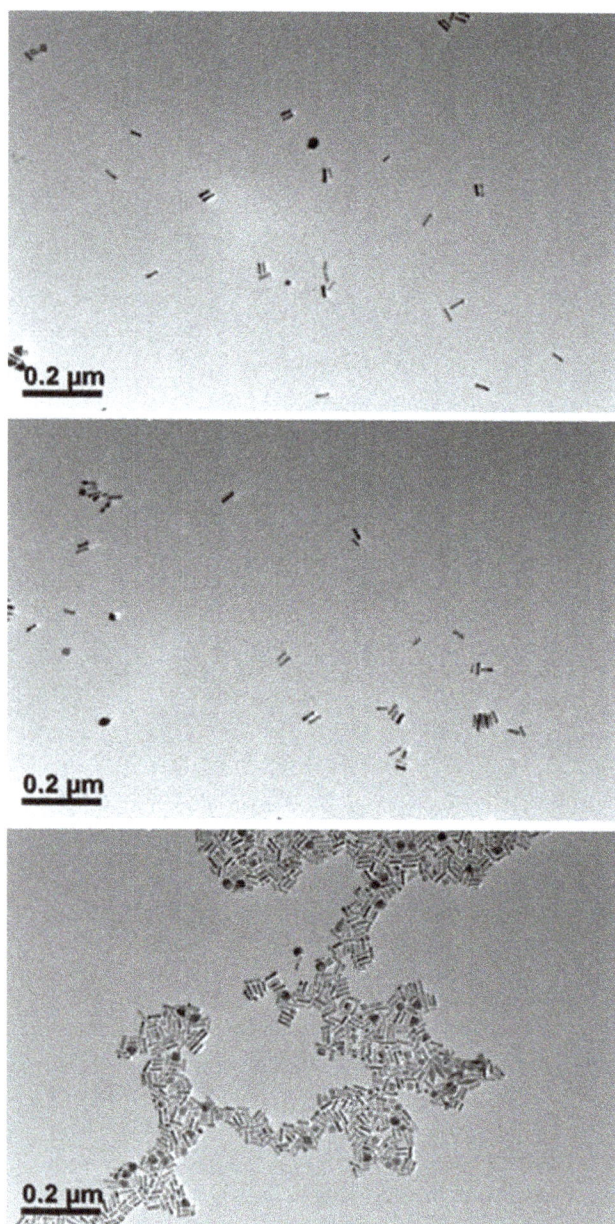

Fig. 3 TEM images of GNRs (**a**) in the presence of *ssDNA* (**b**) and the mixture of *ssDNA* and *Pb²⁺* (**c**)

that GNRs are well dispersed in the aqueous medium in the absence and presence of DNA. However, GNRs obviously aggregated when Pb^{2+} was added into the GNRs/DNA solution in image C, which is accordance with the UV–Visible absorption measurements. Therefore, this colorimetric method was again demonstrated to be feasible and reasonable for the detection of Pb^{2+}.

Identification of G-quadruplex formation

CD is a sensitive technology able to study the configuration inversion of DNA, which could report the structural variations intrinsically and kinetically [48]. To further confirm

the formation of G-quadruplex, CD measurement is utilized to monitor the conformation change of DNA probe in the different cases. Figure 4a shows the CD spectra for the titration of the DNA with increasing amounts of Pb^{2+}. The CD spectra of ssDNA at room temperature exhibited a positive band around 265 nm. Upon the addition of 2 μM Pb^{2+} to the ssDNA, a dramatic change in the CD spectrum was observed. The maximum at 265 nm was gradually increased. Meanwhile, a small positive peak appears near 320 nm, which indicated G-quadruplex formation [15, 49–52]. As the Pb^{2+} concentration increased to 5 μM, we observed a concentration-dependent enhancement of the positive peak around 265 and 320 nm and the negative peak around 240 nm. This CD spectrum suggests the coexistence of the parallel G4 structure with a small amount of the antiparallel one. As we know, K^+ is highly able to stabilize the G4 structure. Therefore, the conformation change of DNA probe induced by K^+ has also been studied. It is clear from Fig. 4b that the CD spectra of DNA probe slightly changed, and no new peak was observed around 320 nm. The DNA probe used in this work is named T30695. Wang et al. previously reported that the DNA melting experiments of K^+-T30695 and Pb^{2+}-T30695 have different stability under the same conditions. The stability of K^+-T30695 is much lower than that of Pb^{2+}-T30695 [15]. CD measurements

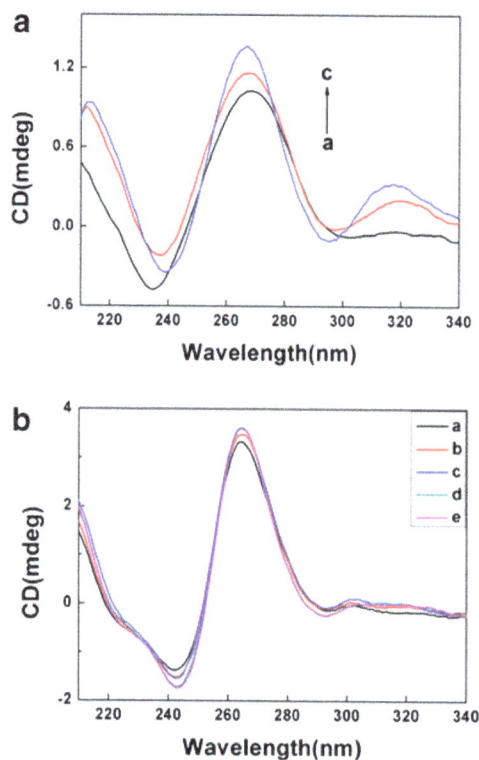

Fig. 4 *CD* spectra for characterizing the structural conversion of DNA (2 μM) in the absence and presence of Pb^{2+} (**a**) and K^+ (**b**). **a** (*a*) without Pb^{2+}; (*b*) 2 μM Pb^{2+}; (*c*) 5 μM Pb^{2+}. **b** (*a*) without K^+; (*b*) 5 μM K^+; (*c*) 10 μM K^+; (*d*) 20 μM K^+; (*e*) 50 μM K^+. All solutions were prepared in 5 mM Tris–acetate (pH 7.4)

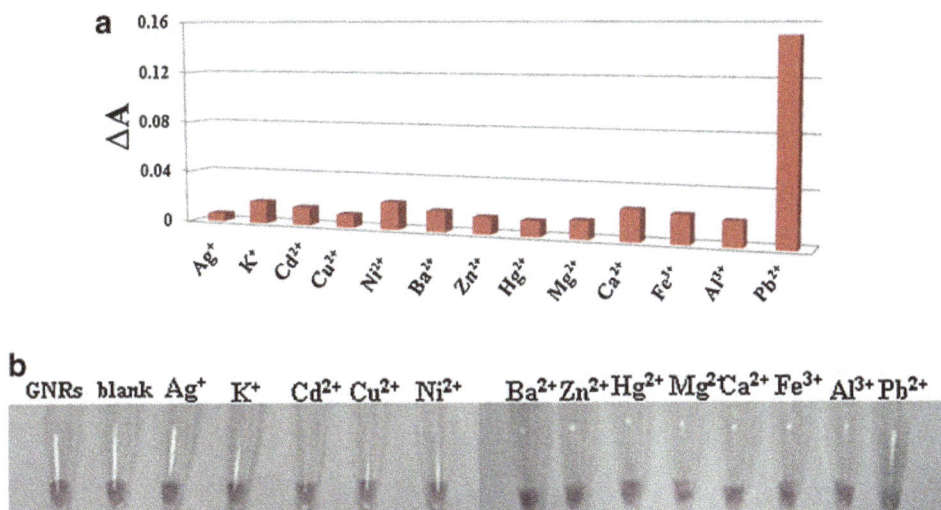

revealed that the formation of G-quadruplex in the presence of Pb^{2+} led to the change of longitudinal band of GNRs.

Selectivity and sensitivity

Selectivity was an important issue to estimate the performance of a sensor. So we tested the selectivity of the proposed method by comparing the absorbance changes of GNRs/DNA caused by Pb^{2+} and a variety of environmentally relevant metal ions, including Ag^+, K^+, Cd^{2+}, Cu^{2+}, Ni^{2+}, Ba^{2+}, Zn^{2+}, Hg^{2+}, Mg^{2+}, Ca^{2+}, Fe^{3+}, and Al^{3+}. Figure 5a illustrates the absorption intensity change ($\Delta A = A_0 - A$) where A_0 and A are the absorption intensity of GNRs/ DNA in the absence and presence of different metal ions, respectively. It is obvious that all the metal ions except Pb^{2+} exhibited little variations in the extinction intensity, which is important and helpful in validation of the method to meet the selectivity requirements of the Pb^{2+} assay in environmental and biological fields. These results clearly reveal that our detection method has high selectivity against other interfering metal ions. As shown in Fig. 5b, the color of

GNRs/DNA solution changed from red to blue when 2 μM Pb^{2+} was added, which indicated the aggregation of GNRs. However, other metal ions have no effect on the GNRs/ DNA solution, even when the concentration of metal ions reached 5 μM. Therefore, this colorimetric method was again demonstrated to be feasible and reasonable for the selective detection of Pb^{2+}.

To evaluate the sensitivity of Pb^{2+} detection, the different concentrations of Pb^{2+} were added into GNRs/DNA solution, respectively. As expected, the decrease of absorbance of GNRs can quantitatively reflect the amount of lead ion added. From Fig. 6, we can see that a dramatic decrease in the absorbance was observed with the increasing of Pb^{2+} concentration. Curve A in Fig. 6 shows the UV–Vis spectra of GNRs/ ssDNA. However, longitudinal absorption of GNRs showed a gradual decrease with the concentration of Pb^{2+} increasing from 5 nM to 3 μM. Upon the addition of Pb^{2+}, the decline of GNRs absorption and the blueshift of the longitudinal band was obversed, which can be ascribed to the strengthened interaction between DNA and GNRs due to the formation of G-quadruplex in the presence of Pb^{2+} because G-quadruplex has higher charge density than the ssDNA [53, 54]. As shown in the inset of Fig. 6, the proposed method exhibited a good linear response ($R = 0.9932$) of absorbance change against the logarithm of Pb^{2+} ion concentration over the range from 5 nM to 1 μM, indicating that Pb^{2+} can be sensitively detected by this method. The detection limit of Pb^{2+} was 3 nM, which was

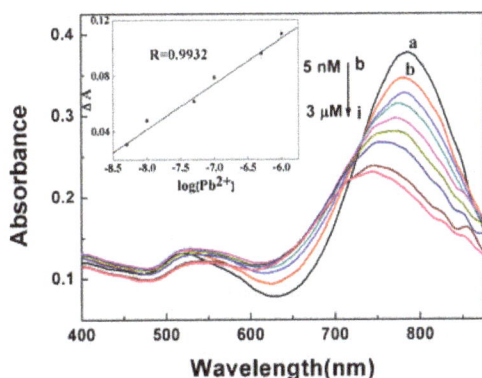

Fig. 6 UV–Vis absorption spectra of GNRs/ssDNA in the absence and presence of Pb^{2+} ion. From curves *a* to *i*, the concentration of Pb^{2+} is 0, 5, 10, 50, 100, and 500 nM and 1, 2, and 3 μM, respectively. (*Inset*) The calibration plots for Pb^{2+} measurements (from 5 nM to 1 μM)

Table 1 Recovery experiments of Pb^{2+} in tap water samples

Samples	Adde0nM)	The proposed method mean[a]±SD[b] (nM)	Recovery (%)
Tap water 1	30	30.08[a]±1.47[b]	100.3
Tap water 2	60	62.12[a]±1.12[b]	103.5

[a] Mean values of three determinations

[b] Standard deviation

much lower than the EPA standard for the maximum allowable level 15 μg/L (72 nM) in drinking water.

Analysis of water

To demonstrate the application potential of our proposed method in environmental analysis, we applied it to analyze real tap water samples. The tap water sample was collected after discharging tap water for ~20 min. Standard addition method was used to valuate the practicality of developed approach. All the water samples were spiked with Pb^{2+} at different concentration levels. The Pb^{2+} concentrations were calculated using standard curves prepared within the same day by our new approach. From Table 1, we can conclude that it is feasible for Pb^{2+} detection in real tap water samples.

Conclusion

In summary, a simple, direct, and cost-effective method has been developed for rapid detection of Pb^{2+} by using GNRs as colorimetric probe. The experimental results show that Pb^{2+} can be detected quickly and accurately with high sensitivity and selectivity against other heavy metal ions. Under the optimal conditions, this method was highly sensitive (LOD=3 nM) and selective toward Pb^{2+} ions, with a linear detection range from 5 nM to 1 μM. From the summary in Table 1 (supporting information), we can conclude that the major advantages of this method are its simplicity, selectivity, and high sensitivity. It is of great theoretical and practical importance for the detection of heavy metal ions.

Acknowledgments This work was financially supported by the National Natural Science Foundation of China (21075079), Program for New Century Excellent Talents in University (NCET-10-0557), and the Program for Changjiang Scholars and Innovative Research Team in 403 University (IRT 404 1070).

References

1. Laterra J, Bressler JP, Indurti RR, Belloni-Olivi L, Goldstein GW (1992) Proc Natl Acad Sci USA 89:10748–10752
2. Kim HN, Ren WX, Kim JS, Yoon JY (2012) Chem Soc Rev 41:3210–3244
3. Zocche JJ, Leffa DD, Damiani AP, Carvalho F, Mendonca RA, Santos CE, Boufleur LA, Dias JF, Andrade VM (2010) Environ Res 110:684–691
4. Huang KW, Cheng-Ju YCJ, Tseng WL (2010) Biosens Bioelectron 25:984–989
5. Chen YY, Chang HT, Shiang YC, Hung YL, Chiang CK, Huang CC (2009) Anal Chem 81:9433–9439
6. Liu HW, Jiang SJ, Liu SH (1999) Spectrochim Acta Part B 54:1367–1375
7. Bowins RJ, Mcnutt RH (1994) J Anal At Spectrom 9:1233–1236
8. Wagner EP, Smith BW, Winefordner JD (1996) Anal Chem 68:3199–3203
9. Neuhause RE, Panne U, Niessner R, Petrucci GA, Cavalli P, Omenetto N (1997) Anal Chim Acta 346:37–48
10. Weidenhamer JD (2007) J Chem Educ 84:1165–1166
11. Saito S, Danzaka N, Hoshi S (2006) J Chromatogr A 1104:140–144
12. Wang L, Jin Y, Deng J, Chen GZ (2011) Analyst 136:5169–5174
13. Song PS, Xiang Y, Xing H, Zhou ZJ, Lu Y (2012) Anal Chem 84:2916–2922
14. Zhao XH, Kong RM, Zhang XB, Meng HM, Liu WN, Tan WH, Shen GL, Yu RQ (2011) Anal Chem 83:5062–5066
15. Li T, Dong SJ, Wang EK (2010) J Am Chem Soc 132:13156–13157
16. Bui MPN, Li CA, Han KN, Pham XH, Seong GH (2012) Analyst 137:1888–1894
17. Fu XB, Qu F, Li NB, Luo HQ (2012) Analyst 137:1097–1099
18. Lee YF, Deng TW, Chiu WJ, Wei TY, Roy P, Huang CC (2012) Analyst 137:1800–1806
19. Liu JW, Lu Y (2003) J Am Chem Soc 125:6642–6643
20. Liu JW, Lu Y (2004) Chem Mater 16:3231–3238
21. Liu JW, Lu Y (2004) J Am Chem Soc 126:12298–12305
22. Liu JW, Lu Y (2004) Anal Chem 76:1627–1632
23. Liu JW, Lu Y (2005) J Am Chem Soc 127:12677–12683
24. Wang ZD, Lee JH, Lu Y (2008) Adv Mater 20:3263–3267
25. Mazumdar D, Liu JW, Lu G, Zhou JZ, Lu Y (2010) Chem Commun 46:1416–1418
26. Li T, Wang EK, Dong SJ (2010) Anal Chem 82:1515–1520
27. Juste JP, Santos IP, Liz-Marzán LM, Mulvaney P (2005) Coord Chem Rev 249:1870–1901
28. Keul HA, Möller A, Bockstaller MR (2007) Langmuir 23:10307–10315
29. Maltzahn GV, Centrone A, Park JH, Ramanathan R, Sailor MJ, Hatton TA, Bhatia SN (2009) Adv Mater 21:3175–3180
30. Wang HF, Huff TB, Zweifel DA, He W, Low PS, Wei A, Cheng JX (2005) Proc Natl Acad Sci USA 102:15752–15756
31. Huang XH, Neretina S, El-Sayed MA (2009) Adv Mater 21:4880–4910
32. Paraba HJ, Jung C, Lee JH, Park HG (2010) Biosens Bioelectron 26:667–673
33. York J, Spetzler D, Xiong FS, Frasch WDD (2008) Lab Chip 8:415–419
34. Li CZ, Male KB, Hrapovic S, Luong JHT (2005) Chem Commun 31:3924–3926
35. Sim HR, Wark AW, Lee HJ (2010) Analyst 135:2528–2532
36. Dujardin E, Hsin LB, Wang CRC, Mann S (2001) Chem Commun 14:1264–1265
37. Li X, Qian J, Jiang L, He SL (2009) Appl Phys Lett 94:063111
38. Liang GX, Pan HC, Li Y, Jiang LP, Zhang JR, Zhu JJ (2009) Biosens Bioelectron 24:3693–3697
39. Ma ZF, Tian L, Wang TT, Wang CG (2010) Anal Chim Acta 673:179–184
40. Jin Y, Chen GZ, Wang YX (2011) Gold Bull 44:163–169
41. Gou XC, Liu J, Zhang HL (2010) Anal Chim Acta 668:208–214
42. Pan BF, Cui DX, Ozkan C, Xu P, Huang T, Li Q, Chen H, Liu FT, Gao F, He R (2007) J Phys Chem C 111:12572–12576
43. Chang JY, Wu HM, Chen H, Ling YC, Tan WH (2005) Chem Commun 8:1092–1094
44. Nehl CL, Hafne JH (2008) J Mater Chem 18:2415–2419
45. Sudeep PK, Joseph ST, Thomas KG (2005) J Am Chem Soc 127:6516–6517
46. Sun ZH, Ni WH, Yang Z, Kou XS, Li L, Wang JF (2008) Small 4:1287–1292

47. Jain PK, Eustis S, El-Sayed MA (2006) J Phys Chem B 110:18243–18253
48. Paramasivan S, Rujan I, Bolton PH (2007) Methods 43:324–331
49. Li CL, Liu KT, Lin YW, Chang HT (2011) Anal Chem 83:225–230
50. Jing NJ, Rando RF, Pommier Y, Hogan ME (1997) Biochemistry 36:12498–12505
51. Smirnov I, Shafer RH (2000) J Mol Biol 296:1–5
52. Li T, Wang EK, Dong SJ (2009) J Am Chem Soc 131:15082–15083
53. Chen GZ, Jin Y, Wang L, Deng J, Zhang CX (2011) Chem Commun 47:12500–12502
54. He W, Huang CZ, Li YF, Xie JP, Yang RG, Zhou PF, Wang J (2008) Anal Chem 80:8424–8430

12

Electron transfer processes on Au nanoclusters supported on graphite

Jie Shen · Juanjuan Jia · Kirill Bobrov ·
Laurent Guillemot · Vladimir A. Esaulov

Abstract Electron transfer processes play an important role in surface chemistry. This paper presents results of a study of changes in resonant electron transfer processes, as a function of gold cluster sizes, on the example of electron transfer between Li^+ ions scattered on Au clusters on highly oriented pyrolytic graphite (HOPG). The gold nanoclusters were grown on lightly sputtered HOPG surface in order to obtain a wide coverage distribution of clusters. The growth of clusters was monitored by scanning tunneling microscopy. We found that electron transfer is much more probable on small clusters, whose lateral size is of the order of 2 to 3 nm and height in the 1-nm range, than on bulk Au or thin Au films. A comparison with Au clusters grown on the semiconducting titania did not reveal significant differences with HOPG.

Keywords Au · Clusters · Electron transfer · Charge exchange · Ion scattering

Introduction

Supported metal nanoparticles have attracted much attention because of their use in many fields such as heterogeneous catalysis, microelectronics, photonics, etc. The size, shape, and nature of the support affect the properties of the nanoparticles and can play a crucial role in determining their use in a particular application [1–3]. The size effects have been discussed in terms of morphology and electronic structure and questions of the interaction with the substrate have been addressed [1–3]. As a function of increasing size, the electronic structure of the clusters evolves from the extreme case of the atomic structure of its constituent atoms to the formation of a system with molecular character, which progressively evolves to that of a band structure of the bulk metal. The molecular aggregate is characterized by an energy gap between occupied and unoccupied levels. The increase in size of the system progressively leads to the disappearance of this gap and transition towards the valence band structure of the metal. According to existing literature, the number of atoms that are necessary to induce this transition may range from several tens to several hundred [1–12]. These evolutions have been studied in a series of elegant scanning tunnel microscopy (STM) studies for various metal clusters on various substrates [4–12].

In recent years, special attention has been paid to gold clusters since the discovery made by Haruta et al. [13] that nanosized gold clusters on titania (TiO_2) exhibit unique catalytic properties. Maximum catalytic activity for these clusters was found to coincide with the metal to nonmetal transition occurring in clusters with a diameter of approximately 3.0 nm, as determined by STS cluster band gap measurements [9–11]. This cluster diameter also coincides with a cluster growth transition from the nucleation of flat, two-dimensional clusters, to their agglomeration into three-dimensional structures, as measured by STM.

In surface reactivity, electron transfer processes between atoms and molecules and the surface play a crucial role. Most surface science experiments, which deal with the study of either the kinetics of adsorption/desorption (e.g., in temperature programmed desorption) or with characterization of adsorbates or products of reactions in situ, do not provide information on the time-dependent dynamics of the electron transfer or on the effect of the surface on the electronic states of the approaching gas-phase particle (atom or molecule). Our objective is to obtain this information in experiments that involve atom or ion beam scattering in which the energy and charge state of particles are monitored. We can, thus, obtain quantitative information in

J. Shen · J. Jia · K. Bobrov · L. Guillemot · V. A. Esaulov (✉)
Institut des Sciences Moléculaires d'Orsay, ISMO,
Centre National de la Recherche Scientifique (CNRS),
Université Paris Sud, UMR 8214,
Bâtiment 351, 91405 Orsay, France
e-mail: vladimir.esaulov@u-psud.fr

controlled conditions that can serve as a basis for theoretical modeling. In this study, we therefore focused on this aspect in the interaction of ions with clusters, since information on electron transfer processes can be directly obtained. Here, we focus on resonant neutralization of Li^+ ions. Because the ionization potential of Li is small and comparable with the workfunction of many metals, we deal with electron transfer near the Fermi level also relevant in chemical reactions.

When dealing with reactivity of clusters, besides size effects, the question of the role of the support has been put forth. Our studies therefore focus both on the changes of electron transfer processes as a function of cluster size and for different types of substrates. Recently, we investigated [14] how electron transfer processes are affected as a function of growth of clusters on the example of neutralization of Li^+ ions on Au clusters grown on TiO_2. It was found that significantly more efficient neutralization occurs on small clusters, with neutralization decreasing as the cluster size grew. Similar results have been obtained by another group [15]. Here, we extend this work to the case of Au clusters grown on highly oriented pyrolytic graphite (HOPG) to investigate in particular if any strong changes would be observed when going from the semiconducting TiO_2 to the conducting graphite.

Experiment

The scattering experiments are performed in an UHV system [16]. The system is equipped with a differentially pumped ion gun used for low-energy ion scattering LEIS and Ar sputtering, and a Li ion gun which uses a getter source. Time of flight measurements were made using a channel plate detector set at 45° with respect to the Li gun, so that the scattering angle of ions corresponds to 135°. The detector is placed at the end of a 124 cm long flight tube.

Our measurements on neutralization probabilities involve low Li^+ beam flux, pulsed beam time of flight measurements. This precludes ion implantation effects. This was, however, regularly checked by performing the same type of measurement, e.g., at a given energy or a given coverage in different conditions, i.e., in the beginning or end of a series of measurements or for different coverage "steps." We, thus, exclude that our results are affected by Li implantation.

The HOPG single crystal is a 1-mm thick, 10×10 mm^2 plate. The sample was cleaved in air with a scotch tape and then attached to a Ta plate, and mounted on a *XYZ* rotary manipulator. It was heated through a combination of radiation and electron bombardment using a tungsten filament positioned behind it. Before the measurements, it was degassed by heating to 600 °C.

The chamber is equipped with a Knudsen cell metal evaporation source, and a quartz crystal microbalance. The deposition rate was generally varied between 0.01 and 0.1 eq mL (where 1 eq. mL=1 monolayer; corresponds to 1.4×10^{15} atom/cm^2) as in our previous work [14].

In order to get an idea of the characteristics of the growth of gold clusters, some STM measurements were performed. The STM experiments were carried out in a separate setup [17], equipped with a variable-temperature STM (Omicron VT-STM) using polycrystalline W tips. The STM chamber is coupled with a second one, equipped with a LEED/Auger system and an Ar sputter gun. The same evaporator was used on both setups.

Results and discussion

Before presenting the results of the study of neutralization on Au clusters, we first briefly describe the results of the STM investigation of the characteristics of cluster formation on HOPG. On pristine HOPG, it is well known that clusters do not form on defect-free HOPG planes, but rather cluster along step edges. This was also verified in our experiments. In order to induce growth over the whole surface of graphite, it is necessary to induce defects that act as nucleation sites. Here,

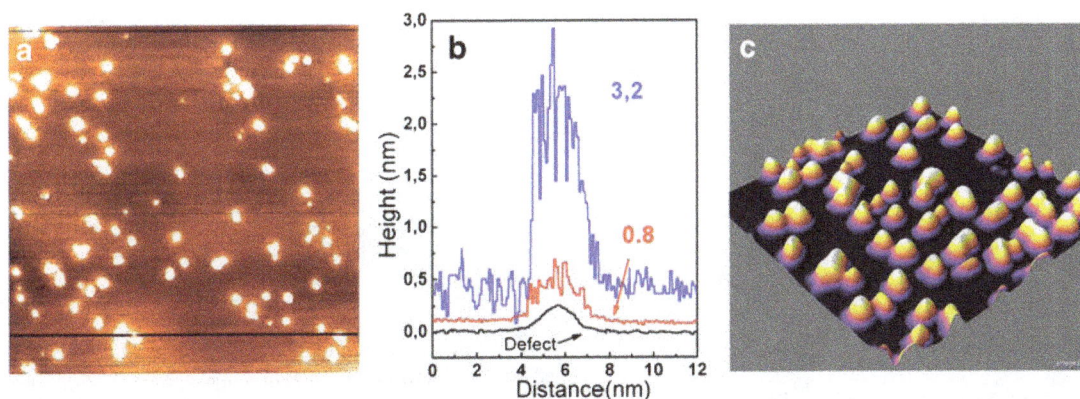

Fig. 1 **a** STM image of a slightly bombarded HOPG surface showing defects (*white spots*). **b** Line profiles of a single defect after 0.8 eq ML Au deposition and after 3.2 eq ML Au deposition. **c** 3D view of the surface after 3.2 eq ML Au deposition

we bombarded HOPG with an Ar ion beam to induce defect formation, examined characteristics of the defects formed, and then deposited Au and examined the gold clusters.

Figure 1a shows an STM image of a pristine HOPG surface bombarded for a few seconds with a 1.5 keV Ar$^+$ beam for 5 s. The white spots correspond to ion-induced damages presumably in single collision events. We see a fairly evenly distributed defect size distribution. Figure 1b shows a line profile along one of the defects in the STM image. In the STM image, the defects look like small protrusions, with an apparent mean height of ~0.2 nm and a lateral dimension (FWHM) of about 2 nm.

When gold is evaporated onto this surface, one observes an increase in size of these structures. Figure 1b shows the evolution of cluster heights as deposition proceeds. Since one cannot monitor the height evolution of a specific cluster, the profiles are taken from arbitrary clusters that are representative of these evolutions. After deposition of 0.8 eq ML of Au, the mean height and width of the structures is about 0.5 and 2.2 nm. For 3.2 eq ML Au deposition, the height increases to about 2 nm and the width is of the order of 2.5 nm. Thus, the clusters grow significantly in height but in the lateral dimensions, it did not increase very significantly.

As the Ar ion beam dose increases, we first observe the formation of a series of isolated, but clustered, defects with an average lateral dimension of the defect cluster in the 8 to 15 nm range. In this case, after 3.2 eq ML Au deposition, we observe clusters that are about 3-nm high and with similar lateral dimensions. In this case, gold clusters are first formed on the individual isolated defects and then coalesce into the larger cluster. At much higher ion fluence, the number of clustered defects increases. The size of the Au clusters is the same as mentioned above for the intermediate Ar ion fluence.

The experiments on Li ion neutralization involve a time of flight scattering study, which allows us to *separate scattering on gold clusters and substrate* as shown in Fig. 2a and also to obtain separately spectra for scattered ion and neutrals (Fig. 2b) and hence determine the *cluster-specific neutralization probability*.

Results of measurements performed as a function of Au evaporation onto a sputtered. HOPG surface are shown in Fig. 3. We observe that generally, *more efficient neutralization*

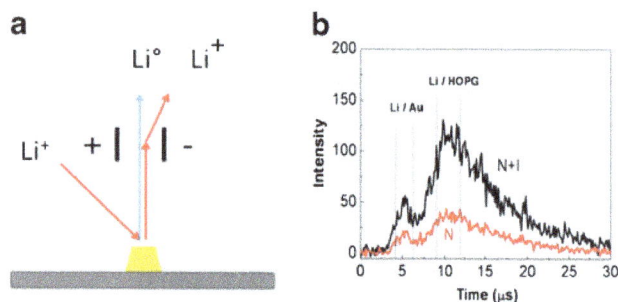

Fig. 3 Neutral fraction dependence on gold clusters on sputtered HOPG as a function of increasing Au coverage. The *horizontal bar* on the right indicates the value of the neutral fraction obtained on a Au(111) surface obtained after a very large evaporation. The *red triangles* summarize data (14) for scattering on Au clusters on TiO$_2$

occurs in the initial stages of deposition as compared to the case of large deposition doses and thin film formation. Initially, at the lowest deposition stages, a slight increase of neutralization occurs followed by a slow decrease for more than 2 eq ML deposition. Because in these experiments we do not perform ion scattering on STM previewed clusters on the same setups, we can only correlate trends between microscopy and scattering experiments, for similar Au deposition fluxes. By comparison with the results of the STM data, it would appear that neutralization is most efficient for cluster heights of the order or less than 1 nm, which would correspond to clusters of few atomic layers. This result would qualitatively concord with the observation that the reactivity of clusters is highest for clusters of about 2 to 3 atomic layers [6–12]. For large Au deposition, the results tend to that of a thin Au film, for which as may be seen, neutralization is much smaller.

The data for neutralization on gold clusters on HOPG is compared with that for the TiO$_2$ substrate studied previously [14]. We observe a fairly similar neutralization for the HOPG case. Note that this comparison is based on equivalent Au coverage, but in the TiO$_2$ case, it was also noted from AFM

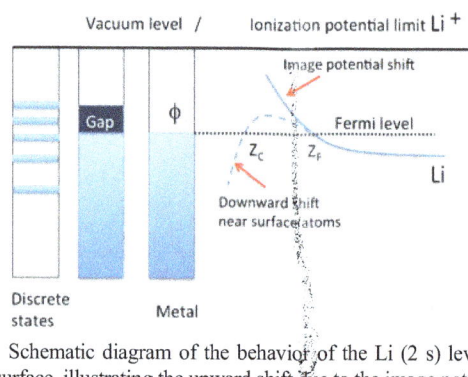

Fig 2 a Schematic diagram of scattering configuration. **b** TOF spectrum of Li scattering on Au clusters on HOPG. *Vertical lines* indicate the areas for integration to derive the neutral fraction

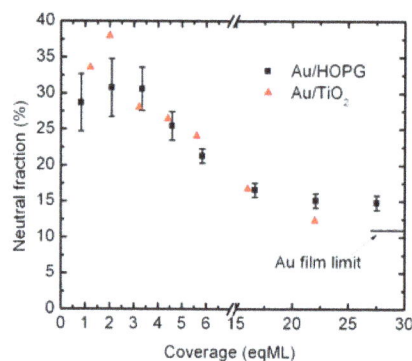

Fig. 4 Schematic diagram of the behavior of the Li (2 s) level near a metal surface, illustrating the upward shift due to the image potential and downward trend near the surface as predicted in recent calculations. On the *left* a schematic view of a metal band structure, a situation with a bandgap and one with only discretized states

images that cluster sizes corresponding to greatest neutralization were for clusters of about 2 to 3 nm lateral dimensions.

As may be seen in both studies, we found that as Au deposition increased, the Li neutral fraction tended to a gold thin film limit, which in the case of TiO_2 had been identified with a Au(111) surface.

At present, it is difficult to give a reliable interpretation of these results. The results of this study need to be put into the perspective of a complex problem related in general to alkali ion neutralization on metal surfaces. While alkali neutralization on bulk metal surfaces seemed well understood [18–20], recent experiments [21–24] revealed very wide discrepancies with predictions of these "standard" jellium-like models of metals using a rate equation approach to describe neutralization. It was indeed found, as may also be seen in the Au thin film limit of Fig. 3, that on high workfunction surfaces, neutralization is still occurring, whereas the usual models would preclude this. Indeed, in these models, it is assumed that near the surface Li(2 s) level is upward shifted due to image potential effects, and Li is ionized as schematized in Fig. 4. Electron capture then occurs at large distances from the surface, when the Li (2 s) level lies below the Fermi level for atom surface distances greater than Z_F. For Au(111) [22, 24] with a workfunction of 5.4 eV, this would only occur at very large distances where the interaction with the surface is very weak, and therefore significant neutralization was not expected.

Initially it was thought [22] that the higher neutralization may be related to such features [25, 26] as projected band gap and surface states for (111) surfaces. However, this anomalously large neutralization appeared to be quite general and not restricted to a given type of surface [16, 24]. More recent theoretical studies show that near the metal surface, the Li(2 s) level actually lies below the Fermi level [24, 27] for some distance less than Z_C, and hence at small distances Li can be neutralized. Therefore, as opposed to the "standard" picture, one does not deal with the neutralization of an ion, as it recedes from the Au surface, but rather a more complex situation involving also neutral atoms that, as they recede from the surface, are first ionized and then neutralized. In the case of a very thin cluster, one could ask the question if the ion would feel the substrate and whether the effective neutralization may be different, although in the case of the present experiment, no significant differences between TiO_2 and HOPG can be noted.

Secondly, it appeared that adiabatically [24], at small velocities, the charge on Li tends to unity (neutral atom), and hence at low velocities neutralization is efficient [24], as opposed to what was expected in the standard descriptions. In a very recent calculation, this seems to be related to a much larger Li (2 s) level width [28] than predicted by earlier DFT calculations [19, 20, 23]. Another possible problem may be that a rate equation approach for determining the level populations, used in such descriptions, may not be suitable for a situation where the atomic level stays close to the Fermi level

for large atom-surface distances. Finally, the latest descriptions do not take into account properly the specifics of the band structure of the metals mentioned above.

Thus, it is clearly difficult to make definitive statements regarding the cluster case. Clearly, the electronic structure of the cluster should play a role. It is possible that, as suggested in some works [9–12], cluster reactivity may correspond to metal–non-metal transitions and opening up of a bandgap in the cluster electronic structure. This as well as appearance of quantized states, due to confinement [5], would obviously affect electron capture and loss probabilities; capture or loss would, e.g., be inhibited in the bandgap region, and the description of it would need to be different. In the case of *discrete states*, non-resonant velocity-dependent charge transfer processes [29–31] may play a role as for dielectric surfaces and should be treated using a *molecular description*, as this has been done also for ionic solids [32, 33]. Finally, it has also been suggested [14] that perhaps defect sites on the cluster or interaction with adatoms, atoms at kinks, and boundaries may somehow play a role. In earlier experiments [14], a test of such effects was made by roughening a Au(111) surface by prolonged bombardment, but no difference with the pristine surface was observed.

To summarize, in this work, we observed that Li ion neutralization on Au clusters grown on sputtered HOPG was more efficient than on bulk gold surfaces and was most efficient on small clusters. The results are fairly close to those obtained on Au clusters on TiO_2. We have also noted a very similar trend on clusters on chains of clusters on pristine HOPG and on clusters grown on alumina films, but these results are beyond the scope of this brief paper. Further theoretical efforts on the description of alkali ion neutralization on bulk metals and for the case of supported clusters would be most welcome.

Acknowledgments Jie Shen and Juanjuan Jia thank the China Scholarship Council which made their stay at Orsay possible. The authors are grateful to Drs. M.L Martiarena, E. Goldberg, J.-P. Gauyacq, and A. Borisov for interesting discussions on this subject.

References

1. Henry CR (1998) Surf Sci Rep 31:235
2. Freund H-J (2002) Surf Sci 500:271
3. Campbell CT (1997) Surf Sci Rep 27:1
4. Barke I, Hövel H (2003) Phys Rev Lett 90:166801
5. Hövel H, Barke I (2003) New J Phys 5:31
6. Nilius N, Kulawik M, Rust H-P, Freund H-J (2004) Surf Sci 572:347
7. Valden M, Lai X, Goodman DW (1998) Science 281:1647
8. Choudhary TV, Goodman DW (2002) Top Catal 21:25
9. Lai X, St. Clair TP, Valden M, Goodman DW (1998) Prog Surf Sci 59:25
10. Lai X, St. Clair TP, Goodman DW (1999) Faraday Discuss 114:279

11. Luo K, St. Clair TP, Lai X, Goodman DW (2000) J Phys Chem B 104:3050
12. Lai X, Goodman DW (2000) J Mol Catal A Chem 162:33
13. Haruta M, Yamada N, Kobayashi T, Iijima S (1989) J Catal 115:301
14. Canario AR, Esaulov VA (2006) J Chem Phys 124:224710
15. Liu GF, Sroubek Z, Yarmoff JA (2004) Phys Rev Lett 92:216801
16. Hamoudi H, Dablemont C, Esaulov VA (2008) Surf Sci 602:2486
17. Guillemot L, Bobrov K (2007) Surf Sci 601:871–875
18. Behringer ER, Andersson DR, Cooper BH, Marston JB (1996) Phys Rev B 54:14765
19. Borisov AG, Teillet-Billy D, Gauyacq JP, Winter H, Dierkes G (1996) Phys Rev B 54:17166
20. Niedfeldt K, Nordlander P, Carter EA (2006) Phys Rev B 74:115109
21. Wiatrowski M, Lavagnino L, Esaulov VA (2007) Surf Sci 601:L39
22. Canario AR, Kravchuk T, Esaulov VA (2006) New J Phys 8:227
23. Canário AR, Borisov AG, Gauyacq JP, Esaulov VA (2005) Phys Rev B 71:121401
24. Chen L, Shen J, Jia J, Kandasamy T, Bobrov K, Guillemot L, Fuhr JD, Martiarena ML, Esaulov VA (2011) Phys Rev A 84:052901
25. Borisov AG, Mertens A, Wethekam S, Winter H (2003) Phys Rev A 68:012901
26. Schmitz A, Shaw J, Chakrabory HS, Thumm U (2010) Phys Rev A 81:042901
27. Garcia EA, Romero MA, Gonzalez C, Goldberg EC (2009) Surf Sci 603:597
28. Meyer C, Bonetto F, Vidal R, Garcıa EA, Gonzalez C, Ferron J, Goldberg EC (2012) Phys Rev A86:032901
29. Shao H, Langreth DC, Nordlander P (1994) Phys Rev B 49:13948
30. Maazouz M, Guillemot L, Esaulov VA, O'Connor DJ (1998) Surf Sci 398:49–59
31. Borisov AG, Esaulov VA (2000) J Phys Condens Matter 12:R177
32. Ustaze S, Verucchi R, Lacombe S, Guillemot L, Esaulov VA (1997) Phys Rev Lett 79:3526
33. Deutscher SA, Borisov AG, Sidis V (1999) Phys Rev A 59:4446

13

Understanding of the oxygen activation on ceria- and ceria/alumina-supported gold catalysts: a study combining $^{18}O/^{16}O$ isotopic exchange and EPR spectroscopy

Pandian Lakshmanan · Frédéric Averseng · Nicolas Bion · Laurent Delannoy · Jean-Michel Tatibouët · Catherine Louis

Abstract Gold supported on ceria or ceria–alumina mixed oxides are very active catalysts for total oxidation of a variety of molecules. The key step of the oxygen activation on such catalysts is still a matter of debate. Gold–ceria (Au/CeO_2) and gold–ceria–alumina ($Au/CeO_2/Al_2O_3$) catalysts were prepared by deposition–precipitation of gold precursor with urea as in former works where their efficiency to catalyze the oxidation of propene and propan-2-ol was demonstrated. To understand the phenomenon of oxygen activation over this class of catalysts, efficient techniques generally used to characterize the interaction between oxygen and cerium-based oxides were applied; the oxygen storage capacity (OSC) measurement, the $^{18}O_2/^{16}O_2$ isotopic exchange study (OIE), as well as characterizations by in situ Raman and electron paramagnetic resonance (EPR) spectroscopies. Each of the techniques allowed showing the impact of the gold nanoparticles on the activation of dioxygen, on the kinetic governing the gas-phase/ solid oxygen atom exchange, and on the nature and the location of the adsorbed oxygen species. Gold nanoparticles were shown to increase drastically the OSC values and the rate of oxygen exchange. OIE study demonstrated the absence of pure equilibration reaction ($^{16}O_{2(g)}+{}^{18}O_{2(g)} \leftrightarrow 2\ {}^{16}O^{18}O_{(g)}$), indicating that gold did not promote the dissociation of dioxygen. Peroxo adspecies were observed by Raman spectroscopy only in the presence of gold. On the contrary, EPR spectroscopy indicated that the concentration of superoxo adspecies was lower for oxide-supported gold samples than for bare oxides. The combination of techniques allowed reinforcing the hypothesis that the gold nanoparticules promote the activation of dioxygen by generating extremely mobile diatomic-oxygenated species at the gold/ceria interfacial perimeter. This specific gold–ceria interaction, which leads to the increase in oxygen mobility, is probably also responsible for the higher catalytic performance of Au/CeO_2 and $Au/CeO_2/Al_2O_3$ in oxidation reaction compared to bare supports.

Keywords Gold · Heterogeneous catalysis · $^{18}O_2/^{16}O_2$ isotopic exchange · EPR · Raman · Ceria · Oxygen storage capacity · Oxygen species

Introduction

Since the exceptional discovery of Haruta who demonstrated that, when smaller than 5 nm, the gold nanoparticules can display very high catalytic CO oxidation activity even at 203 K [1, 2], hundreds of papers on the topic of catalysis by gold have been annually published with a progression which remains exponential. Then, numerous catalytic formulations have been developed, and the efficiency of nanostructured gold deposited on reducible transition metal oxide supports was particularly reported [3]. Among the reducible oxide supports, ceria was particularly investigated because of its capability to act as an oxygen reservoir by storing/releasing oxygen through a redox process which involves the Ce^{4+}/Ce^{3+} couple. Nanocrystalline ceria-supported gold catalysts were then studied in various catalytic reactions including the water–gas shift reaction [4, 5], the VOC combustion [6–10], the selective oxidation of alcohol [11], the selective oxidation of CO in excess of H_2 (CO-PROX) [12], and the most investigated reaction remaining the low-temperature CO oxidation [13, 14]. The morphology of

P. Lakshmanan · F. Averseng · L. Delannoy · C. Louis
Laboratoire de Réactivité de Surface, UMR 7197 CNRS, Université Pierre et Marie Curie-UMPC, 4 place Jussieu, 75252 Paris Cedex 05, France

P. Lakshmanan · N. Bion (✉) · J.-M. Tatibouët
CNRS UMR 7285, Institut de Chimie des Milieux et Matériaux de Poitiers (IC2MP), University of Poitiers, 4 rue Michel Brunet, 86022 Poitiers Cedex, France
e-mail: nicolas.bion@univ-poitiers.fr

the ceria crystallites, the size of the gold clusters (particles), and the oxidation state of the active gold atoms are the parameters usually discussed because of their determining importance for the catalytic oxidation activity [15]. The nature of the intermediate species is also studied. For instance, it has been clearly demonstrated that one of the key elements for explaining the high reactivity of nanometer-sized gold nanoparticles in CO oxidation was the abundance of low-coordinated Au atoms in the small particles, where CO can be preferentially adsorbed [16]. On the contrary, the nature of the intermediate species in the oxygen activation step is still a matter of debate [17]. One reason is the fact that the adsorbed oxygen species are not experimentally easy to identify in the conditions of the reaction. Some recent papers showed that the anion photoelectron spectroscopy [18] and infrared multiple photon dissociation [19] allowed determining which oxygenated species are involved in the activation process, but these spectroscopic studies have not been applied to ceria-supported gold catalysts. Computational method is a way to get around the experimental limitations, and very impressive demonstrations in terms of reactive intermediate species and mechanism scheme were reported [17, 20]. Nevertheless, the use of the theoretical modeling did not permit to prevent controversial conclusions.

In this work, we used a combination of characterization techniques involving oxygen storage capacity measurement (OSC), oxygen isotopic exchange (OIE), and Raman and electron paramagnetic resonance (EPR) spectroscopies. These techniques were extensively used on cerium-based oxides to understand the mechanism of dioxygen activation [21–23] and to explain the property of these oxides to store and release oxygen [24, 25]. Thus, we proposed to apply the same techniques on ceria-supported gold catalysts. In addition, gold supported on mixed oxide consisting of cerium oxide supported on alumina with two CeO_2 loadings (5 to 10 wt.%) were also studied. The latter catalytic formulations showed interesting catalytic performances in the oxidation of two types of VOCs as follows: propene [7] as model of hydrocarbon and 2-propanol [26] as model of alcohol. Ceria–alumina was preferred to ceria support because of the poor resistance of the latter against thermal sintering.

Experimental part

Materials

Alumina, AluC Degussa (110 m^2 g^{-1}), was used as the support to load cerium oxide (5 and 10 wt.% with respect to alumina) by impregnation in excess of aqueous solution of $Ce(NO_3)_3$, $6H_2O$ (Aldrich, 99.9 %), followed by calcination at 500 °C. One weight percent of gold was loaded by deposition–precipitation with urea on the various CeO_2–Al_2O_3 samples as well as on pure alumina and ceria of high surface area (HSA-5 Rhodia,

200 m^2 g^{-1}) and low surface area, sintered HSA-5 by aging at 800 °C under a flow of humid air for 12 h (abbreviated as CeO_2 LSA 50 m^2 g^{-1}). The catalysts are named as $Au/xCeO_2/Al_2O_3$ (x=wt% of CeO_2), Au/Al_2O_3, and Au/CeO_2. The details of experimental procedures regarding catalyst preparation and characterization by elemental analyses, N_2 physisorption at 77 K, XRD, X-ray photoelectron spectroscopy (XPS), and combined TEM–energy-filtered transmission electron microscopy (EFTEM) techniques were largely described in our previous study [7].

Oxygen storage capacity

OSC measurements were carried out at 673 K using a U-form reactor connected to a gas chromatograph equipped with a Porapak column and a thermal conductivity detector. The experimental set-up and the protocol of experiment were reported elsewhere [27]. The samples were placed into the reactor and heated under a continuous flow of helium (30 cm^3 min^{-1}) up to 673 K for 30 min before pretreatment with ten pulses (0.246 cm^3) of pure O_2 at atmospheric pressure followed by a purge with pure He for 10 min. Alternate pulses of CO and O_2 were undertaken three times in order to check the reproducibility of the measurement. The amounts of unconverted CO and O_2 as well as of produced CO_2 were quantified. The OSC was calculated from the average value of CO_2 production (after CO pulse) and was expressed in μmol O g^{-1} or μmol O m^{-2} taking into account the mass or the BET surface area of the samples.

The number of surface oxygen atoms involved in the OSC process was calculated considering that only ceria participated and assuming a preferential (100) orientation of the ceria surface. Calculations have been already described in a previous publication [27].

Oxygen isotopic exchange

OIE experiments were performed in a set-up already described elsewhere [27, 28]. A U-form reactor was placed in a closed recycle system which was connected on one side to a mass spectrometer (Pfeiffer Vacuum, QMS 200) for the monitoring of the gas-phase composition and on the other side to a vacuum pump. A recycling pump removed limitations due to gas-phase diffusion. OIE experiments were undertaken on 20 mg of catalyst, subjected to a $^{16}O_2$ activation step at 873 K under atmospheric pressure for 1 h prior to cooling to the desired temperature, at which point the system was degassed and the isotopic mixture charged. The study of homo-exchange (also called equilibration reaction) was performed using an equimolar mixture of $^{16}O_2$ and $^{18}O_2$ (99.9 % purity, supplied by Isotec), whereas for the study of the hetero-exchange, the mixture was replaced by pure $^{18}O_2$. The masses 32, 34, and 36 m/z were monitored as a function of time to follow the exchange. The m/z values of 28 and 44 were also recorded to check the absence of air or CO_2. The atomic fraction of ^{18}O in the gas phase (α_g), the rate of exchange (R_e), and the number of O atoms exchanged (N_e) were calculated as described in previous references [29]. Typically:

$$\alpha_g = \frac{P_{36} + \frac{1}{2}P_{34}}{P_{36} + P_{34} + P_{32}} \qquad (1)$$

Where P_{36}, P_{34}, and P_{32} were the partial pressures of $^{18}O_2$, $^{18}O^{16}O$, and $^{16}O_2$, respectively;

$$R_e = -N_g \frac{d\alpha_g}{dt} \qquad (2)$$

Where N_g was the number of ^{18}O atoms in gas phase at the beginning of the reaction;

$$N_e = N_g(1 - \alpha_g) \qquad (3)$$

Finally, the number of exchangeable atoms could be calculated when equilibrium between the gas-phase and the solid was reached by using:

$$N_s = \frac{N_e}{\alpha^*} = N_g \left[\frac{1 - \alpha^*}{\alpha^*} \right] \qquad (4)$$

Where α^* was the value of α_g at equilibrium.

Raman spectroscopy

The Raman study was performed using a KAISER RXN1 spectrometer equipped with a NIR (785 nm) laser diode (25< power <50 mW). The powder sample was put in a built-in cell that allowed heating and spectra acquisition under gas flow. Once in the cell, the sample was flushed under a flow of O_2 (50 cm^3 min^{-1}) and heated up to 693 K (5 K min^{-1}) for 1 h. Then, the O_2 flow was switched to Ar (same flow rate), and the cell was cooled down to room temperature (RT), at which point a spectrum was registered. The sample was finally flushed 30 min with O_2 flow (150 cm^3 min^{-1}) at RT, then possibly flushed with an Ar flow (150 cm^3 min^{-1}).

Electron paramagnetic resonance spectroscopy

The EPR spectra were recorded on a JEOL FA-300 series EPR spectrometer at ~9.3 GHz (X-band) using a 100 kHz field modulation and a 2.5 G standard modulation width. The spectra were recorded at 77 K using an insertion Dewar containing liquid nitrogen. Computer simulation of the spectra was performed using the EPRsim32 program [30].

The sample was introduced into a cell consisting of a U-shape reactor with a porous disk for thermal treatment and connected to an EPR tube. The cell can be closed using vacuum valves. After 10 min in dynamic primary vacuum ($\approx 10^{-2}$ mbar) at RT, 250 mbar O_2 was introduced into the cell, which was then heated at 773 K (5 K min^{-1}) for 2 h. After decreasing the temperature down to 723 K, the sample was evacuated under dynamic primary then secondary vacuum ($\approx 10^{-4}$ mbar, 10 min) before rapid cooling to RT. At RT, increasing pressures of O_2 were introduced into the cell (from 1.5 up to 50 mbar). After internal transfer of the sample into the EPR tube, the spectra were recorded after 30 min cooling at 77 K (acquisition time: 8 min). The "raw" double integration of the EPR signal (arbitrary units) was performed using the cwEsr software (3.3.36E XB version) provided by JEOL. The double integration per gram of sample, which is proportional to the amount of paramagnetic species in the absence of dipolar interaction, was calculated considering the amount of sample in the EPR tube for each measurement.

Results and discussion

The characterization of the samples studied in this work has been already described in a previous work [7]. The main physicochemical parameters are summarized in Table 1. The BET surface area of pure alumina support slightly decreased as the CeO$_2$ loading increased, while the high BET surface area of commercial ceria was maintained in the Au/CeO$_2$ catalyst to 200 m^2 g^{-1}. In Au/xCeO$_2$/Al$_2$O$_3$ (x=5 and 10 wt.%), mainly 2-D patches and 3-D nanoparticles of CeO$_2$ (ca 8 nm) were detected by XRD and EFTEM in accordance with similar observation made by Martínez-Arias [31]. The average sizes of the gold particles visible by EFTEM also reported in Table 1 did not depend on the support but were generally smaller after reduction than after calcination treatment (note that gold particles are visible only on alumina because of the poor contrast between gold and ceria). Finally, XPS and CO oxidation model reaction showed that whatever the mode of activation (thermal treatment under H$_2$ or O$_2$), all

Table 1 BET surface area, Au loading, and Au particle size of the catalysts

Catalysts	S$_{BET}$	Au	Au particle size (nm)	
	(m^2 g^{-1})	(%)	After calcination	After reduction
Au/Al$_2$O$_3$	110	0.88	2.4 (0.56)[a]	2.0 (0.49)
Au/5CeO$_2$/Al$_2$O$_3$	99	0.89	2.6 (0.43)	2.2 (0.47)
Au/10CeO$_2$/Al$_2$O$_3$	92	0.89	2.6 (0.57)	2.0 (0.45)
Au/CeO$_2$	200	0.97	n.m.[b]	n.m.

[a] Standard deviation is given in brackets

[b] n.m. not measurable

gold was metallic after reduction under H_2, but gold remained unreduced on ceria or ceria patches after calcination under O_2, while it was metallic on alumina.

Oxygen storage capacity

OSC measurements were performed at 673 K. The values obtained for the various samples are reported in Table 2, in which one can distinguish the OSC of the bare supports (denoted "without Au") and the OSC of the samples containing gold (denoted "with Au"). As it could be anticipated, the Al_2O_3 support did not display any OSC activity at 673 K. The activity, i.e., the CO_2 production, remained null for Au/Al_2O_3, confirming the reduced state of gold when deposited on alumina after an oxidation stage at 673 K. Indeed, oxidized gold would have been reduced by the CO pulses, and CO_2 would have been produced. OSC was detectable for $5CeO_2/Al_2O_3$ and $10CeO_2/Al_2O_3$ but was not really different for the two samples. The OSC of bare ceria was much higher, 162 μmol O g^{-1}, corresponding to 0.81 μmol O m^{-2}. In order to assess the proportion of oxygen atoms involved in this process, the latter value was compared with the theoretical one. Taking into account that only one oxygen atom out of four is involved in the Ce^{4+}/Ce^{3+} reduction step, the theoretical OSC for ceria can be estimated as 5.7 μmol O m^{-2}. The ratio between the experimental and theoretical OSC values gives the number of surface oxide layers (N_L) participating to the process. For the bare ceria (200 m^2 g^{-1}), the results in Table 2 show that at 673 K, N_L is restricted to a small fraction of the surface (N_L=0.14). The calculation performed for the ceria of low LSA area shows that N_L is the same (N_L=0.15) and therefore that the OSC process is only dependent on the surface area for a given oxide. In contrast to what was expected, the N_L values were higher when ceria was supported on alumina.

We can notice a beneficial effect of the presence of gold nanoparticles on the OSC values for the various supports

(Table 2). On CeO_2, after removing the CO_2 produced by the reduction of oxidized gold, the OSC value was three times as high in the presence of Au. It is worth noting that on CeO_2 LSA, the increase of the OSC by the presence of gold was much lower, showing that the OSC is not only dependent on the ceria surface but also on the Au/CeO_2 interaction. Flytzani-Stephanopoulos et al. also reported the improvement of the OSC property of ceria due to the presence of gold [32]. The impact of Au is more pronounced for the ceria–alumina mixed oxide supports, for which the OSC process involves the participation of bulk oxygen atoms (N_L >1 indicates that the OSC is not limited to the surface) and for which the OSC seems to depend on the CeO_2 content (N_L=1.81 and 2.19 for $Au/5CeO_2/Al_2O_3$ and $Au/10CeO_2/Al_2O_3$, respectively).

Isotopic oxygen exchange ($^{18}O_2/^{16}O_2$ IE)

We first investigated the homo-exchange reaction ($^{16}O_{2(g)}+$ $^{18}O_{2(g)} \leftrightarrow 2\ ^{16}O^{18}O_{(g)}$) by introducing an equimolar mixture of $^{18}O_2/^{16}O_2$ in the Au/CeO_2 system, and the oxygen isotopic exchange evolution was followed in a temperature-programmed experiment. When temperature increased, a decrease of the $^{18}O_2$ partial pressure and an increase of the partial pressures of $^{16}O^{18}O$ and $^{16}O_2$ were observed (Fig. 1). A decrease of the ^{18}O atomic fraction (α_g) was observed in the gas phase when the exchange was taking place. This behavior is not consistent with a homo-exchange reaction during which α_g should remain constant. This result is therefore the indication of a hetero-exchange process in which ^{16}O lattice oxygen atoms, supplied by ceria, participates to the reaction. A very similar evolution of the oxygen isotopomer partial pressures was observed on pure CeO_2 (Fig. S1), meaning that the hetero-exchange is due to ceria. Such a result emphasizes the inability of the gold nanoparticles to dissociatively adsorb molecular dioxygen or shows that the dissociation and the exchange with lattice oxygen atoms occur simultaneously.

In the objective to study whether gold nanoparticles influence the mobility of ceria oxygen atoms or not, we performed

Table 2 OSC of $xCeO_2/Al_2O_3$ and CeO_2 at 673 K without or with the presence of 1 wt.% Au

Supports	Without Au		With Au	
	OSC (μmol O g^{-1})	N_L[a]/ CeO_2	OSC (μmol O g^{-1})	N_L/ CeO_2
Al_2O_3	0	0	0	0
$5CeO_2/Al_2O_3$	15	0.53	51	1.81
$10CeO_2/Al_2O_3$	17	0.36	115	2.19
CeO_2 (200 m^2 g^{-1})	162	0.14	474	0.42
CeO_2 LSA (49 m^2 g^{-1})	41	0.15	74	0.26

[a] N_L number of layers; N_L=1 when the surface is entirely involved in the OSC process, N_L>1 when bulk oxygen atoms are involved in the OSC process

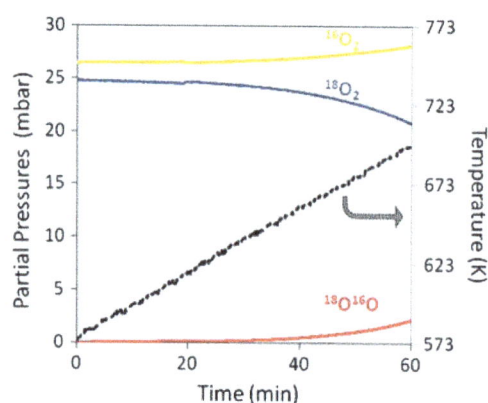

Fig. 1 Evolution of the oxygen isotopomer partial pressures during a tempertaure-programmed homo-exchange experiment on Au/CeO_2 catalyst

an isothermal hetero-exchange experiment, introducing pure $^{18}O_2$ (ca 50 mbar) in the reactor cell at 723 K. The results, in terms of initial rate of exchange (R_e) and number of atoms exchanged (N_e) after 20 min reaction, are reported in Table 3. Again, in this set of experiments, we compared the bare supports (columns denoted "without Au") with the gold catalysts ("with Au"). It clearly appears that the gold nanoparticles exalt the isotopic exchange activity. Since it is well known that Al_2O_3 support is not able to exchange oxygen at 723 K [33], these results show that the exchange is promoted by the interaction between Au nanoparticles and CeO_2. The R_e values, which were low for $5CeO_2/Al_2O_3$ and $10CeO_2/Al_2O_3$, increased in the presence of gold to a value quite similar for both, ca 6–7×10^{17} at O g^{-1} s^{-1}. The similarity between both solids was confirmed in comparing the number of exchanged atoms (N_e values): 5.43 and 5.53 at O nm^{-2} were exchanged for Au/$5CeO_2/Al_2O_3$ and Au/$10CeO_2/Al_2O_3$, respectively, suggesting that the proportion of gold particles in interaction with ceria was close in the two cases. The R_e value normalized per gram of sample was largely higher on bare CeO_2 (55.3×10^{17} at O g^{-1} s^{-1}). Despite this high value, the presence of Au led to an increase of Re (77.8×10^{17} at O g^{-1} s^{-1}). If one examines the N_e values, their increase was not as drastic as for the R_e values (13.7 at O nm^{-2} and 14.7 at O nm^{-2} for CeO_2 and Au/CeO_2, respectively). This can be explained by an equilibrium rapidly reached in the ^{18}O concentration between the gas phase and the solid (depending on the mass of catalyst and the initial pressure of $^{18}O_2$), which minimized the differences that could be detected after 20 min of exchange. Finally, we studied the effect of the ceria surface area with the results obtained on CeO_2 LSA. Contrary to the conclusion previously made regarding the OSC measurement (Table 2), it clearly appears that the oxygen exchange activity is not dependent on the surface area only, since the N_e value, normalized per surface unit, was five times as small for CeO_2 LSA as compared to CeO_2. The beneficial effect of Au was also noticed on CeO_2 LSA, since the R_e and N_e values were almost four times higher in the

presence of the metal. Nevertheless, the activity remained smaller than that of the high surface area ceria.

In order to ascertain the conclusion that the interaction between Au particles and CeO_2 surface was the key element to explain the strong activity in oxygen exchange, we prepared two other gold catalysts with higher gold loading, i.e., containing 4 wt.% Au deposited on $10CeO_2/Al_2O_3$ and CeO_2. Note that the size of the Au particles was maintained around 2.5 nm. The higher gold loading resulted in higher R_e and N_e values (see values between parentheses in Table 3). A closer inspection of the evolutions of the $^{16}O_2$ (m/z=32) and $^{18}O^{16}O$ (m/z=34) partial pressures as a function of time (Fig. 2) provided possible reasons for the beneficial impact of gold in the oxygen exchange reaction. At the beginning of the reaction, both $^{18}O^{16}O$ and $^{16}O_2$ molecules appeared simultaneously on $10CeO_2/Al_2O_3$ (Fig. 2a) and on ceria support (Fig. 2c). This suggests that simple exchange (Eq. 4) and multiple exchange (Eq. 5) reactions occurred on these supports as already reported for ceria in previous studies [27, 29] as follows:

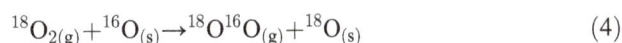

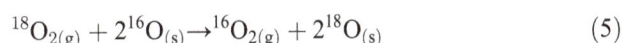

$$^{18}O_{2(g)} + {}^{16}O_{(s)} \rightarrow {}^{18}O^{16}O_{(g)} + {}^{18}O_{(s)} \tag{4}$$

$$^{18}O_{2(g)} + 2{}^{16}O_{(s)} \rightarrow {}^{16}O_{2(g)} + 2{}^{18}O_{(s)} \tag{5}$$

The presence of gold made the $^{16}O_2$ isotopomer the main molecule present at the very beginning of the experiments (Fig. 2b, d), indicating that Au favored the multiple exchange. The promotion of the multiple hetero-exchange mechanism by gold when supported on Ce-doped ordered mesoporous alumina materials has already been reported by Fonseca et al. [34]. Taking into account that the multiple exchange must involve binuclear oxygen species as intermediate, it can be inferred from Fig. 2 that the interfaces between the gold nanoparticles and the ceria crystallites are the preferential location for dioxygen activation, in the form of either diatomic peroxide or superoxide species. The correlation between multiple exchange and peroxide or superoxide intermediate species was suggested by Winter to explain the complex exchange mechanism observed over basic oxides [35]. Corma et al. provided spectroscopic evidences that superoxide species and peroxide adspecies at the one-electron defect site were the active species in the CO oxidation reaction for gold supported on nanocrystalline CeO_2 [36].

To investigate the nature of the binuclear oxygen species adsorbed on the catalytic surface, characterizations by Raman and electron paramagnetic resonance spectroscopies were performed in focusing the study on the impact of the presence of gold nanoparticles.

Identification of oxygen active species

In an attempt to detect undissociated oxygen species by Raman spectroscopy, the catalyst pretreatments were performed in

Table 3 $^{18}O_2/^{16}O_2$ isotopic exchange experiments on pre-oxidized samples at 723 K

Supports	Without Au		With Au	
	R_e (10^{17} at O g^{-1} s^{-1})	N_e at 20 min (at O nm^{-2})	R_e (10^{17} at O g^{-1} s^{-1})	N_e at 20 min (at O nm^{-2})
$5CeO_2/Al_2O_3$	1.29	1.06	7.70	5.43
$10CeO_2/Al_2O_3$	1.96	1.61	6.30 (25)[a]	5.53 (10.50)
CeO_2	55.30	13.70	77.80 (183)	14.70 (16.40)
CeO_2 LSA	1.40	2.72	5.40	9.92

[a] Values between parentheses are the results for Au(4 wt.%)/$10CeO_2$/Al_2O_3 and Au(4 wt.%)/CeO_2

Fig. 2 Evolution of the oxygen isotopomer partial pressures as a function of time during an hetero-exchange experiment at 723 K on **a** 10CeO₂/Al₂O₃, **b** Au(4 wt.%)/10CeO₂/Al₂O₃, **c** CeO₂, and **d** Au(4 wt.%)/CeO₂ catalysts

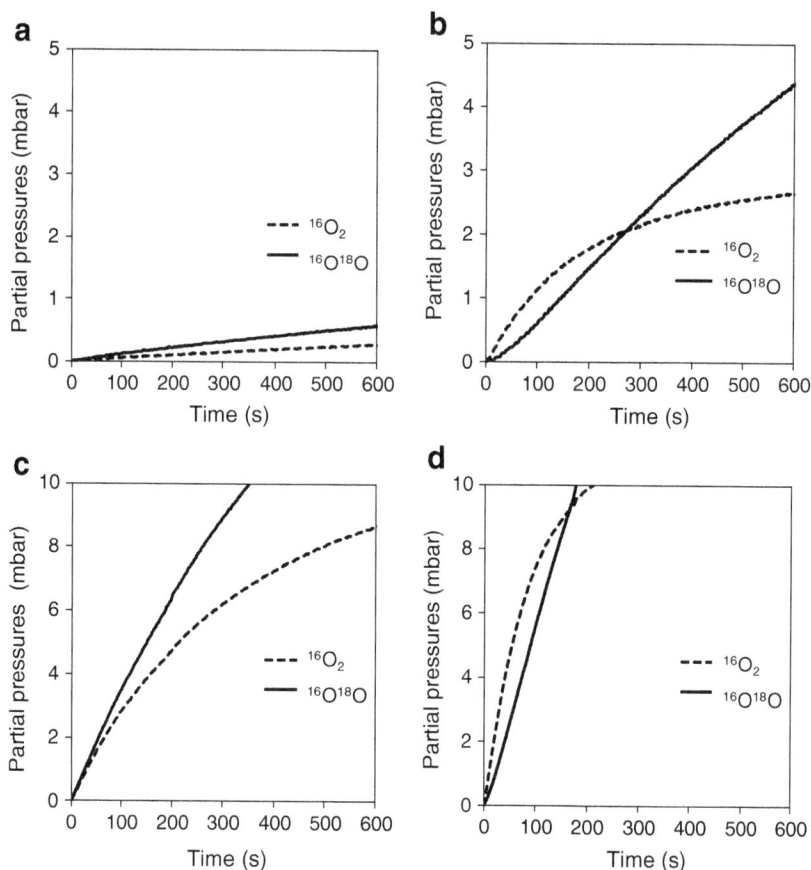

conditions as close as possible as those used for the isotopic exchange study (see Experimental part). Raman spectroscopy is a sensitive and specific technique capable to discriminate oxygen species (O_2, O_2^- or O_2^{2-}) adsorbed on CeO_2, as reported in earlier works [22, 37]. The vibrational frequencies of these adsorbed oxygen species strongly decrease with their

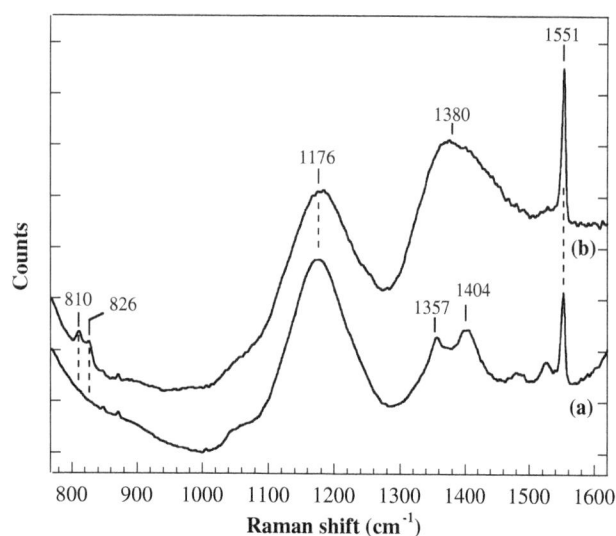

electronic charge; in the 820–890 cm⁻¹ range for O_2^{2-} (peroxo), 1,120–1,140 cm⁻¹ for O_2^- (superoxo), and 1,480–

Fig. 3 Raman spectra after calcination at 693 K Ar, then O_2 flow of **a** CeO_2, and **b** Au/CeO₂

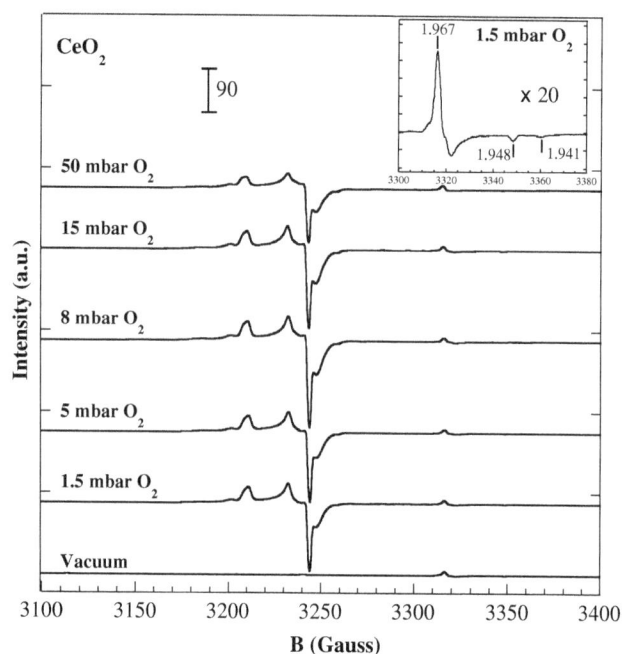

Fig. 4 EPR spectra of CeO_2 sample (recorded after 30 min at 77 K) under increasing PO_2

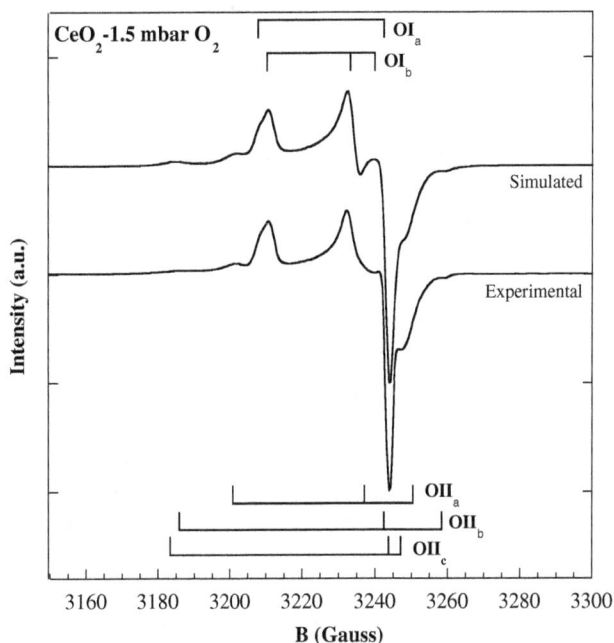

Fig. 5 Experimental (recorded after 30 min at 77 K) and simulated EPR spectra of CeO$_2$ sample, under 1.5 mbar O$_2$

$1,570 \text{ cm}^{-1}$ for physically adsorbed O$_2$ [22, 37]. Moreover, the resulting bands are sharp compared to the ones related to ceria lattice vibrations, which make them quite easy to identify.

First of all, due to the strong fluorescence of the Al$_2$O$_3$-containing materials, only CeO$_2$ and Au/CeO$_2$ materials could be studied. Under O$_2$ flow at RT, apart from the vibration bands characteristic of ceria (Fig. S2), i.e., an intense band at 458 cm^{-1} attributed to 1st order F_{2g} lattice vibrational mode of O_h and weaker bands at 253 and 1,176 cm^{-1} attributed to second order modes, arising from a mixing of A$_{1g}$, E$_g$, and F$_{2g}$

lattice vibrational modes [38] (additional bands were visible at 1,357, 1,404, and 1,380 cm^{-1}, assigned to residual carbonate/carboxylate species), the spectra of CeO$_2$ and Au/CeO$_2$ (Fig. 3) revealed the presence of a sharp band at 1,551 cm^{-1} that grew under O$_2$ flow. It can be reasonably attributed to physically adsorbed O$_2$, since it easily disappeared by flushing with inert gas. One can note that apart from the vibration bands characteristic of ceria described above, a shoulder around 590 cm^{-1} was also visible for both CeO$_2$ and Au/CeO$_2$ (Fig. S2). It is related to the presence of oxygen vacancies [39]. The intensity of the 590 cm^{-1} shoulder relative to the F2g mode band (458 cm^{-1}) is about twice as high for Au/CeO$_2$ (9×10^{-2}) as for CeO$_2$ (4×10^{-2}), revealing a much more oxygen-defective material in the presence of gold. These additional vacancies are probably located at the perimeter of the interface between gold and ceria and would explain the exaltation of the multiple exchange mechanism observed by isotopic exchange technique for gold-containing catalysts [40]. In addition, tiny bands at 810 and 826 cm^{-1} could be observed on Au/CeO$_2$ only (not on CeO$_2$), in the frequency range of adsorbed O$_2^{2-}$ peroxo species. No O$_2^-$ (superoxo) species could be observed. It may be noted that the observation of such superoxo species often requires low temperature (93 K), which could not be reached with our spectrometer. In order to ascertain the presence of O$_2^-$ (superoxo) species, EPR experiments were also performed.

To avoid the interaction between the radical oxygen species and the gas phase O$_2$ which is known to induce signal broadening, it was necessary to perform the EPR experiments at low temperature, i.e., at the liquid nitrogen temperature, 77 K, which is the usual temperature used for the identification of the oxygen radicals. The EPR study was first performed over CeO$_2$ and Au/CeO$_2$. Again, the pretreatments had to be

Table 4 Simulated EPR parameters of O$_2^-$ species adsorbed on CeO$_2$, Au/CeO$_2$, 10CeO$_2$/Al$_2$O$_3$, and Au/10CeO$_2$/Al$_2$O$_3$ under 1.5 and 50 mbar O$_2$

Sample	O$_2^-$ species Assignment*	g$_z$ (g$_{||}$)	g$_x$ (g$\perp$)	g$_y$	Proportion (%) 1.5 mbar O$_2$	Proportion (%) 50 mbar O$_2$
CeO$_2$	OI$_a$	2.033	2.011	2.011	26.5	23.5
	OI$_b$	2.031	2.017	2.012	30.5	20.5
	OII$_a$	2.037	2.014	2.007	26.5	27.0
	OII$_b$	2.047	2.011	2.001	3.5	5.5
	OII$_c$	2.048	2.010	2.008	13.0	23.5
Au/CeO$_2$						
	OI$_a$	2.033	2.011	2.011	48.0	45.5
	OI$_b$	2.031	2.016	2.011	14.0	16.0
	OII$_a$	2.036	2.012	2.007	17.0	18.0
	OII$_b$	2.048	2.011	2.001	1.5	6.5
	OII$_c$	2.049	2.009	2.008	19.5	14.0
10CeO$_2$/Al$_2$O$_3$	OCA	2.027	2.017	2.011	$\approx$100	
Au/10CeO$_2$/Al$_2$O$_3$	OCA	2.027	2.017	2.011	$\approx$100	

*According to notations of Martínez-Arias et al. [31]

Table 5 Overall intensity (double integration) of O_2^- species adsorbed on CeO_2, Au/CeO_2, $10CeO_2/Al_2O_3$, and Au/$10CeO_2/Al_2O_3$ upon increasing O_2 pressures

PO_2 (mbar)		1.5	5	8	15	50
Overall intensity	CeO_2	6 590	7 140	9 200	9 060	6 740
	Au/CeO_2	4 070	4 940	6 020	6 220	4 000
	$10CeO_2/Al_2O_3$	51 300	36 000	29 500	8 350	200
	Au/$10CeO_2/Al_2O_3$	22 300	8 700	3 800	550	≈ 0

adapted to the technical constraints, but were performed in conditions as close as possible to those of isotopic exchange (see Experimental part). After treatment under vacuum at 723 K, the EPR spectra of CeO_2 and Au/CeO_2 samples revealed a weak anisotropic signal at g≈1.97 (Figs. 4 and S3) composed of two species ($g_{\perp1,2}$≈1.97, $g_{\|1}$=1.948, and $g_{\|2}$=1.941, insert of Fig. 4). The signal obtained for Au/CeO_2 was slightly broader because of the presence of a third species (1.98<$g_{\perp3}$<1.96 and $g_{\|3}$=1.937–1.936) (Fig. S3). Its intensity (double integration) was similar for CeO_2 and Au/CeO_2 samples. Though this signal at g≈1.97 was currently observed on ceria-based samples, its attribution is still under debate. It was tempting to assign it to Ce^{3+}; however, according to the literature, such paramagnetic ion could be observed at very low temperature only (T <20 K) [41, 42]. Some authors proposed that this signal originated from Ce^{3+} in peculiar geometry/symmetry [43, 44], while other authors proposed that it arises from quasi-free electrons with some orbital mixing with the empty f-orbitals of Ce^{4+} ions [23, 45].

Upon addition of O_2, the signal at g≈1.97 remained unchanged, which further strengthened the hypothesis of a paramagnetic species trapped in the bulk of the CeO_2 particles (no possible dipolar interaction with O_2 that could affect the signal). Moreover, a complex new signal of high intensity appears in the 3,150 G<B<3300 G region, typical of O_2^- species (Fig. 4). It had almost the same shape for CeO_2 and Au/CeO_2 (Fig. S3), and the simulation of the signal (Figs. 5 and S4) indicated the presence of five different O_2^- species in both samples with slightly different proportions (Table 4). The OI's and OII's species have been previously observed on similar materials [46–48]. These species were attributed to different adsorption modes of O_2^- on the surface, OI with EPR-equivalent oxygen atoms (adsorption parallel to the surface), and OII with nonequivalent oxygen atoms (end-on or other asymmetric adsorption).

Upon increasing O_2 pressure, the shape of the O_2^- signal was barely affected, but its intensity increased, then decreased when PO_2>8 mbar and 15 mbar for CeO_2 and Au/CeO_2, respectively, as attested by the values of the double integration reported in Table 5. The decreasing intensity above a critical O_2 pressure probably resulted from dipolar interactions between O_2^- and nearby O_2 molecules [49, 50]. This points out that the O_2^- signal intensity depends not only on the number of paramagnetic species, but also on this broadening effect, thus comparisons of the intensities are meaningful for the smallest O_2 pressures,

only. The interesting point regarding these experiments is that the intensity (double integration) of the overall O_2^- signal is smaller for Au/CeO_2 than for CeO_2 (Table 4), indicating that no extra O_2^- species were generated by the presence of gold.

In the case of $10CeO_2/Al_2O_3$ and Au/$10CeO_2/Al_2O_3$, upon O_2 addition, the signal of O_2^- was very different and much more intense at low PO_2 (more than five times) (Fig. 6) than that for CeO_2 and Au/CeO_2 samples (Figs. 4 and S3). Again, the shape of the O_2^- signal was similar for $10CeO_2/Al_2O_3$ and Au/$10CeO_2/Al_2O_3$ (Fig. S5). It was also much less complex than the former ones (Figs. 4 and S3), and could be accurately simulated considering only one O_2^- species, thereafter referred as OCA in the literature [31] (Table 4). It has been proposed that the OI and OII species formed on 3-D CeO_2 nanoparticles, while the OCA species formed on 2-D platelets of CeO_2 [31]. This assignment fits with the fact that CeO_2 consists of 3-D particles, and that, as mentioned at the beginning of the section "Results and discussion", previous EFTEM images of xCeO_2/Al_2O_3 revealed the presence of both 3-D and 2-D (platelets) CeO_2 particles on alumina [7]. The intensity of the OCA species strongly decreased and broadened for O_2 pressures >1.5 mbar, and almost disappeared at PO_2≈50 and 15 mbar, for $10CeO_2/Al_2O_3$ and Au/$10CeO_2/Al_2O_3$, respectively, which

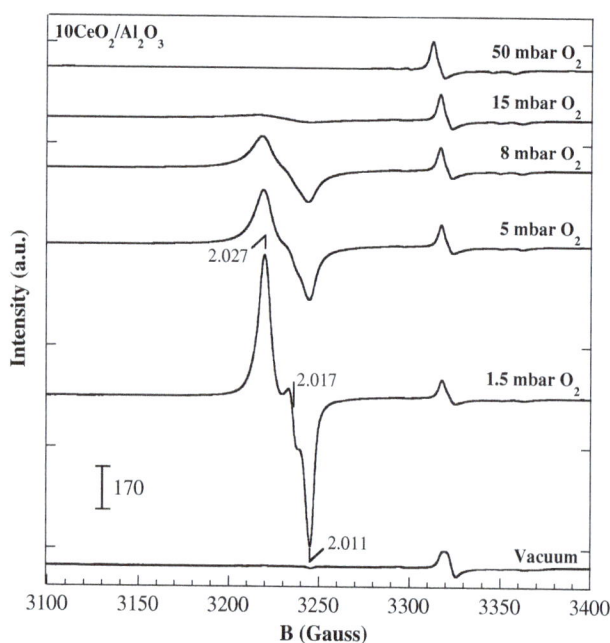

Fig. 6 EPR spectra of $10CeO_2/Al_2O_3$ sample (recorded after 30 min at 77 K) under increasing PO_2

was not observed with CeO_2 and Au/CeO_2. Such different behavior towards O_2 dipolar broadening has been reported earlier [51] on CeO_2/SiO_2 and CeO_2/Al_2O_3 materials, and the authors assigned it to the difference in O_2 accessibility to O_2^- species, thus locating O_2^- in the bulk (CeO_2/SiO_2: little to no broadening effect) or on the surface (CeO_2/Al_2O_3: strong broadening effect) of CeO_2. Again, the double integration of the O_2^- signal was somewhat smaller for $Au/10CeO_2/Al_2O_3$ than for $10CeO_2/Al_2O_3$, whatever the pressure of O_2 (Table 5).

To summarize, the comparison of the intensity of EPR signals of O_2^- species in CeO_2 and Au/CeO_2 pointed out a lower amount of O_2^- species in the presence of Au, for which the OSC property and the oxygen isotopic exchange activity was exalted, which ruled out the hypothesis that O_2^- species would play an active role in the corresponding processes. However, the formation of a noticeable amount of O_2^{2-} species adsorbed on Au/CeO_2 and not on CeO_2 strengthened the hypothesis of an activation of dioxygen molecule via Au nanoparticles through the formation of a peroxo species. This interpretation matches the conclusion brought by Pal et al. [18] who reported the results obtained by combining photoelectron spectroscopy and computational method, showing that the O–O bond was more activated (more elongated) in the peroxo form than that in the superoxo one, and suggesting that peroxo mode of chemisorptions plays a crucial role in the dioxygen activation.

Conclusion

In this work, we studied the oxygen mobility and the nature of the oxygenated species responsible for the chemisorption of dioxygen on Au/CeO_2 and $Au/xCeO_2/Al_2O_3$ (x=5 and 10 wt.%) catalysts. We stressed the role of the Au nanoparticles in the step of dioxygen activation. To achieve this goal, oxygen storage capacity measurements and $^{18}O/^{16}O$ isotopic exchange reaction were undertaken to study the oxygen mobility, while Raman spectroscopy and electron paramagnetic resonance spectroscopies were employed to determine the nature and the location of the adsorbed oxygen species. All the characterizations were performed on catalysts previously activated in conditions as close as possible to the thermal pretreatment used prior to the reactions of oxidation studied in our previous work [7].

Au nanoparticles were shown to exalt both the OSC property of CeO_2 and the exchange rate of oxygen between the gas phase and the lattice oxygen atoms of CeO_2, whether the Au nanoparticles were supported on CeO_2 or on CeO_2/Al_2O_3. Complementary results obtained by modifying the gold loading and the ceria surface indicated that the improvement in the oxygen mobility was dependent on the Au/CeO_2 interfacial perimeter. Moreover, the close inspection of the evolutions of the isotopomer partial pressures during the isotopic exchange experiments led to the conclusion that the interfaces between the gold nanoparticles and the ceria crystallites were the preferential location for dioxygen activation via adsorption of binuclear species.

Further characterization by Raman and EPR spectroscopies were performed to determine the nature of the dioxygen species. The study of CeO_2 and Au/CeO_2 by in situ Raman under oxygen revealed the presence of peroxo species on Au/CeO_2 and of additional vacancies probably located at the perimeter of the interface between gold and ceria. EPR revealed the presence of superoxo species on both samples, but their concentration was lower in the presence of gold. The same observation was made for $10CeO_2/Al_2O_3$ and $Au/10CeO_2/Al_2O_3$. As a consequence, the efficiency of Au/CeO_2 and $Au/xCeO_2/Al_2O_3$ catalysts for oxidation reaction could be explained by the activation of dioxygen molecule at the ceria/Au nanoparticles interfacial perimeter involving peroxo species.

Acknowledgments The authors thank Jean-Marc Krafft, engineer at LRS for the Raman measurements, and Pantea Baripour, master student at LRS for the preliminary experiments of EPR. The authors also acknowledge the Agence Nationale pour la Recherche for financial support (ANR-BLANC07-2 183612).

References

1. Haruta M, Kobayashi T, Sano H, Yamada N (1987) Novel Gold Catalyst for the Oxidation of Carbon Monoxide at a Temperature far below 0 °C. Chem. Lett. 405–408
2. Haruta M, Yamada N, Kobayashi T, Iijima S (1989) Gold catalysts prepared by co-precipitation for low-temperature oxidation of hydrogen and of carbon monoxide. J Catal 115:301–309
3. Bond GC, Louis C, Thompson DT (2006) Catalysis by gold. Imperial College Press, London
4. Fu Q, Saltsburg H, Flytzani-Stephanopoulos M (2003) Active non-metallic Au and Pt species on ceria-based water-gas shift catalysts. Science 301:935–938
5. Leppelt R, Schumacher B, Plzak V, Kinne M, Behm RJ (2006) Kinetics and mechanism of the low-temperature water-gas shift reaction on Au/CeO_2 catalysts in an idealized reaction atmosphere. J Catal 244:137–152
6. Delannoy L, Fajerwerg K, Lakshmanan P, Potvin C, Méthivier C, Louis C (2010) Supported gold catalysts for the decomposition of VOC: Total oxidation of propene in low concentration as model reaction. Appl Catal B Environ 94:117–124
7. Lakshmanan P, Delannoy L, Richard V, Méthivier C, Potvin C, Louis C (2010) Total oxidation of propene over $Au/xCeO_2$-Al_2O_3 catalysts: Influence of the CeO_2 loading and the activation treatment. Appl Catal B Environ 96:117–125
8. Solsona B, Garcia T, Murillo R, Mastral AM, Ndifor EN, Hetrick CE, Amiridis MD, Taylor SH (2009) Ceria and gold/ceria catalysts for the abatement of polycyclic aromatic hydrocarbons: an in situ DRIFTS study. Top Catal 52:492–500

9. Scire S, Minico S, Crisafulli C, Satriano C, Pistone A (2003) Catalytic combustion of volatile organic compounds on gold/cerium oxide catalysts. Appl Catal B Environ 40:43–49

10. Andreeva D, Petrova P, Sobczak JW, Ilieva L, Abrashev M (2006) Gold supported on ceria and ceria-alumina promoted by molybdena for complete benzene oxidation. Appl Catal B Environ 67:237–245

11. Enache DI, Knight DW, Hutchings GJ (2005) Solvent-free oxidation of primary alcohols to aldehydes using supported gold catalysts. Catal Lett 103:43–52

12. Bion N, Epron F, Moreno M, Mariño F, Duprez D (2008) Preferential oxidation of carbon monoxide in the presence of hydrogen (PROX) over noble metals and transition metal oxides: advantages and drawbacks. Top Catal 51:76–88

13. Carrettin S, Concepcion P, Corma A, Nieto JML, Puntes VF (2004) Nanocrystalline CeO₂ increases the activity of an for CO oxidation by two orders of magnitude. Angew Chem Int Ed 43:2538–2540

14. Widmann D, Leppelt R, Behm RJ (2007) Activation of a Au/CeO₂ catalyst for the CO oxidation reaction by surface oxygen removal/oxygen vacancy formation. J Catal 251:437–442

15. Guan Y, Ligthart DAJM, Pirgon-Galin O, Pieterse JAZ, van Santen RA, Hensen EJM (2011) Gold stabilized by nanostructured ceria supports: nature of the active sites and catalytic performance. Top Catal 54:424–438

16. Hvolbæk B, Janssens TVW, Clausen BS, Falsig H, Christensen CH, Nørskov JK (2007) Catalytic activity of Au nanoparticles. Nano Today 2:14–18

17. Boronat M, Corma A (2010) Oxygen activation on gold nanoparticles: separating the influence of particle size, particle shape and support interaction. Dalton Trans 39:8538–8546

18. Pal R, Wang L-M, Pei Y, Wang L-S, Zeng XC (2012) Unraveling the mechanisms of O₂ activation by size-selected gold clusters: transition from superoxo to peroxo chemisorption. J Am Chem Soc 134:9438–9445

19. Woodham AP, Meijer G, Fielicke A (2013) Charge separation promoted activation of molecular oxygen by neutral gold clusters. J Am Chem Soc 135:1727–1730

20. Green IX, Tang W, Neurock M, Yates JT Jr (2011) Spectroscopic observation of dual catalytic sites during oxidation of CO on a Au/TiO₂ catalyst. Science 333:736–739

21. Yao HC, Yao YFY (1984) Ceria in automotive exhaust catalysts.1. oxygen storage. J Catal 86:254–265

22. Pushkarev VV, Kovalchuk VI, d'Itry JL (2004) Probing defect sites on the CeO₂ surface with dioxygen. J Phys Chem B 108:5341–5348

23. Oliva C, Termignone G, Vatti FP, Forni L, Vishniakov AV (1996) Electron paramagnetic resonance spectra of CeO₂ catalyst for CO oxidation. J Mater Sci 31:149–158

24. Li C, Domen K, Maruya K, Onishi T (1990) Oxygen-exchange reactions over cerium oxide—An FT-IR study. J Catal 123:436–442

25. Li C, Domen K, Maruya K, Onishi T (1989) Dioxygen adsorption on well-outgassed and partially reduced cerium oxide studied by FT-IR. J Am Chem Soc 111:7683–7687

26. Lakshmanan P, Delannoy L, Louis C, Bion N, Tatibouët JM (2013) Au/xCeO₂/Al₂O₃ catalysts for VOC elimination: oxidation of 2-propanol. Catal Sci Technol.

27. Madier Y, Descorme C, Le Govic AM, Duprez D (1999) Oxygen mobility in CeO₂ and Ce$_x$Zr$_{(1-x)}$O₂ Compounds: study by CO transient oxidation and ¹⁸O/¹⁶O isotopic exchange. J Phys Chem 103:10999–11006

28. Ojala S, Bion N, Rijo Gomes S, Keiski RL, Duprez D (2010) Isotopic oxygen exchange over Pd/Al₂O₃ catalyst: study on C¹⁸O₂ and ¹⁸O₂ exchange. ChemCatChem 2:527–533

29. Duprez D (2006) Oxygen and hydrogen surface mobility in supported metal catalyst. Study by ¹⁸O/¹⁶O and ²H/¹H exchange. In: Hargreaves JSJ, Jackson SD, Webb G (eds) Isotopes in Heterogeneous Catalysis. Imperial College Press, London, pp 133–181

30. Spalek T, Pietrzyk P, Sojka Z (2005) Application of the genetic algorithm joint with the Powell method to nonlinear least-squares fitting of powder EPR spectra. J Chem Inf Model 45:18–27

31. Martínez-Arias M, Fernández-García M, Salamanca LN, Valenzuela RX, Conesa JC, Soria J (2000) Structural and redox properties of ceria in alumina-supported ceria catalyst supports. J Phys Chem B 104:4038–4046

32. Fu Q, Kudriavtseva S, Saltsburg H, Flytzani-Stephanopoulos M (2003) Gold–ceria catalysts for low-temperature water-gas shift reaction. Chem Eng J 93:41–53

33. Martin D, Duprez D (1996) Mobility of surface species on oxides. 1. isotopic exchange of ¹⁸O₂ with ¹⁶O of SiO₂, Al₂O₃, ZrO₂, MgO, CeO₂, and CeO₂-Al₂O₃. Activation by noble metals. Correlation with oxide basicity. J Phys Chem 100:9429–9438

34. Fonseca J, Royer S, Bion N, Pirault-Roy L, do Carmo Rangel M, Duprez D, Epron F (2012) Preferential CO oxidation over nanosized gold catalysts supported on ceria and amorphous ceria-alumina. Appl Catal B Environ 128:10–20

35. Winter ERS (1968) Exchange reactions of oxides. Part IX. J. Chem. Soc. A 2889–2902

36. Guzman J, Carrettin S, Corma A (2005) Spectroscopic evidence for the supply of reactive oxygen during CO oxidation catalyzed by gold supported on nanocrystalline CeO₂. J Am Chem Soc 127:3286–3287

37. Choi YM, Abernathy H, Chen H-T, Lin MC, Lu M (2006) Characterization of O₂–CeO₂ interactions using in situ Raman spectroscopy and first-principle calculations. ChemPhysChem 7:1957–1963

38. Weber WH, Hass KC, McBride JR (1993) Raman study of CeO₂. Second-order scattering, lattice dynamics, and particle-size effects. Phys Rev B 48:178–185

39. McBride JR, Hass KC, Poindexter BD, Weber WH (1994) Raman and x-ray studies of Ce$_{1-x}$RE$_x$O$_{2-y}$, where RE=La, Pr, Nd, Eu, Gd, and Tb. J Appl Phys 76:2435–2441

40. Widmann D, Leppelt R, Behm RJ (2007) CO Oxidation activity activation of a Au/CeO₂ catalyst for the CO oxidation reaction by surface oxygen removal/oxygen vacancy formation. J Catal 251:437–442

41. McLaughlan SD, Forrester PA (1966) Orthorhombic and trigonal electron-spin-resonance spectra of Ce³⁺ ions in CaF₂. Phys Rev 151:311–314

42. Barrie JD, Momoda LA, Dunn B, Gourier D, Aka G, Vivien D (1990) ESR and optical spectroscopy of Ce³⁺-β-alumina. J Sol St Chem 86:94–100

43. Dufaux M, Che M, Naccache C (1969) Electron paramagnetic resonance study of oxygen adsorption on supported molybdenum and cerium oxides. Comptes Rendus Acad Sci Paris C86:2255–2257

44. Fierro JLG, Soria J, Sanz J, Rojo JM (1987) Induced changes in ceria by thermal treatments under vacuum or hydrogen. J Sol St Chem 66:154–162

45. Gideoni M, Steinberg M (1972) Study of oxygen sorption on cerium(IV) oxide by electron-spin resonance. J Sol St Chem 4:370–373

46. Mendelovici L, Tzehoval H, Steinberg M (1983) The adsorption of oxygen and nitrous-oxide on platinum ceria catalyst. Appl Surf Sci 17:175–188

47. Soria J, Martínez-Arias A, Conesa JC (1995) Spectroscopic study of oxygen-adsorption as a method to study surface-defects on CeO₂. J Chem Soc Faraday Trans 91:1669–1678

48. Zhang X, Klabunde KJ (1992) Superoxide (O₂⁻) on the surface of heat-treated ceria-intermediates in the reversible oxygen to oxide transformation. Inorg Chem 31:1706–1709

49. Povich MJ (1975) Electron-spin resonance oxygen broadening. J Phys Chem 79:1106–1109

50. Pake GE, Tuttle TR (1959) Anomalous loss of resolution of paramagnetic resonance hyperfine structure in liquids. Phys Rev Lett 3:423–425

51. Aboukais A, Zhilinskaya EA, Lamonier JF, Filimonov IN (2005) Colloid Surf A Physicochem Eng Asp 260:199–207

14

Au as an efficient promoter for electrocatalytic oxidation of formic acid and carbon monoxide: a comparison between Pt-on-Au and PtAu alloy catalysts

Qiang Zhang · Ruirui Yue · Fengxing Jiang ·
Huiwen Wang · Chunyang Zhai · Ping Yang · Yukou Du

Abstract The absence of unpaired d-electrons of gold leads to its lack of reactivity and paucity of catalytic activity. Synergistic activity of bimetallic PtAu has been proved, and its structure greatly influences on the electrocatalytic activity toward formic acid and carbon monoxide oxidation. Here, a comparison between Pt-modified Au (designated as Pt-on-Au) and PtAu alloy catalysts has been studied. The Pt-on-Au catalyst was prepared by electrodeposition of Pt on the pre-prepared Au, while PtAu alloy was obtained by co-electrodeposition. As a whole, both types of PtAu catalysts were found to be more active toward formic acid electrooxidation compared to pure Pt, exhibiting maximum activity on Pt-on-Au catalyst with Pt to Au atomic ratio of 1:10.22. Moreover, the Pt/Au atomic ratio directly relates to the oxidation pathway of formic acid and carbon monoxide oxidation. The results may be ascribed to much less CO_{ads} on the surface than single Pt catalyst due to the effect of Au nanoparticles. CO stripping voltammograms present the obvious variation between Pt-on-Au and PtAu alloy catalysts. Meanwhile, the electrocatalytic activities of bimetallic PtAu are evaluated by electrochemical impedance spectroscopy and Tafel analysis.

Keywords Pt-on-Au catalyst · PtAu alloy · Electrocatalytic activity · Formic acid · CO stripping voltammograms

Introduction

In order to solve the problems of air pollution, increasing energy demands, as well as limited fuel reserves caused by traditional fuel consuming, energy storage devices including fuel cells become more and more concerned [1–7]. In particular, direct formic acid fuel cell (DFAFC) has attracted great attention with its advantages of nontoxicity and low fuel crossover as compared to methanol, which makes DFAFC a promising candidate for power source in portable electronic devices [8, 9]. As we have known, a catalyst is a very important part of fuel cells, and platinum is the most common anode catalyst for formic acid (FA) oxidation. However, Pt is prone to poisoning by CO-like intermediates, leading to a significant decrease of catalytic performance [10–12]. It is due to the fact that the indirect pathway by dehydration is predominant in the oxidation reaction of FA on pure Pt, but not by the dehydrogenation pathway [13].

In order to improve the catalytic performance of FA oxidation by dehydrogenation pathway, the development of bimetallic Pt–M (M = Au, Pd, Ru, Bi, etc.) catalysts has been recognized as one of the most effective strategies. The facts show that a Pt-based catalyst has improved the catalytic performance of FA oxidation compared with pure Pt [14–19]. Among various Pt-based catalysts, bimetallic Pt–Au is one of the best catalysts for formic acid oxidation, which is attributed to its enhanced activity [11, 20, 21] and stability against dissolution by raising the Pt oxidation potential [20, 22–26]. Moreover, the incorporation of Au into Pt could lead to the segregation of Pt sites and further reduce the number of adsorption sites for CO, thereby yielding an improvement in the activity of FA oxidation [13, 20, 27].

Recently, various bimetallic Pt–Au catalysts with different structures, such as PtAu alloy [21, 28], Pt-modified Au (Pt-on-Au) [29, 30], and Au@Pt [31, 32], show excellent catalytic activity toward FA oxidation. Park et al. [33] found that the uniform Pt-on-Au nanoparticles showed higher electrocatalytic activities than the pure Pt electrocatalyst in the area- and mass-specific current densities. It is attributed to the enhancement effect of Au atoms and the high Pt utilization in the FA electrooxidation reaction. Ding et al. [29]

Q. Zhang · R. Yue · F. Jiang (✉) · H. Wang · C. Zhai · P. Yang ·
Y. Du (✉)
College of Chemistry, Chemical Engineering and Materials
Science, Soochow University, Suzhou 215123,
People's Republic of China
e-mail: jiangfx82@163.com
e-mail: duyk@suda.edu.cn

Table 1 Preparation variables and composition of Pt-on-Au catalysts

Catalysts	Precursor concentration/mmol L^{-1}		Q_{dep}/C		Atomic ratio[a]
	HAuCl$_4$	H$_2$PtCl$_6$	Au	Pt	Pt/Au
M-1:20.01	3.0	0.77	5×10^{-2}	5×10^{-5}	1:20.01
M-1:10.22	3.0	0.77	5×10^{-2}	1×10^{-4}	1:10.22
M-1:7.16	3.0	0.77	5×10^{-2}	2×10^{-4}	1:7.16
M-1:5.28	3.0	0.77	5×10^{-2}	5×10^{-4}	1:5.28
M-1:2.19	3.0	0.77	5×10^{-2}	1×10^{-3}	1:2.19
Pure Pt	–	0.77	–	1×10^{-2}	–

[a] The atomic ratio on the surface of catalyst obtained by EDX

fabricated the monolayer Pt-modified nanoporous Au catalyst with ultralow Pt loading, great tolerance to poisoning, and high stability for FA electooxidation. They suggested that the outmost Au layer may not only inhibit Pt oxidation but also hold a themodynamic stabilization effect. Xu and co-workers [21] studied the electrocatalytic activity of PtAu alloy nanoparticles with 1:1 atomic ratio and demonstrated that the PtAu/C catalyst exhibited a higher activity for FA oxidation reaction than commercial Pt/C. Liu et al. [34] developed a 3D porous AuPt alloy foam film catalyst with superior electrocatalytic activity toward FA oxidation. Huang et al. [35] prepared a PtAu alloy catalyst by electrodeposition method for FA electrooxidation with the dehydration pathway. Additionally, the surface composition of atomic Pt/Au ratio and the preparation method are significantly related to the electrocatalytic activity of bimetallic PtAu catalyst for FA oxidation reaction [13, 34–36]. In general, the high electrocatalytic activity of bimetallic PtAu catalyst has been ascribed to the special electronic effect [29, 36, 37], ensemble effect [11, 20, 30], and synergistic effect [34, 38, 39]. Many research findings have revealed that both Pt-on-Au and PtAu alloy catalysts exhibited high electrocatalytic activity toward FA oxidation with much less CO$_{ads}$ than that

on Pt catalyst surface. However, to the best of our knowledge, there are no reports on the comparison of electrocatalytic activity of Pt-on-Au and PtAu alloy obtained by electrodeposition toward FA and CO oxidation.

In this work, we synthesized two types of bimetallic PtAu catalysts by electrodeposition of Pt on pre-prepared Au nanoparticle (designated as Pt-on-Au) and co-deposition of Pt and Au (designated as PtAu alloy). The electrocatalytic activities of bimetallic PtAu catalysts with a different Pt/Au ratio toward FA and CO oxidation were investigated systematically by cyclic voltammetry, Tafel plots, and electrochemical impedance spectroscopy (EIS).

Materials and methods

Materials

H$_2$PtCl$_6$ and HAuCl$_4$ were purchased from Sinopharm Chemicals Reagent Co., Ltd., China. All chemicals (HCOOH and H$_2$SO$_4$) were of analytical grade. All aqueous solutions were prepared with double-distilled water.

Table 2 Preparation variables and composition of PtAu alloy catalysts

Catalysts	Precursor concentration[a]/mmol L^{-1}		Q_{dep}/C	Atomic ratio[b]
	HAuCl$_4$	H$_2$PtCl$_6$	PtAu[c]	Pt/Au
A-5.08:1	2.5	0.5	5×10^{-3}	5.08:1
A-3.44:1	2.25	0.75	5×10^{-3}	3.44:1
A-0.94:1	1.5	1.5	5×10^{-3}	0.94:1
\A-1:2.04	1.0	2.0	5×10^{-3}	1:2.04
A-1:2.70	1.0	2.0	5×10^{-3}	1:2.07
A-1:3.85	0.75	2.25	5×10^{-3}	1:3.85

[a] A mixed solution containing HAuCl$_4$ and H$_2$PtCl$_6$

[b] The atomic ratio on the surface of catalyst obtained by EDX

[c] The total deposited charge for PtAu alloy

Fig. 1 SEM images of M-1:10.22 (**a**) and A-1:2.04 (**b**)

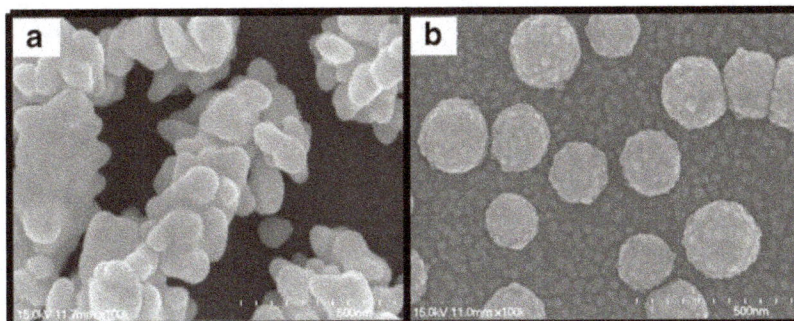

Apparatus

The electrochemical experiments were carried out in a conventional three-electrode cell using a CHI660B electrochemical workstation (Shanghai Chenhua Instrumental Co., Ltd., China). A glassy carbon electrode (GCE, 3 mm in diameter) was used as the working electrode. Before use, GCE surface was polished with 0.3 μm alumina slurry and then rinsed with doubly distilled water in ultrasonic bath. The counter electrode and the reference electrode were platinum wire and saturated calomel electrode, respectively, which were carefully cleared before the experiment. Electrolyte solutions were deaerated by a dry nitrogen stream and maintained with a slight overpressure of nitrogen during the electrochemical measurements. All of the electrochemical measurements were carried out at room temperature. Scanning electron microscope (SEM, S-4700, Japan) equipped with an energy-dispersive X-ray analyzer (EDX, S-4700, Japan) was used to determine the morphology and composition of catalysts. X-ray diffraction (XRD) measurements and X-ray photoelectron spectroscopy (XPS) were performed on an X'Pert-Pro MPD X-ray diffractometer using CuKa radiation (50 kV) and on an ESCALab220i-XL electron spectrometer from VG Scientific using 300-W AlKa radiation, respectively.

Preparation of the catalysts

The procedures of Pt-on-Au nanoparticles were prepared by a two-step electrodeposition. First, the Au-modified GCE (Au/GCE) were obtained by the electrodeposition of Au on GCE at −0.2 V in a solution consisting of 3.0 mM $HAuCl_4$ and 0.5 M H_2SO_4 and then was rinsed with double-distilled water for the following experiment use. The pre-prepared Au/GCE as working electrode was immersed into 0.5 M H_2SO_4 solution containing 0.77 mM H_2PtCl_6 for the electrodeposition of Pt. Pt nanoparticles were deposited under an applied potential of −0.2 V by a potentiostatic method. The resultant Pt-modified Au/GCE (Pt-on-Au) was rinsed thoroughly with double-distilled water several times. The Pt–Au catalysts with different atomic ratios on the surface of catalysts by controlling the deposited charge of Pt are designated as M-1:20.01, M-1:10.22, M-1:7.16, M-1:5.28, and M-1:2.19 in Table 1. For the preparation of PtAu alloy catalysts, it was completed by co-deposition at a constant applied potential of −0.2 V in a mixed H_2PtCl_6/$HAuCl_4$ solution containing 0.5 M H_2SO_4. The different atomic ratios of PtAu alloy were prepared by changing the concentration of H_2PtCl_6 and $HAuCl_4$, and the corresponding samples were designed as A-5.08:1, A-3.44:1, A-0.94:1, A-1:2.04, A-

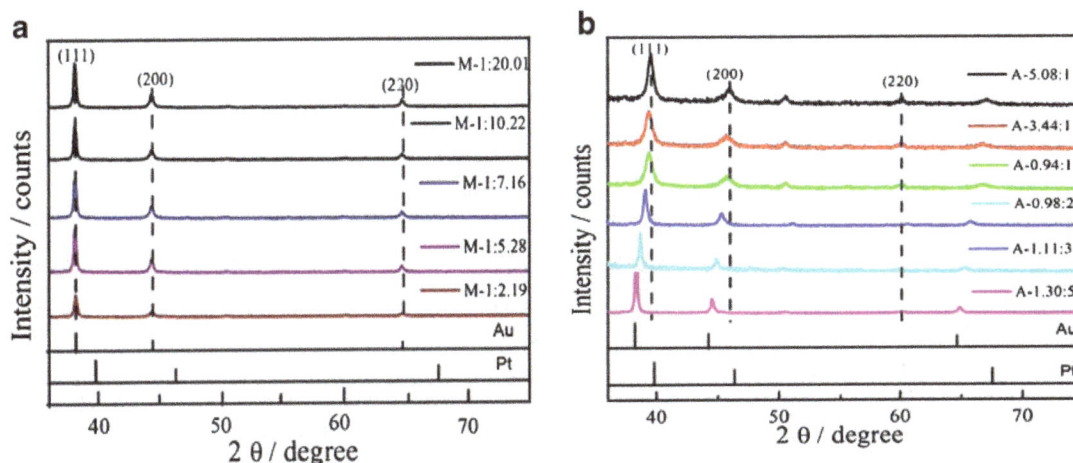

Fig. 2 XRD patterns of Pt-on-Au and PtAu alloy nanoparticles: **a** Pt-on-Au nanoparticles, **b** PtAu alloy nanoparticles

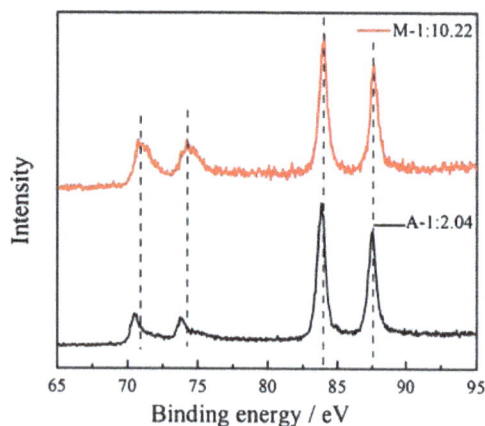

Fig. 3 XPS spectra of Pt_{4f} and Au_{4f} for the M-1:10.22 and A-1:2.04

1:3.27, and A-1:3.85 in Table 2, respectively. For a comparison of voltammetric features and electrocatalytic activities, the pure Pt catalyst on GCE (Pt/GCE) was also prepared under similar conditions as stated earlier.

Pt or Au loading was evaluated by the charge integrated during the deposition process (Q_{dep}) with an assumption of 100 % current efficiency according to Eq. (1):

$$W = \frac{\eta Q_{dep} M}{FZ} \tag{1}$$

Here, W is the mass of deposited Pt or Au, η is current efficiency (assuming 100 % current efficiency here), Q_{dep} is the total charge passed through the electrodes during the deposition process, M is the molecular weight, F is the Faraday constant (96,485 C mol^{-1}), and Z is the number of electrons transferred (taken as four for the Pt and three for the Au formation).

Results and discussion

Morphology and structure characterization

The morphology and structural features of the two types of PtAu catalysts were investigated by SEM and XRD. Figure 1 shows a SEM micrograph for the M-1:10.22(A) and A-1:2.04(B) on GC electrode. As can be seen from Fig. 1a, the M-1:10.22 catalyst formed with Pt particles growing layer by layer on the surface of pre-deposited Au substrate exhibits rough surface morphology and coral-like structure. However, the A-1:2.04 shown in Fig. 1b displays a quasi-spherical particle structure with small pricks uniformly dispersing on the surface, which is the nucleation site for the alloy deposition. Figure 2 shows the XRD patterns of the two types of as-prepared PtAu catalysts. As is known, the peaks of pure gold nanoparticles ($2\theta=38.2°$, 44.4°, and 64.6°) and pure platinum nanoparticles ($2\theta=39.8°$, 46.2°, and 67.5°) are assigned to the (111), (200), and (220) planes, respectively, indicating the typical face-centered-cubic crystal structure. As shown in Fig. 2a, besides the diffraction peaks of Au, no obvious reflection peaks of Pt can be observed due to the low loading of Pt on the Au surface, also indicating that the Pt nanoparticles are uniformly deposited on Au surface without agglomeration. In Fig. 2b, it is worth noting that the (111) peaks of PtAu alloy occur between (111) peaks of pure Au and Pt nanoparticles, which move to a lower 2θ value tending towards the corresponding peak of pure Au as the Pt/Au ratio decreases. It suggests that a single-phase alloy of PtAu has been formed by electrochemical co-deposition. In addition, the EDX analysis in Tables 1 and 2 has also confirmed the presence of Pt and Au in the as-prepared Pt-on-Au and PtAu alloy catalysts.

To investigate the atomic composition and properties of M-1:10.22 and A-1:2.04, XPS was then used to characterize

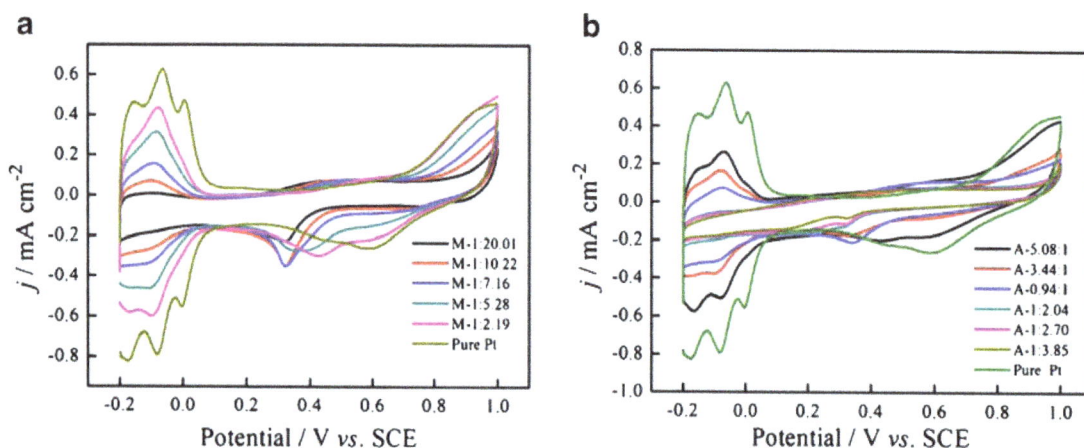

Fig. 4 **a** CVs of Pt/GC and Pt-on-Au electrodes and **b** Pt/GC and PtAu alloy electrodes recorded in 0.5 M H_2SO_4 aqueous solution. Scan rate, 50 mV s^{-1}

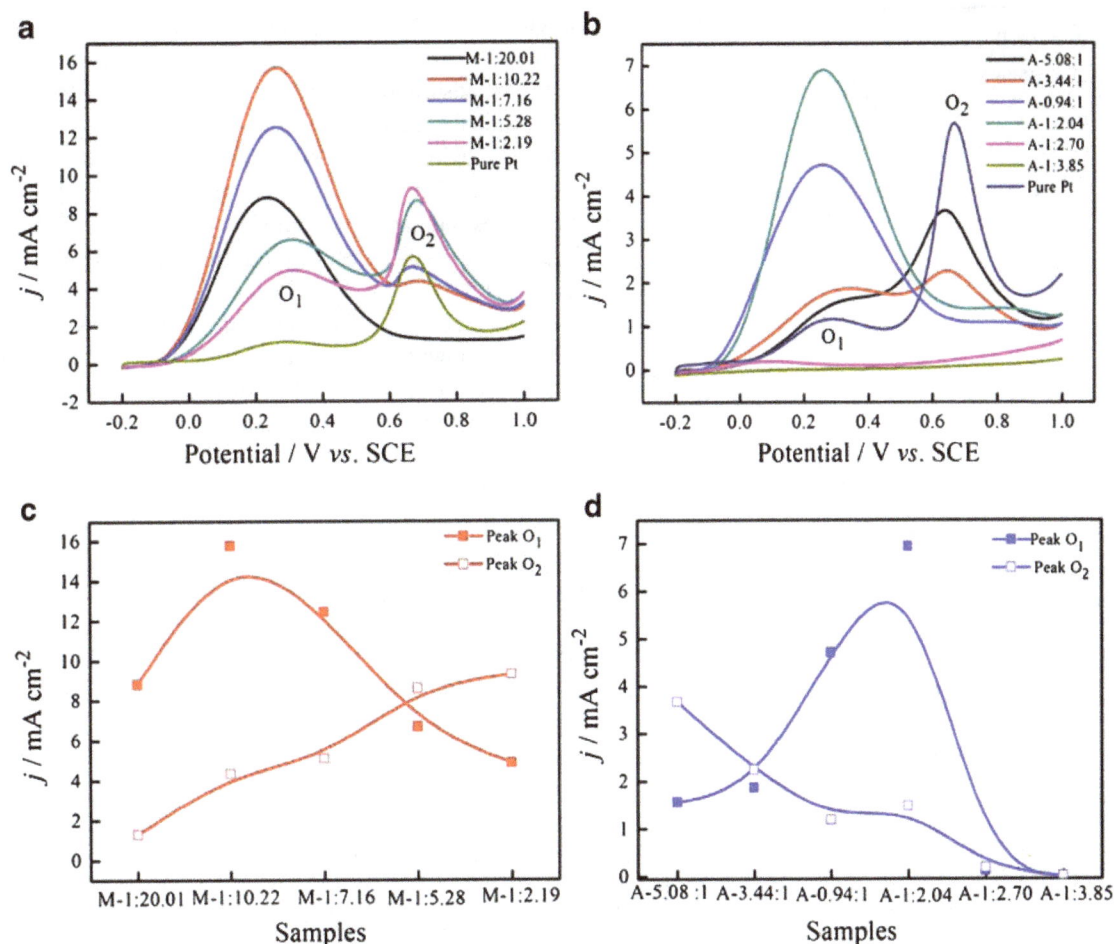

Fig. 5 **a** CVs of Pt/GC and Pt-on-Au electrodes. **b** PtAu alloy and Pt/GC electrodes in 0.5 M HCOOH + 0.5 M H_2SO_4 solution only the positive-going potential scans at 50 mV s^{-1} and current densities of

peak O$_1$ and the peak O$_2$ in HCOOH oxidation on Pt-on-Au electrodes (**c**) and on PtAu alloy electrodes (**d**) as a function of sample compositions

these samples. As shown in Fig. 3, for M-1:10.22, the peaks at 74.1 and 70.9 eV correspond to the 4f$_{5/2}$ and 4f$_{7/2}$ of

metallic Pt, which shift slightly to lower binding energy as compared to bulk Pt (74.5 and 71.2 eV) [40], indicating that

Table 3 Parameters of FA oxidation and CO stripping voltammograms on the Pt-on-Au and PtAu alloy catalysts, respectively

Catalysts	FA oxidation			CO stripping		
	j_{O1} mA cm^{-2}	j_{O2} mA cm^{-2}	j_{O1}/j_{O2}	E_I V	E_{II} V	Q_{CO} C
M-1:20.01	8.82	1.30	6.78	0.671	0.84	0.013
M-1:10.22	15.76	4.33	3.64	0.682	–	0.062
M-1:7.16	12.44	5.10	2.44	0.688	–	0.052
M-1:5.28	6.67	8.59	0.78	0.693	–	0.043
M-1:2.19	4.86	9.29	0.52	0.689	1.10	0.067
A-5.08 :1	1.57	3.67	0.43	0.71	–	0.097
A-3.44 :1	1.87	2.25	0.83	0.73	1.10	0.034
A-0.94 :1	4.71	1.19	3.97	0.76	1.07	0.010
A-1:2.04	6.94	1.48	4.69	0.84	1.04	0.008
A-1:2.70	0.13	0.18	0.72	–	1.01	–
A-1:3.85	–	–	–	–	1.01	–
Pure Pt	1.13	5.56	0.20	0.660	0.97	0.136

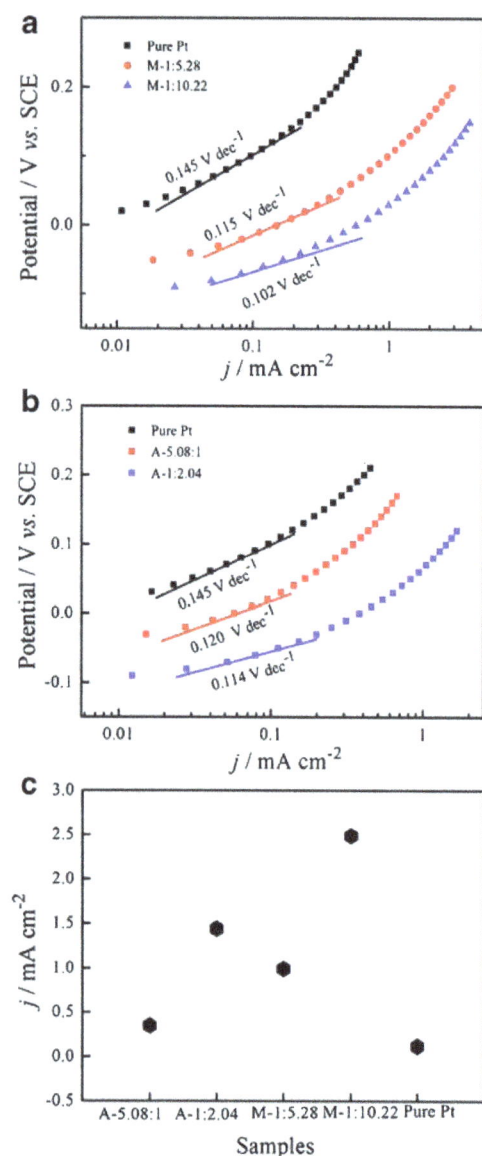

Fig. 6 Quasi-steady-state polarization curves for the oxidation of HCOOH in 0.5 M HCOOH + 0.5 M H_2SO_4 electrolyte recorded on **a** Pt-on-Au and Pt electrodes, **b** PtAu alloy, and Pt electrodes (**c**) area-specific current densities at 0.1 V. Scan rate, 1.0 mV s^{-1}

the electronic structure of Pt is modified by Au [41]. While for Au in M-1:10.22, its $4f_{7/2}$ and $4f_{5/2}$ peaks can be observed at 84.00 and 87.70 eV, the same as that of bulk Au due to the scattering dispersion of Pt nanoparticles on the Au substrate [40]. However, as exhibited in Fig. 3, both the peaks of Au_{4f} and of Pt_{4f} in A-1:2.04 shift to lower binding energies as compared to those of bulk Au and Pt, which indicates that the PtAu in A-1:2.04 catalyst is an alloy [42].

Voltammetric analysis

Figure 4 shows the cyclic voltammograms (CVs) of pure Pt, Pt-on-Au (Fig. 4a), and PtAu alloy (Fig. 4b) on GCE in

deaerated 0.5 M H_2SO_4 solution under a scan rate of 50 mV s^{-1}. For Pt/GCE, the curve presents a well-defined hydrogen adsorption/desorption ($H_{ads/des}$), the broad double layer, and the cathodic reduction peak of Pt oxide in the potential region of −0.2~0.1, 0.1~0.28, and 0.28~0.9 V, respectively. The CVs of Pt-on-Au and PtAu alloy show the shapes similar to the pure Pt. Moreover, the peaks of $H_{ads/des}$ are smaller than that on pure Pt and gradually increase with the increase of Pt loadings. It can be observed that the reduction peaks of Pt oxide on both PtAu catalysts shift to a lower potential as the decrease of Pt loading ascribed to the size effect of Pt islands [43]. Additionally, it is noted that the reduction peak of Pt oxide on Pt-on-Au consists of two overlapping peaks when the ratio of Pt:Au is larger than 1:5.28, indicating the formation of two different Pt particles/structures/agglomerates [44, 45]. A similar phenomenon can be seen on the PtAu alloy catalysts with the Pt/Au ratio exceeding 0.94:1. When the Pt composition on/in Au becomes larger, the $H_{ads}/_{des}$ peaks and the reduction peaks of Pt oxide markedly increase. Kristian [46] pointed out that the Pt entities on the Au surface have more negative potential for the reduction of Pt oxide at lower Pt/Au ratios since it makes obtaining electrons more difficult for Pt atoms compared to higher Pt/Au ratios and pure Pt.

Electrooxidation of formic acid

Figure 5a presents the positive scan CVs of Pt/GCE and Pt-on-Au catalysts toward FA oxidation in a mixed solution of 0.5 M HCOOH and 0.5 M H_2SO_4. For Pt/GCE, the typical feature of the FA electrooxidation is observed according to the dual-pathway mechanism. As can be seen, a weak peak (O_1) current density at about 0.3 V is related to the direct oxidation of FA via the dehydrogenation mechanism, while the other peak (O_2) at 0.70 V is due to the oxidation of intermediate CO generated from the dehydration of FA [47]. Obviously, the peak current density at O_1 (j_{O1}) on Pt-on-Au is larger, and the peak current density at O_2 (j_{O2}) relatively becomes lower than that on pure Pt toward FA oxidation (in Fig. 5a and Table 3), which is in agreement with the results of previous reports [29, 30]. Especially, the M-1:10.22 catalyst of Pt-on-Au shows the highest j_{O1} (15.76 mA cm^{-2}), which is ~14 times higher than that of Pt/GCE. With further decrease of the Pt/Au ratios to 1:20.01, the first anodic peak (O_1) decreases drastically and the second peak (O_2) almost vanishes. It indicates that FA electrooxidation mainly follows the dehydrogenation pathway on the M-1:10.22 and M-1:20.01 Pt-on-Au catalysts.

The ratio of j_{O1}/j_{O2} (in Table 3) was further used to evaluate the effect of Pt/Au composition on FA oxidation pathway. The j_{O1}/j_{O2} on Pt-on-Au is larger than that on pure Pt (0.20) and gradually increases as the Pt loadings on Au surface decrease. It is due to the dehydration pathway on pure

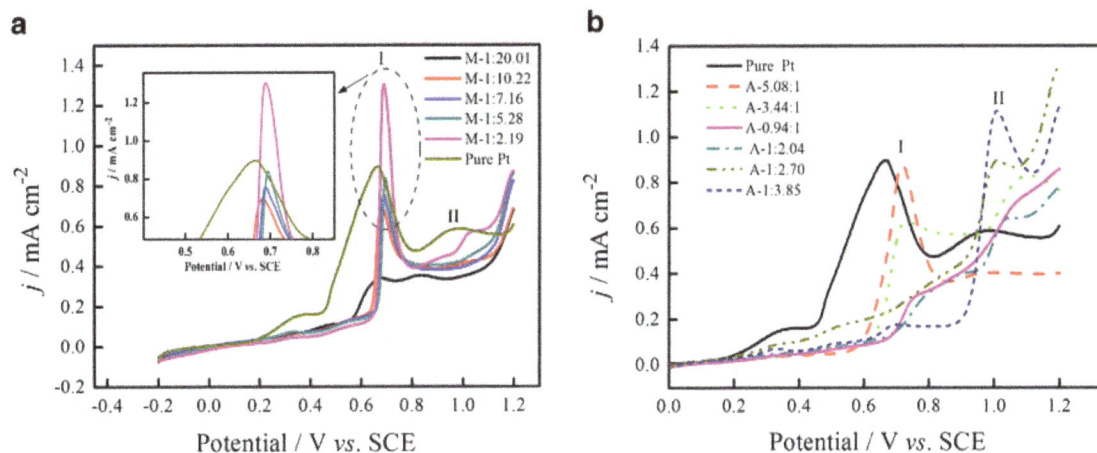

Fig. 7 **a** CO_{ads} stripping voltammograms on M-1:20.01, M-1:10.22, M-1:7.16, M-1:5.28, M-1:2.19, and pure Pt, respectively. **b** CO_{ads} stripping voltammograms on pure Pt, A-5.08:1, A-3.44:1, A-0.94:1, A-1:2.04, A-1:2.70, and A-1:3.85, respectively, recorded in 0.5 M H_2SO_4 at a scan rate of 20 mV s^{-1}

Pt or Pt-rich catalyst, leading to the formation of CO intermediates. When the Pt/Au ratio is below 1:7.16, the pathway of FA oxidation transforms into dehydrogenation. It is in agreement with the observation by Kristian et al. [30] who attributed such to a proposed ensemble effect for Pt-on-Au.

For comparison, Fig. 5b shows the positive scan CVs of PtAu alloy catalysts for the electrooxidation of FA. A similar phenomenon for FA electrooxidation can be observed on PtAu alloy. Interestingly, the highest j_{O1} and j_{O1}/j_{O2} were obtained on the A-1:2.04 PtAu alloy (6.94 mA cm^{-2}) (Fig. 5b and Table 3), which is different from Pt-on-Au. However, on A-1:2.70 and A-1:3.85, the FA oxidation peak at both O_1 and O_2 almost vanishes due to the increase of Au. Although a similar dehydrogenation pathway for FA oxidation can be observed on Pt-on-Au (M-1:10.22) and PtAu alloy (A-1:2.04) catalysts, the different Pt/Au composition suggests that the electronic structure of PtAu has a significant influence for the FA oxidation pathway, which needs more and further work to study.

As explained earlier, the height of the peak O_1 and O_2 and their ratio give an indication along which path the reaction is dominant during the electrooxidation of formic acid on the catalyst. Figure 5c, d shows a systematic dependence of these two peaks on the surface composition of Pt-on-Au and PtAu alloy, respectively. At Pt-on-Au surface, peak O_1 is the highest at the ratio of Pt/Au 1:10.22 and decreases with an increase in the ratio of Pt/Au, while peak O_2 height ascends with further increase of Pt loading. For PtAu alloy catalyst, the trend of peak O_1 is similar to that of Pt-on-Au catalyst; however, the height of peak O_2 starts to descend and would probably reach zero with a decrease in the ratio of Pt/Au.

To further investigate the kinetics of FA oxidation, Tafel measurements were performed under a steady-state condition. As seen from Fig. 6a, b, the slope of Tafel plots for the Pt-on-Au and PtAu alloy shows a much lower value than that of Pt/GCE (145 mV dec^{-1}). Moreover, the M-1:10.22 Pt-on-Au and A-1:2.04 PtAu alloy catalysts show the lowest values of 102 and 114 mV dec^{-1}, respectively. As we have known, the Tafel slope is related to the CO coverage on catalyst, and a lower CO coverage means a lower Tafel slope value [10, 33]. In addition, Fig. 6c shows the area-specific current densities at 0.1 V based on the Tafel plots. The activities of Pt-on-Au and PtAu alloy catalysts are much higher than Pt/GC. Also, the current density for FA oxidation on M-1:10.22 Pt-on-Au is about two times higher than that of A-1:2.04 PtAu alloy. Compared with Pt/GCE, the lower slopes and the higher current density of both PtAu catalysts are attributed to the low amount of adsorbed CO or intermediates on the Pt sites. Meanwhile, this result indicates that the Au atoms of Pt-on-Au nanoparticles are more powerful in improving the electrocatalytic activity toward FA oxidation than those of PtAu alloy particles [33].

CO_{ads} stripping voltammetry

The CO_{ads} stripping voltammetry measurements were performed on Pt-on-Au (Fig. 7a) and PtAu alloy catalysts on GCE (Fig. 7b) in 0.5 M H_2SO_4 solution. As shown in Fig. 7, the pure Pt on GCE presents a typical CO_{ads} oxidation peak at ~0.65 V [43]. For the Pt-on-Au catalysts, the CO stripping peak shifts positively to a higher potential and becomes narrower compared with that on pure Pt. The CO stripping curve of M-1:20.01 Pt-on-Au evolves into a small and broad peak consisting of two poorly separated peaks (a peak at 0.67 V and a shoulder at 0.83 V), indicating that two types of Pt sites exist on the Au substrate [48]. With the increase of Pt loading on Au surface, the CO stripping peaks appear at a lower potential (at ~0.7 V) and becomes larger. Yu and coworkers [49] reported similar results and illustrated that the ratio of the isolated single Pt atom within the Pt adatom population decided the CO adsorption on the Pt-modified

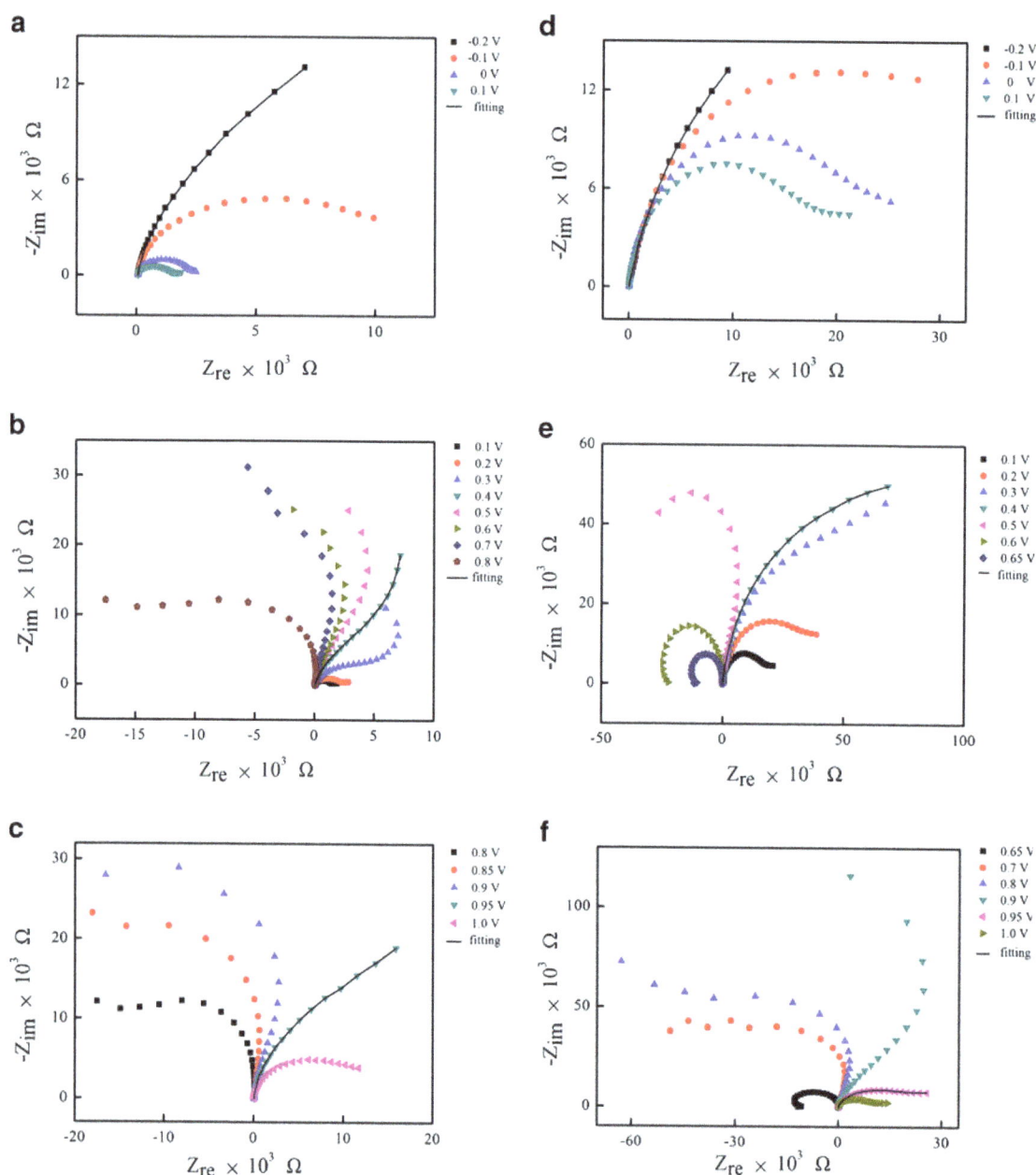

Fig. 8 Nyquist plots of HCOOH electrooxidation on M-1:10.22 (a–c) and on A-1:2.04 (d–f) electrodes in 0.5 M HCOOH + 0.5 M H_2SO_4 solution at electrode potentials from −0.2 V to 1.0 V

Au surface. The Pt atoms on the Au surface are isolated with lower coverage of Pt, and consequently, the CO stripping peak shifts negatively owing to the weak Pt–CO bond energy compared with pure Pt. Conversely, Pt–Pt neighboring atoms gradually increase with increasing Pt coverage; accordingly, the adsorption of CO is much stronger than that on bulk Pt, leading the CO stripping peak to shift positively.

The CO_{ads} stripping voltammograms of PtAu alloy (Fig. 7b) are somewhat different from those on Pt-on-Au catalysts. As can be seen from Fig. 7b, the CO stripping peak appears at 0.71 V (peak I) at the Pt/Au ratio of 5.08:1, which is 60 mV higher than that of pure Pt. With the ratio of Pt/Au up to 3.44:1,

the current density of peak I decreases; however, a new peak (peakII) appears at a more positive potential. Peak I becomes smaller and finally vanishes at a lower Pt/Au ratio, indicating a less or no poisoning on PtAu alloy catalyst by CO_{ads} with the increasing Au composition [46]. It can be clearly observed that peaks I and II lose their intensity rapidly on the PtAu alloy catalysts (Pt/Au=0.94:1 and 1:2.04). However, upon further decreasing the ratio of Pt/Au to 1:2.70 and 1:3.85, peak I was not observed, while the intensity of peak II obviously becomes larger. Kristian and co-workers [46] has pointed out that the up-shift in the d-band center of the Pt atoms results in increasing Pt–CO bond strength.

EIS studies for the composite catalysts

Electrochemical impedance spectroscopy as a sensitive and powerful technique has been used to test the catalytic activity of electrocatalyst for FA oxidation. Figure 8 presents the Nyquist plots of formic acid electrooxidation on M-1:10.22 Pt-on-Au and A-1:2.04 PtAu alloy catalysts at various potentials in 0.5 M HCOOH + 0.5 M H_2SO_4 at 0.3 V. When the applied potential is below 0.1 V, the impedance arcs of M-1:10 Pt-on-Au are located in the first quadrant, and the diameter of the arcs decreases with increasing electrode potential, as shown in Fig. 8a. This indicates that more active sites are available for formic acid oxidation [50], which coincidentally increases the oxidation current density of formic acid with increasing potential, as shown in Fig. 5. However, with the electrode potential increasing further, the arc diameter increases correspondingly, and interestingly, the arc reverses to the second quadrant (Fig. 8b, c). Such interesting impedance behaviors have also been reported in the studies of methanol or formic acid electrooxidation on Pt or Pt-based alloy electrodes [50–52]. At a positive potential more than 0.95 V, the impedance curves return to normal behaviors in the first quadrant and the arc diameter decreases with the potentials. At these positive potentials, surface-adsorbed CO was almost removed completely, which is indicative of the diminishing charge transfer resistance (R_{ct}). From Fig. 8d–f, similar features can be also found on A-1:2.04 PtAu alloy, except that the arc diameter is substantially larger than that on M-1:10.22 Pt-on-Au catalyst, indicating a higher R_{ct} for formic acid oxidation on A-1:2.04 PtAu alloy catalyst. This coincides with the results of the aforementioned CV in Fig. 5 wherein the catalytic activity of M-1:10.22 is higher than that of A-1:2.04 for formic acid electrooxidation.

Conclusions

In the present work, PtAu catalysts with various Pt/Au ratios obtained by electrodeposition of Pt precursor on pre-prepared Au nanoparticles (Pt-on-Au) and by simultaneous co-electrodeposition of Au and Pt (PtAu alloy) have been investigated as an anodic electrocatalyst toward formic acid and carbon monoxide oxidation. The two types of PtAu catalysts show a higher electrocatalytic activity toward formic acid oxidation compared to pure Pt on GCE. The Au atoms in PtAu catalysts promote the activity of formic acid electrooxidation. The highest activities under potentiodynamic and quasi-steady-state condition were obtained on M-1:10.22 Pt-on-Au and on A-1:2.04 PtAu alloy, respectively. The results of CO_{ads} stripping experiment show that the CO stripping peak position shifts positively on the two types of PtAu nanoparticles compared with that of the Pt catalyst. Moreover, EIS data indicate that the performance of M-1:10.22 Pt-on-Au for formic acid electrooxidation is much better than that of A-1:2.04 PtAu alloy.

Acknowledgments　This work was supported by the National Natural Science Foundation of China (Grant Nos. 51073114, 20933007, 21173261, 51073074, and 50963002), the Academic Award for Young Graduate Scholar of Soochow University, the Opening Project of Xinjiang Key Laboratory of Electronic Information Materials and Devices (XJYS0901-2010-01), and the Priority Academic Program Development of Jiangsu Higher Education Institutions (PAPD).

References

1. Raoof JB, Ojani R, Sahar RN (2010) Electrochemical synthesis of bimetallic Au@Pt nanoparticles supported on gold film electrode by means of self-assembled monolayer. Int J Hydrog Energy 641:71–77
2. Habibi B, Delnavaz N (2010) Electrocatalytic oxidation of formic acid and formaldehyde on platinum nanoparticles decorated carbon-ceramic substrate. Int J Hydrog Energy 35:8831–8840
3. Raoof JB, Karimi MA, Hosseini SR, Mangelizade S (2011) Enhanced electrocatalytic activity of nickel particles electrodeposited onto poly (m-toluidine) film prepared in presence of CTAB surfactant on carbon paste electrode for formaldehyde oxidation in alkaline medium. Int J Hydrog Energy 36:13281–13287
4. Zhu M, Lu Y, Du Y, Li J, Wang X, Yang P (2011) Photocatalytic hydrogen evolution without an electron mediator using a porphyrin–pyrene conjugate functionalized Pt nanocomposite as a photocatalyst. Int J Hydrog Energy 36:4298–4304
5. Belousov OV, Belousova NV, Sirotina AV, Solovyov LA, Zhyzhaev AM, Zharkov SW, Mikhlin YL (2011) Formation of bimetallic Au–Pd and Au–Pt nanoparticles under hydrothermal conditions and microwave irradiation. Langmuir 27:11697–11703
6. Yao Z, Zhu M, Jiang F, Du Y, Wang C, Yang P (2012) Highly efficient electrocatalytic performance based on Pt nanoflowers modified reduced graphene oxide/carbon cloth electrode. J Mater Chem 22:13707–13713
7. Maye MM, Kariuki NN, Luo J, Han L, Njoki P, Wang L, Lin Y, Naslund HR, Zhong C (2004) Electrocatalytic reduction of oxygen: gold and gold–platinum nanoparticle catalysts prepared by two-phase protocol. Gold Bull 37:217–223
8. Rice C, Ha S, Masel RI, Waszczuk P, Wieckowski A, Barnard T (2002) Direct formic acid fuel cells. J Power Sources 111:83–89
9. Wang X, Hu J, Hsing IM (2004) Electrochemical investigation of formic acid electro-oxidation and its crossover through a Nafion® membrane. J Electroanal Chem 562:73–80
10. Habibi B, Gahramanzadeh R (2011) Fabrication and characterization of non-platinum electrocatalyst for methanol oxidation in alkaline medium: nickel nanoparticles modified carbon-ceramic electrode. Int J Hydrog Energy 36:1913–1923
11. Park S, Xie Y, Weaver MJ (2002) Electrocatalytic pathways on carbon-supported platinum nanoparticles: comparison of particle-size-dependent rates of methanol, formic acid, and formaldehyde electrooxidation. Langmuir 18:5792–5798
12. Chang SC, Leung LWH, Weaver MJ (1990) Metal crystallinity effects in electrocatalysis as probed by real-time FTIR spectroscopy: electrooxidation of formic acid, methanol, and ethanol on ordered low-index platinum surfaces. J Phys Chem 94:6013–6021

13. Chen G, Li Y, Wang D, Zheng L, You G, Zhong C, Yang L, Cai F, Cai J, Chen B (2011) Carbon-supported PtAu alloy nanoparticle catalysts for enhanced electrocatalytic oxidation of formic acid. J Power Sources 196:8323–8330

14. Rigsby MA, Zhou W, Lewera A, Duong HT, Bagus PS, Jaegermann W, Hunger R (2008) Experiment and theory of fuel cell catalysis: methanol and formic acid decomposition on nanoparticle Pt/Ru. J Phys Chem C 112:15595–15601

15. Mazumder V, Lee Y, Sun S (2010) Recent development of active nanoparticle catalysts for fuel cell reactions. Adv Funct Mater 20:1224–1231

16. Zhang H, Jin M, Xia Y (2012) Enhancing the catalytic and electrocatalytic properties of Pt-based catalysts by forming bimetallic nanocrystals with Pd. Chem Soc Rev 41:8035–8049

17. Kang S, Lee J, Lee JK, Chung S, Tak Y (2006) Influence of bimodification of Pt anode catalyst in direct formic acid fuel cells. J Phys Chem B 110:7270–7274

18. Lee H, Habas SE, Somorjai GA, Yang P (2008) Localized Pd overgrowth on cubic Pt nanocrystals for enhanced electrocatalytic oxidation of formic acid. J Am Chem Soc 130:5406–5407

19. Schmidt TJ, Behm RJ (2000) Formic acid oxidation on pure and bimodified Pt(111): temperature effects. Langmuir 16:8159–8166

20. Choi JH, Jeong KJ, Dong Y, Han J, Lim TH, Lee JS, Sung YE (2006) Electro-oxidation of methanol and formic acid on PtRu and PtAu for direct liquid fuel cells. J Power Sources 163:71–75

21. Xu J, Zhao T, Liang Z (2008) Carbon supported platinum–gold alloy catalyst for direct formic acid fuel cells. J Power Sources 185:857–861

22. Jia J, Cao L, Wang Z (2008) Platinum-coated gold nanoporous film surface: electrodeposition and enhanced electrocatalytic activity for methanol oxidation. Langmuir 24:5932–5936

23. Zhang Y, Huang Q, Zou Z, Yang J, Vogel W, Yang H (2010) Enhanced durability of Au cluster decorated Pt nanoparticles for the oxygen reduction reaction. J Phys Chem C 114:6860–6868

24. Luo J, Njoki PN, Lin Y, Mott D, Wang L, Zhong C (2006) Characterization of carbon-supported AuPt nanoparticles for electrocatalytic methanol oxidation reaction. Langmuir 22:2892–2898

25. Kim S, Jung C, Kim J, Rhee CK, Choi SM, Lim TH (2010) Modification of Au nanoparticles dispersed on carbon support using spontaneous deposition of Pt toward formic acid oxidation. Langmuir 26:4497–4505

26. Zhang J, Sasaki K, Sutter E, Adzic RR (2007) Stabilization of platinum oxygen-reduction electrocatalysts using gold clusters. Science 315:220–222

27. Habrioux A, Vogel W, Guinel M, Guetaz L, Servat K, Kokoh B, Alonso-Vante N (2009) Structural and electrochemical studies of Au–Pt nanoalloys. Phys Chem Chem Phys 11:3573–3579

28. Malaknaz ME, Mehran M, Bineta K, Louis N, Patricia K, Hynd R (2010) Bimetallic Au–Pt nanoparticles synthesized by radiolysis: application in electro-catalysis. Gold Bull 43:49–56

29. Wang R, Wang C, Cai W, Ding Y (2010) Ultralow-platinum-loading high-performance nanoporous electrocatalysts with nanoengineered surface structures. Adv Mater 22:1845–1848

30. Kristian N, Yan Y, Wang X (2008) Highly efficient submonolayer Pt-decorated Au nano-catalysts for formic acid oxidation. Chem Commun 0:353–355

31. Ren B, Lian X, Li J, Fang P, Lai Q, Tian Z (2008) Spectroelectrochemical flow cell with temperature control for investigation of electrocatalytic systems with surface-enhanced Raman spectroscopy. Faraday Discuss 140:155–165

32. Liu C, Wei Y, Liu C, Wang K (2012) Pt–Au core/shell nanorods: preparation and applications as electrocatalysts for fuel cells. J Mater Chem 22:4641–4644

33. Park IS, Lee KS, Choi JH, Park HY, Sung YE (2007) Surface structure of Pt-modified Au nanoparticles and electrocatalytic activity in formic acid electro-oxidation. J Phys Chem C 111:126–133

34. Liu J, Cao L, Huang W, Li Z (2011) Preparation of AuPt alloy foam films and their superior electrocatalytic activity for the oxidation of formic acid. ACS Appl Mater Interfaces 3:3552–3558

35. Huang J, Hou H, You T (2009) Highly efficient electrocatalytic oxidation of formic acid by electrospun carbon nanofiber-supported Pt_xAu_{100} $_{-x}$bimetallic electrocatalyst. Electrochem Commun 11:1281–1284

36. Zhang G, Zhao D, Feng Y, Zhang B, Su D, Liu G, Xu B (2012) Catalytic Pt-on-Au nanostructures: why Pt becomes more active on smaller Au particles. ACS Nano 6:2226–2236

37. Luo M, Wang C, Hu G, Lin W, Ho C, Lin Y, Hsu Y (2009) Active alloying of Au with Pt in nanoclusters supported on a thin film of Al_2O_3/NiAl(100). J Phys Chem C 113:21054–21062

38. Hu Y, Zhang H, Wu P, Zhang H, Zhou B, Cai C (2011) Bimetallic Pt–Au nanocatalysts electrochemically deposited on graphene and their electrocatalytic characteristics towards oxygen reduction and methanol oxidation. Phys Chem Chem Phys 13:4083–4094

39. Mott D, Luo J, Njoki PN, Lin Y, Wang L, Zhong C (2007) Synergistic activity of gold–platinum alloy nanoparticle catalysts. Catal Today 122:378–385

40. Moulder JF, Stickle WF, Sobol PE, Bomben KD (1995) Handbook of x-ray photoelectron spectroscopy. Physical Electronics, Inc., Eden Prairie, MN

41. Xu YY, Dong YN, Shi J, Xu ML, Zhang ZF, Yang XK (2011) Au@Pt core-shell nanoparticles supported on multiwalled carbon nanotubes for methanol oxidation. Catal Commun 13:54–58

42. Yi CW, Luo K, Wei T, Goodman DW (2005) The composition and structure of Pd-Au surfaces. J Phys Chem B 109:18535–18540

43. Arenz M, Mayrhofer KJJ, Stamenkovic V, Blizanac BB, Tomoyuki T, Ross PN, Markovic NM (2005) The effect of the particle size on the kinetics of CO electrooxidation on high surface area Pt catalysts. J Am Chem Soc 127:6819–6829

44. Xia Y, Liu J, Huang W, Li Z (2012) Electrochemical fabrication of clean dendritic Au supported Pt clusters for electrocatalytic oxidation of formic acid. Electrochim Acta 70:304–312

45. Scheijen FJE, Beltramo GL, Hoeppener S, Housmans THM, Koper MTM (2008) The electrooxidation of small organic molecules on platinum nanoparticles supported on gold: influence of platinum deposition procedure. J Solid State Electrochem 12:483–495

46. Kristian N, Yu Y, Gunawan P, Xu R, Deng W, Liu X, Wang X (2009) Controlled synthesis of Pt-decorated Au nanostructure and its promoted activity toward formic acid electro-oxidation. Electrochim Acta 54:4916–4924

47. Obradović MD, Rogan JR, Babić BM, Tripković AV, Gautam ARS, Radmilović VR, Gojković SL (2012) Formic acid oxidation on Pt–Au nanoparticles: relation between the catalyst activity and the poisoning rate. J Power Sources 197:72–79

48. Du B, Tong Y (2005) A coverage-dependent study of Pt spontaneously deposited onto Au and Ru surfaces: direct experimental evidence of the ensemble effect for methanol electro-oxidation on Pt. J Phys Chem B 109:17775–17780

49. Yu Y, Lim KH, Wang JY, Wang X (2012) CO adsorption behavior on decorated Pt@Au nanoelectrocatalysts: a combined experimental and DFT theoretical calculation study. J Phys Chem C 116:3851–3856

50. Yue R, Jiang F, Du Y, Xu J, Yang P (2012) Electrosynthesis of a novel polyindole derivative from 5-aminoindole and its use as catalyst support for formic acid electrooxidation. Electrochim Acta 77:29–38

51. Chen W, Kim J, Sun S, Chen S (2006) Electro-oxidation of formic acid catalyzed by FePt nanoparticles. Phys Chem Chem Phys 8:2779–2786

52. Chen W, Kim J, Sun S, Chen S (2007) Composition effects of FePt alloy nanoparticles on the electro-oxidation of formic acid. Langmuir 23:11303–11310

15

Theoretical insights on the effect of reactive gas on the chemical ordering of gold-based alloys

Hazar Guesmi

Abstract Alloy catalysts typically operate under high-pressure and high-temperature conditions, and these reactive environments may substantially influence the alloy surface composition. Theoretical studies of catalytic properties are often investigated on model systems where no account is taken for the possibility that the surface composition can be modified after the gas exposure. This is a serious drawback that may prevent reliable description of the catalyst reactivity that mainly depends on the configuration of the surface. Nowadays, modelling the equilibrium structure of metal surfaces and alloys in a reactive environment is still a barely studied subject and remains an extremely challenging task. Recent methodological advances and their applications, mainly on gold-based alloy systems, are presented and discussed in this brief overview.

Keywords Gold · Alloy · Under gas · Structure · Reactivity

Introduction

The desire to synthesise efficient catalysts with well-defined, controllable properties and structures at the nanometre scale has generated great interest in bimetallic nanoclusters [1, 2]. One of the major reasons for such interest is the fact that their chemical and physical properties may be tuned by varying the composition and the size of the clusters, which leads to different atomic ordering.

Gold-based nanoalloys (or gold alloy nanoparticles) are attracting growing attention due to their specific properties as optical [3], catalytic [2] and electro-catalytic [4] materials. In heterogeneous catalysis, bimetallic gold nanoparticles have shown high reactivity in a number of catalytic reactions including the direct synthesis of hydrogen peroxide from H_2 and O_2 [5, 6], synthesis of vinyl acetate [7], selective hydrogenation of butadiene [8] and so forth. In the context of CO reforming, recent experiments on gold nanoparticles with different Ni contents show an improvement of the CO oxidation rate [9]. Palladium [10] and platinum [11] also emerge as good candidates for the enhancement of such reaction.

Surface structures, compositions and segregation properties of nanoalloys are of prior interest as they are important in determining chemical reactivity and especially catalytic activity. Moreover, the adsorbate-induced segregation of metal alloys under the reaction conditions and thus the changes in local atomic composition and surface structure have been predicted and demonstrated to occur for a number of gold alloy systems [12, 13]. In particular, for Au–Pd nanoalloys (that will be more detailed bellow) although the gold surface enrichment is predicted to be thermodynamically favourable under vacuum conditions [14], a reversed segregation of Pd as a more active component to the surface is reported to occur in the presence of adsorbates [15–17]. Concerning the adsorbate-induced surface reconstruction, Yoshida et al. [18] have recently reported the results of visualizing gas molecules interacting with supported nanoparticle catalysts at reaction conditions. Using the newly developed aberration-corrected environmental transmission electron microscopy, these authors succeeded to show how adsorbed CO causes the (100) facets of gold nanoparticles to reconstruct into Au(100)-hex. This phenomenon which could also happen on specific gold-

H. Guesmi
CNRS—Laboratoire de Réactivité de Surface, Université Pierre et Marie Curie (UMR 7197), 3 rue Galilée, 94200 Ivry, France

H. Guesmi (✉)
CNRS—Institut Charles Gerhardt-équipe MACS, Ecole Nationale de Chimie de Montpellier (UMR 5253), 8 rue de l'Ecole Normale, 34296 Montpellier, France
e-mail: hazar.guesmi@enscm.fr

based alloy systems was not yet visualised. Therefore, while a given local surface structure and ensemble may exhibit a desired property under idealized ultrahigh vacuum conditions, it is important to understand whether the particular configuration is stable under the operating environment for a specific application.

Many theoretical methods are recognized as successful tools to study the structure and the chemical ordering of bulk and surface alloys [19, 20]. For nanoalloys [see 21 and references herein], structural searches via empirical potentials [22, 23], tight-binding [24, 25] and methods based on density functional theory (DFT) calculations [26, 27] have been performed. DFT methods, even limited to small bimetallic clusters [28], can be of sufficiently high accuracy, affording the possibility of treating a wide variety of structures and chemical compositions. For instance, by using a quick screen of a large number of bimetallic configurations, DFT methods are able to computationally identify the most stable structure and sometimes to design new catalysts with improved performance [29]. Nevertheless, in the vast majority of theoretical studies, no account is taken for the possibility that the surface composition can be modified during the gas exposure, i.e. in the presence of the adsorbate. In this context, much effort is needed to identify, on one hand, realistic systems in which the structure and chemical order correspond to equilibrium phases (potential energy surface under vacuum) and, on the other hand, the evolution of such phases in the presence of the adsorbates. Up to now, little is known on the latter topic in which many important questions are still open. Therefore, it is important to bring the attention of interested researches and specialists in the field. This paper aims to provide a brief overview of the state of the art of the few theoretical studies devoted to the change occurring on the outermost surface layers of gold-based alloy catalysts during reaction conditions (in situ).

Theoretical approaches

Two well-established DFT-based methodologies have been employed to model the equilibrium structures of alloys in a reactive environment: the DFT atomistic thermodynamics and the cluster expansion approaches. In addition, DFT vibrational frequency calculations of a probe molecule can help to identify the local atomic surface structure of alloys under reactive gas. This latter characterization method can be useful when combined with experimental results. In general, the interplay between in situ techniques and theory is the better way to bridge the pressure gap [30] and to allow extending the comprehension of the surface state of metallic nanoparticles under reaction conditions. This is of particular importance in the case of multi-metallic

catalysts for which reactant-driven changes of the surface composition may occur.

DFT atomistic thermodynamics

In the early 1970s, DFT methods were firstly employed in simulations of surface phonon. Thanks to the fast increase in computer technology, these simple beginnings developed rapidly into more sophisticate modelling that are able to calculate the geometry of molecular and extended systems to experimental accuracy and bond energies to within a few percent errors [31]. To model extended systems representing alloy surfaces in reactive gas, periodic boundary conditions through a supercell approach are commonly used. By considering the alloy surface in contact with a bulk phase and with the gas-phase atmosphere, the free energies of several structures and alloy compositions are calculated to determine the most stable one under specific conditions of temperature and partial pressure of gas-phase species.

Nevertheless, the weakness of this approach is that only ordered alloy structures can be simulated. Furthermore, as small unit cells periodically repeated in space cannot accommodate large deformations that might be induced by the addition of impurities, size mismatch and stresses cannot be considered. Detailed description of the basic features of this approach can be found in [32] for the case of bimetallic alloy in thermodynamic equilibrium with single gas molecule and in [33, 34] for the case of multiple gas species.

Cluster expansion method (or lattice–gas–Hamiltonian)

Cluster expansion (CE) method is a formalism that uses a set of DFT calculations of large number of ordered and disordered fixed structures to extract many-body interaction terms that describe the Hamiltonian of the alloy system [32]. A CE of the Hamiltonian is then termed by a polynomial equation involving the neighbouring energy interactions that, in principle, includes an infinite number of terms and summations. The success of this approach is built upon the fact that, in metal alloys, the strongest interactions are usually short ranged and, therefore, the CE can be truncated. Once the CE is constructed, Monte Carlo simulations at finite temperatures can be performed to study ordered and disordered alloy systems.

In principle, one only needs a finite number of cluster interactions to correctly reproduce the configuration energies. However, this method is limited to the treatments of structures with pre-determined lattice topology.

DFT alloy surface probe

The other way to identify the local atomic surface structure of alloys under reactive gas is to use a probe molecule as CO for

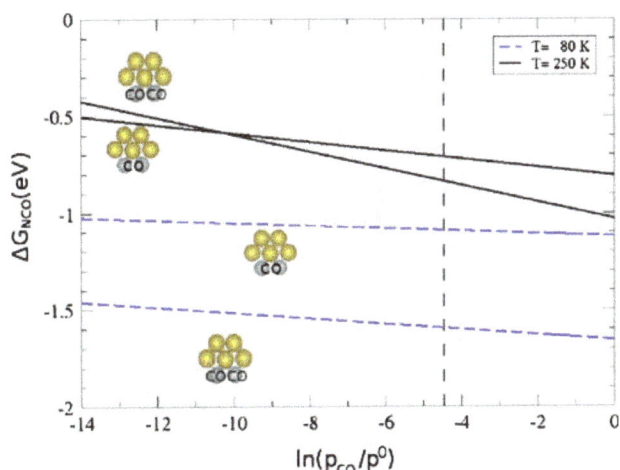

Fig. 1 The excess free energy of single and multiple CO adsorptions on Pd2@Au dimers, as a function of the pressure. At low temperature, the equilibrium configuration for CO adsorption on Pd2@Au is 2*CO molecules over all range of pressures, while at high temperature, a single CO bound to the bridge site is favoured at low pressures and CO pairs are likely only at high CO pressures. The crossing between multiple and single CO adsorption is observed at pCO~ 0.03 Torr. Reproduced with the permission from [44]

example. The approach is based on the statement that IR bands of carbonyl adsorbed species recorded after CO introduction are relevant of specific CO/metal interactions and serve as a fingerprint of the adsorption site [35, 36] as well as the surface chemical bonds [37]. Combined with experimental work, DFT energy and harmonic vibrational calculations allow the assignment of spectra, recorded by in situ or operando techniques, thus providing a more reliable description of surface active sites and the nearest chemical environment.

Gold-based alloy systems

Despite a considerable progress of the application of the previously described DFT methodologies in the identification of equilibrium alloy structures under reactive gas conditions [see

the recent reviews 32, 38], only very few studies at this level of theory have considered Au-based bimetallic nanosystems. In the following, we briefly review the main theoretical insights in this topic by presenting some examples of gold-based alloy systems.

Bimetallic gold–palladium alloy

Au–Pd catalyst is one of the bimetallic systems that have attracted the most significant interest due to its superior performance in various catalytic reactions [1, 2, 5–7, 10]. The interplay between gold and palladium components leads to its superior catalytic reactivity in terms of the so-called ligand and ensemble effects [39]. In addition, the evolution of surface composition during exposure to reaction atmosphere seems to enhance, in many cases, the catalytic properties of the Au–Pd alloy. For instance, Piccolo et al. have reported that the increase in the conversion rate of hydrogenation of butadiene into butenes observed on Au–Pd surfaces during time on stream could partly originate from a modification of the surface composition and precisely from Pd surface enrichment induced by reactant or product adsorption [40]. Recent studies of Goodman's group showed that CO adsorption induces Pd segregation on AuPd(100) surfaces [10, 41]. The authors have reported that such segregation leads to the formation of contiguous Pd sites, at least Pd dimers able to dissociate O_2, responsible for the observed enhancement of low-temperature CO oxidation reaction. Moreover, the segregation of Pd in Au–Pd nanoparticles after other gas exposure such O_2 [16], H_2 [42] and NO [43] was also observed through various techniques.

From this unique behaviour of Pd under reaction conditions, several questions rise: what is the distribution of Pd atoms between the surface and bulk phases, and how this distribution depends on CO gas phase, on the temperature and on the Pd concentration?

According to the work of García and López using a first principle-based thermodynamic model, CO-induced segregation

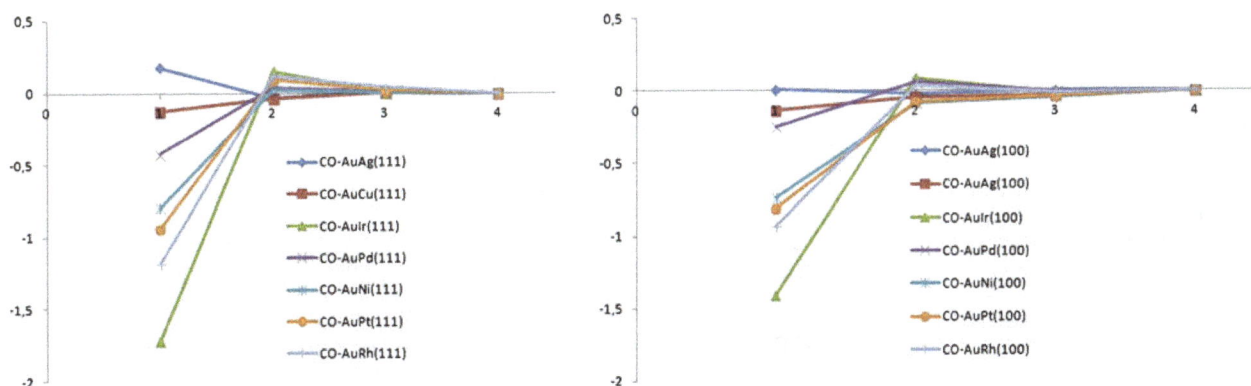

Fig. 2 DFT segregation energies of group 9-10-11 transition metals (TM=Ag, Cu, Ir, Pd, Ni, Pt and Rh), substituted in semi-infinite gold (111) and (100) surfaces, calculated in the presence of CO

Fig. 3 Assignations of DRIFT spectra in the carbonyl region of Au–Pd(20)/Al₂O₃ exposed to CO to the corresponding DFT calculated local geometries

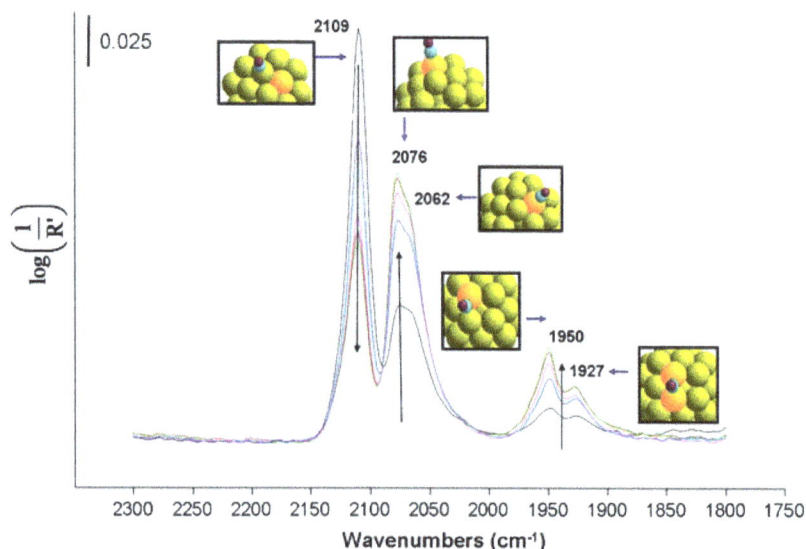

occurs for PdAu (111) at moderate CO pressures, about 10^{-2} Torr (Fig. 1) [44]. By performing DFT energy calculations coupled by thermodynamics together with a simple lattice–gas model, Soto-Verdugo and Metiu [45] have investigated the equilibrium composition of the (111) and (100) diluted Au–Pd alloy exposed to CO as a function of temperature, CO pressure and Pd/Au ratio. These authors have reported that in the presence of CO, the interaction between two Pd remains less favourable than Au–Pd interaction, which explains the non-formation of aggregates within the surface layer and the non-existence of Pd domains in the bulk. From the study of equilibrium composition at the surface of spherical nanoalloys, these authors have suggested that the ability of Pd to segregate to the

surface was controlled by the binding energy of the reactants to the Pd surface atom. Following the same idea, using DFT periodic calculations in the presence of adsorbed CO, we have recently investigated the segregation behaviours of group 9-10-11 transition metals (TM=Ag, Cu, Ir, Pd, Ni, Pt and Rh), substituted in semi-infinite gold surfaces (Dhouib et al., submitted). The ability of TMs to segregate from the gold bulk to the surface was found to increase by increasing their binding strengths with adsorbed molecule. In addition, the investigation of different surface orientations shows a better segregation of TMs to the closed-packed (111) surface compared to the (100) (Fig. 2). Concerning Au–Pd alloy, we have also shown that the segregation behaviour of Pd was oxygen coverage-dependent

Fig. 4 DFT calculated formation energies of Au–Pt surface alloys on Pt(111). The *grey lines* connecting the lowest energy structures show a convex hull, indicating ordered structures exist for the surface. *Red dots* are striped structures with the size of the circles showing the average period of the stripes. The *black line* shows the formation energies of the random surfaces, and the *green square* is the formation energy of the 2D special quasi-random structure generated at 50 % composition. Reproduced with permission from [52]

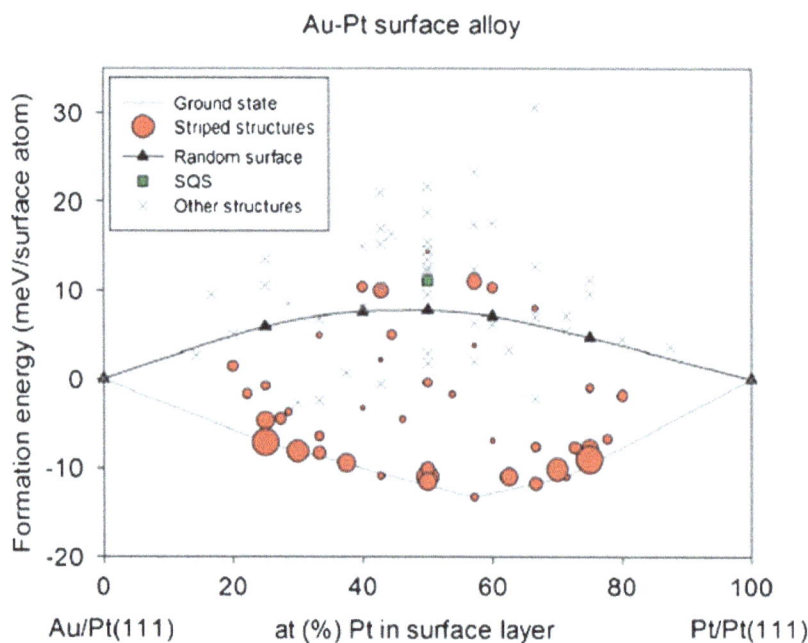

[15]. Indeed, the DFT energy calculations predict that at $T=0$ K, Pd segregation to the surface occurs only in the presence of more than 1/3 ML of chemisorbed oxygen. This is not the case for Au–Pt and Au–Ni alloys where, as soon as one oxygen atom is adsorbed, the stability of the metal impurities is reversed from the bulk to the surface [16]. Both CO and H chemisorption effects on the chemical ordering of bimetallic Au–Pd clusters have been reported by West et al. [46]. From calculated DFT ligand bonding strengths, these authors have predicted that in contrast with conditions under CO atmosphere, core-shell segregation inversion under H_2 could not be favourable.

In a recent work devoted to the calculation of CO–Pd frequencies and IR experimental assignments, we have demonstrated that in gold–palladium nanoalloys (with very low Pd loading), only isolated or dimer palladium could exist on the surface [17]. This combined approach allowed identification of different CO alloy-bonding and quantification of all Pd-type coordinations of the surface in the presence of gas (Fig. 3). Moreover, from energetic and electronic analysis, the DFT results have shown that, under conditions where migration and restructuring can occur, Pd would migrate towards the edge positions on small AuPd nanoparticles.

Finally, for designing new bimetallic catalysts for specific applications, several DFT calculations were devoted to the prediction and to the comparison of the free activation energies of bond breakings and bond formations of intermediates during reaction conditions. For instance, by comparing the dissociation energy barriers of NO on a series of doped gold surfaces, Fajín et al. [47] have recently demonstrated that NO dissociation is only possible in the case of the Ir–Au(110) bimetallic surface but only at high temperature. In a similar context, by investigating the effect of surface structure on the catalytic activity of Au–Pd alloy, Wei et al. [48] have demonstrated by means of DFT reaction energy profile calculations that alloying Pd with gold would facilitate the desorption of O_2, which is generally the rate-determining step for N_2O decomposition reaction.

Bimetallic gold–platinum alloy

Alloying of Pt catalysts has been found to improve activity and selectivity as compared to the respective monomeric system. For example, a Cu–Pt surface alloy shows better activity than Pt for the water–gas shift reaction because the surface alloy better activates H_2O and at the same time binds CO more weakly [49]. A Pt monolayer on Pd(111) has improved activity for the oxygen reduction reaction as compared to a pure Pt catalyst [50]. Similarly to the Au–Pd system, Pt-based alloy system can show surface modifications and segregation in the presence of reactive gas. Using density functional theory, Tenney et al. have shown that it is thermodynamically more favourable for Pt to diffuse to the Au–Pt cluster surface in order to bind to CO [51]. By using CE-based DFT method, Chen and co-workers [52] have studied the surface structure and ordering of a mixed Au–Pt/Pt(111) surface alloy and have investigated the related oxygen binding. The DFT energies of about 90 Au–Pt structures were used as input to train the surface CE Hamiltonian. Even though the Au–Pt system is well known to phase separate in the bulk (miscibility gap), this alloy showed a series of stable low-temperature lateral ordered striped structures (Fig. 4). The existence of such structures was explained by the competition between the energy penalty of Au–Pt bonds at the boundaries between the stripes and the favourable strain relaxation from forming stripes. Because the presence of adsorbates can alter the atomic ordering of the surface atoms, these latter authors have investigated the effect of adsorbed oxygen on these ordered and disordered identified Au–Pt structures. The calculation of molecular oxygen binding was found to be highly correlated with the type of local surface alloy arrangement but information concerning structure change after such oxygen adsorption remains missing.

Conclusions

In the presence of adsorbates, the surface composition may change due to the adsorbate effect on solute segregation tendencies. The understanding of surface segregation is thus of primary importance for controlling the behaviour of bimetallic catalysts. However, due to the complexity of the alloy systems, segregation and surface structure are often difficult to predict. By the consequence, in spite of the considerable progress of the theoretical methodologies for the identification of bulk and surface alloy structures under idealized ultrahigh vacuum conditions, very few studies are devoted to the equilibrium alloy structures under reactive gas conditions. In this paper, the recent slight progresses in this challenging task were briefly reviewed through some gold-based alloy examples.

Acknowledgment The author acknowledges the support of EU (COST-MP0903) and addresses her warm thanks to Dr. Tzonka Minèva for the critical reading of the manuscript.

References

1. Hugon A, Delannoy L, Krafft J-M, Louis C (2010) Supported gold-palladium catalysts for selective hydrogenation of 1,3 butadiene in an excess of propene. J Phys Chem C 114:10823–10835
2. El Kolli N, Delannoy L, Louis C (2013) Bimetallic Au–Pd catalysts for selective hydrogenation of butadiene: influence of the preparation method on catalytic properties. J Cat 297:79–92

3. Alissawi N, Zaporojtchenko V, Strunskus T, Kocabas I, Chakravadhanula VSK, Kienle L, Garbe-Schönberg D, Faupel F (2012) Effect of gold alloying on stability of silver nanoparticles and control of silver ion release from vapor-deposited Ag-Au/polytetrafluoroethylene nanocomposites. Gold Bull 46:3–11

4. Luo J, Wang LY, Mott D, Njoki P, Lin Y, He T, Xu Z, Wanjana B, I-Im Lim S, Zhong CJ (2008) Core@Shell nanoparticles as electrocatalysts for fuel cell reactions. Adv Mater 20:4342–4347

5. Edwards JK, Hutchings GJ (2008) Palladium and gold-palladium catalysts for the direct synthesis of hydrogen peroxide. Angew Chem Int Ed 47:9192–9198

6. Edwards JK, Solsona B, Carley AF, Herzing AA, Kiely CJ, Hutchings GJ (2009) Switching-off hydrogen peroxide hydrogenation in the direct synthesis process. Science 323:1037–1041

7. Chen M, Kumar D, Yi CW, Goodman DW (2005) The promotional effect of gold in catalysis by palladium–gold. Science 310:291–293

8. Hugon A, El Kolli N, Louis C (2010) Advances in the preparation of supported gold catalysts: mechanism of deposition, simplification of the procedures and relevance of the elimination of chlorine. J Catal 274:239–250

9. Chandler BO, Long CG, Gilberson JD, Pursell CJ, Vijayaraghavan G, Stevenson KJ (2010) Enhanced oxygen activation over supported bimetallic Au−Ni catalysts. J Phys Chem C 114:11498–11508

10. Gao F, Wang Y, Goodman DW (2009) CO oxidation over AuPd(100) from ultrahigh vacuum to near-atmospheric pressures: the critical role of contiguous Pd atoms. J Am Chem Soc 131:5734–5735

11. Zhou S, Jackson GS, Eichhorn B (2007) Architectural effects on the catalytic activity of Au-Pt bimetallic nanostructures: alloys and contact aggregates particles for CO tolerant hydrogen activation. Adv Funct Mater 17:3099–3104

12. García-Mota M, López N (2011) The role of long-lived oxygen precursors on AuM alloys (M=Ni, Pd, Pt, Pt) in CO oxidation. Phys Chem Chem Phys 13:5790–5797

13. Tenney SA, He W, Roberts CC, Ratliff JS, Shah SI, Shafai GS, Turkowski V, Rahman TS, Chen DA (2011) CO-induced diffusion of Ni atoms to the surface of Ni-Au clusters on TiO2(110). J Phys Chem C 115:11112–11123

14. Pittaway F, Paz-Borbón LO, Johnston RL, Arslan H, Ferrando R, Mottet C, Barcaro G, Fortunelli A (2009) Theoretical studies of palladium-gold nanoclusters: Pd-Au clusters with up to 50 atoms. J Phys Chem C 113:9141–9152

15. Guesmi H, Louis C, Delannoy L (2011) Chemisorbed atomic oxygen inducing Pd segregation in PdAu(1 1 1) alloy: energetic and electronic DFT analysis. Chem Phys Lett 503:97–100

16. Dhouib A, Guesmi H (2012) DFT study of the M segregation on MAu alloys (M = Ni, Pd, Pt) in presence of adsorbed oxygen O and O2. Chem Phys Lett 521:98–103

17. Zhu B, Thrimurthu G, Delannoy L, Louis C, Mottet C, Creuze J, Legrand B, Guesmi H (2013) Evidence of Pd segregation and stabilization at edges of AuPd nano-clusters in the presence of CO: a combined DFT and DRIFTS study. J Catal.

18. Yoshida H, Kuwauchi Y, Jinschek JR, Sun K, Tanaka S, Kohyama M, Shimada S, Haruta M, Takeda S (2012) Visualizing gas molecules interacting with supported nanoparticulate catalysts at reaction conditions. Sience 335:317–319

19. Treglia G, Legrand B (1987) Surface-sandwich segregation in Pt-Ni and Ag-Ni alloys: two different physical origins for the same phenomenon. Phys Rev B 35:4338–4344

20. Treglia G, Legrand B, Maugain P (1990) Surface segregation in CuNi and AgNi alloys formulated as an area-preserving map. Surf Sci 225:319–330

21. Ferrando R, Jellinek J, Johnston RL (2008) Nanoalloys: from theory to applications of alloy clusters and nanoparticles. Chem Rev 108:845–910

22. Bochicchio D, Ferrando R (2010) Size-dependent transition to high-symmetry chiral structures in AgCu, AgCo, AgNi and AuNi nanoalloys. Nano Lett 10:4211–4216

23. Bochicchio D, Ferrando R (2013) Morphological instability of core-shell metallic nanoparticles. Phys Rev B 87(16):13

24. Moreno V, Creuze J, Berthier F, Mottet C, Tréglia G, Legrand B (2006) Site segregation in size-mismatched nanoalloys: application to Cu-Ag. Surf Sci 600:5011–5020

25. Creuze J, Braems I, Berthier F, Mottet C, Tréglia G, Legrand B (2008) Model of surface segregation driving forces and their coupling. Phys Rev B 78:075413

26. Barcaro G, Fortunelly A, Polak M, Rubinovich L (2011) Patchy multishell segregation in Pd–Pt alloy nanoparticles. Nano Lett 11:1766–1769

27. Rossi G, Rapallo A, Mottet C, Fortunelli A, Baletto F, Ferrando R (2004) Magic poluicosahedral core-shell clusters. Phys Rev Lett 93(10):105503–1055037

28. Heiles S, Logsdail AJ, Sch fer R, Johnston RL (2012) Dopant-induced 2D-3D transition in small Au-containing clusters: DFT-global optimization of 8-atom Au-Ag nanoalloys. Nanoscale 4:1109–1115

29. Norskov JK, Biligaard T, Rossmeisl J, Christensen CH (2009) Towards the computational design of solid catalysts. Nat Chem 1:37–46

30. Molenbroek AM, Helveg S, Topsøe H, Clausen BS (2009) Nanoparticles in heterogeneous catalysis. Top Catal 52:1303–1311

31. Tielens F, Calatayud M (2011) The synergistic power of theory and experiment in the field of catalysis Preface. Catal Today 177:1–2

32. Zafeiratos S, Piccinin S, Teschner D (2012) Alloys in catalysis: phase separation and surface segregation phenomena in response to the reactive environment. Catal Sci Tech 2:1787–1801

33. Sun Q, Reuter K, Scheffler M (2003) Effect of a humid environment on the surface structure of RuO2(110). Phys Rev B: Condens Matter 67:205424–205431

34. Nguyen NL, Piccinin S, de Gironcoli S (2011) Stability of intermediate states for ethylene epoxidation on Ag-Cu alloy catalyst: a first-principles investigation. J Phys Chem C 115:10073–10079

35. Lamberti C, Bordiga S, Geobaldo F, Zecchina A, Otero Areán C (1995) Stretching frequencies of cation–CO adducts in alkali–metal exchanged zeolites: an elementary electrostatic approach. J Chem Phys 103:3158–3166

36. Cairon O, Guesmi H (2011) How does CO capture process on microporous NaY zeolites? A FTIR and DFT combined study. Phys Chem Chem Phys 13:11430–11437

37. Risse T, Carlsson A, Baumer M, Kluner T, Freund HJ (2003) Using IR intensities as a probe for studying the surface chemical bond. Surf Sci 546:L829–L835

38. Tao F, Zhang S, Nguyen L, Zhang X (2012) Action of bimetallic nanocatalysts under reaction conditions and during catalysis: evolution of chemistry from high vacuum conditions to reaction conditions. Chem Soc Rev 41:7980–7993

39. Boscoboinik JA, Calaza FC, Garvey MT, Tysoe WT (2010) Identification of adsorption ensembles on bimetallic alloys. J Phys Chem C 114:1875–1880

40. Piccolo L, Piednoir A, Bertolini JC (2005) Pd-Au single-crystal surfaces: segregation properties and catalytic activity in the selective hydrogenation of 1,3-butadiene. Surf Sci 592:169–181

41. Gao F, Wang YL, Goodman DW (2009) CO adsorption-induced surface segregation and reaction kinetics. J Phys Chem C 13:14993–15000

42. Di Vece M, Bals S, Verbeeck J, Lievens P, Van Tendeloo G (2009) Compositional changes of Pd–Au bimetallic nanoclusters upon hydrogenation. Phys Rev B 80(12):4

43. de Bocarmé TV, Moors M, Kruse N, Atanasov IS, Hou M, Cerezo A, Smith GDW (2009) Surface segregation of Au–Pd alloys in UHV and

reactive environments: quantification by a catalytic atom probe. Ultramicroscopy 109:619–624

44. García M, López N (2010) Temperature and pressure effects in CO titration of ensembles in PdAu(111) alloys using first principles. Phys Rev B 82(7):9

45. Soto-Verdugo V, Metiu H (2007) Segregation at the surface of an Au/Pd alloy exposed to CO. Surf Sci 601:5332–5339

46. West PS, Johnston RL, Barcaro G, Fortunelli A (2010) The effect of CO and H chemisorptions on the chemical ordering of bimetallic clusters. J Phys Chem C 114:19678–19686

47. Fajín JLC, Cordeiro MNDS, Gomes JRB (2013) A DFT study of the NO dissociation on gold surfaces doped with transition metals. J Chem Phys 138:74701–74710

48. Wei X, X-F Yang A-Q, Wang LL, Liu X-Y, Zhang T, Mou C-Y, Li J (2012) Bimetallic Au-Pd alloy catalysts for N2O decomposition: effects of surface structures on catalytic activity. J Phys Chem C 116:6222–6232

49. Nilekar AU, Mavrikakis M (2008) Improved oxygen reduction reactivity of platinum monolayers on transition metal surfaces surf. Sci 602:L89–L94

50. Tenney SA, Ratliff JS, Roberts CC, He W, Ammal SC, Hayden A, Chen DA (2010) Adsorbate-induced changes in the surface composition of bimetallic cluster: Pt-Au on TiO2(110). J Phys Chem C 114: 21652–21663

51. Knudsen J, Nilekar AU, Vang RT, Schnadt J, Kunkes EL, Dumesic JA, Mavrikakis M, Besenbacher F (2007) A Cu/Pt near-surface alloy for water-gas shift catalysis. J Am Chem Soc 129:6485–6490

52. Chen W, Schmidt D, Schneider WF, Wolverton C (2011) Ordering and oxygen adsorption in Au-Pt/Pt(111) surface alloys. J Phys Chem C 115:17915–17924

16

Performance and comparison of gold-based neutron flux monitors

Georg Steinhauser · Stefan Merz · Franziska Stadlbauer · Peter Kregsamer · Christina Streli · Mario Villa

Abstract In two test series, liquid and solid gold-based neutron flux monitor materials were investigated with respect to the effects of neutron absorbers such as chlorine, scattering effects, and the dependence of the enhanced activation caused by the epithermal resonance integral. The liquid monitors were prepared from aqueous solutions of tetraamminegold(III) nitrate and tetrachloroauric(III) acid. The presence of chlorine-35 partly suppresses the activation of gold-197; this effect depends not only on the concentration of the absorber but also on the state of the neutron flux density monitor. Aqueous samples show greater relative losses than solid monitors. Neutron scattering occurs in hydrogen-rich sample matrices which is shown by the fact that cadmium-shielded aqueous samples show an over proportional activation. Hence, fast neutrons must be moderated to the epithermal energies covered by the resonance integral, which is characterized by much greater cross sections for the capture of neutrons. The insight of this study with respect to neutron scattering in hydrogen-rich matrices must be taken into account also for neutron activation analysis; sample and standard must have a similar matrix with respect to its neutron scattering properties, otherwise the effect of increased activation as well as of enhanced self-shielding are underestimated.

Keywords Neutron flux monitor · Neutron flux density · Neutron scattering · Matrix effects · Chlorine

G. Steinhauser (✉) · S. Merz (✉) · F. Stadlbauer · P. Kregsamer · C. Streli · M. Villa
Atominstitut, Vienna University of Technology,
Stadionallee 2,
1020 Vienna, Austria
e-mail: georg.steinhauser@ati.ac.at
e-mail: merz@ati.ac.at

Introduction

One of the most important physical parameters of a nuclear research reactor is its neutron flux density. It is measured by determining the activity of a material—the neutron flux monitor—that is exposed to the neutron beam causing its activation. The activity of the monitor is given by the parameters of the activation equation (Eq. 1).

$$A = N\Phi\sigma\left(1 - e^{-\lambda t}\right), \tag{1}$$

where A = activity [becquerel], N = number of atoms in the monitor [ ], Φ = neutron flux density [per square centimeter per second], σ = nuclear cross section [square centimeter], λ = decay constant [per second], and t = irradiation duration [seconds].

Gold is among the most commonly used materials for neutron flux monitors. It offers several advantages, e.g., it can be produced in great purity; it yields high activities due to its large thermal cross section (98.65 barn; 1 b = 10^{-24} cm^2) for neutron capture in the nuclear reaction ^{197}Au(n,γ)^{198}Au; ^{198}Au has a very convenient half-life of 2.7 days; the cross section of gold and its other nuclear properties are among the best known of all nuclei [1]. Due to its cross section characteristics, in particular its large resonance integral for the capture of epithermal neutrons, it is possible to apply the cadmium difference method [2] to obtain information on the energy spectrum of the neutron source. Gold in pure metallic form or in the form of alloys (e.g., Al-0.1% Au) is thus usually applied in the form of foils, disks, or wires to monitor the flux in a certain irradiation position.

The application of gold-based flux monitors, however, offers some disadvantages as well. First, it is often activated too strongly if applied in larger amounts than in the milligram range (especially in long-term activation experiments) and hence requires long cooling times before it can be measured properly. As an alternative to reducing the activation duration, the power level of the reactor can be diminished for the flux measurement. However, this not only affects the reactor operation and possibly influences the experiments of other users of the reactor; it often also makes the flux measurement appear as an approximation in case the samples and the monitor are activated at different power levels (mainly due to reactor power nonlinearity). Second, the high cross section of gold can lead to significant neutron self-shielding and thus has to be corrected mathematically (e.g., [3]). Third, gold foils or wires not necessarily simulate the sample geometry accordingly, if the neutron flux is determined in preparation of the irradiation of samples in the reactor. Many samples which are commonly irradiated in research reactors, however, have other geometry than the disk, foil, or wire shape (but rather the shape of a more or less cylindrical vial).

The use of gold alloys (such as the aforementioned Al–Au alloys, e.g., IRMM-530RA or IRMM-530RC) may overcome most of these disadvantages. In particular, such materials avoid neutron self-shielding effects as well as strong activation beyond the levels of safe handling and measurability. However, those materials are relatively pricey; they do not necessarily simulate the sample geometry and, due to the aluminum matrix, they can yield other activation products which may cause interferences and thus require certain periods of cooling (important nuclear reactions of Al are $^{27}Al(n,\gamma)^{28}Al$ with a half-life of 2.2 min; ^{27}Al $(n,\alpha)^{24}Na$ with a half-life of 15 h).

In order to overcome these problems, aqueous gold solutions may be regarded as a suitable alternative. Some research reactor operators are reluctant to irradiate liquids because of potential contamination risks. Due to the short half-life of ^{198}Au, however, gold-based neutron flux monitors can be recycled after some weeks of cooling, thus justifying additional efforts to seal the irradiation vials tightly in order to minimize this risk. In any case, the amount of gold can be easily reduced in liquids at the experimenter's wish, hence reducing the total activity of the monitor and minimizing self-shielding effects. Furthermore, a solution can be filled into almost any sample geometry, thus allowing monitoring the neutron flux, exactly where and how the samples are exposed to it. Unfortunately, almost all common and purchasable gold compounds are based on chlorine, namely the tetrachloroaurate anion $[AuCl_4^-]$. Chlorine has a nasty activation product stemming from $^{37}Cl(n,\gamma)^{38}Cl$ ($T_{1/2}=$ 37 min) and is also a strong neutron absorber (^{35}Cl, $\sigma=44$ barn for thermal (n,γ) neutron capture). Hence, Cl compounds are likely to cause self-shielding effects in the sample. Tetrabromoaurates(III) are purchasable as well, but the highly gamma ray emitting activation product ^{82}Br ($T_{1/2}=35.3$ h) makes them even less suitable.

The aim of this study was to compare the performance of various flux monitor materials, in particular to compare liquid and solid gold monitors. We tried to conduct experiments to gain further insight into important effects such as neutron self-shielding, neutron scattering, and neutron energy aspects, which all need to be considered.

Experimental and methods

The experiments were conducted in two series: first, the comparison of various liquid neutron flux monitors and, second, the comparison of liquid and solid materials under different conditions. For all irradiation experiments in this study, the so-called Weber-tube no. 2 of the institute's 250 kW_{th} TRIGA Mark II reactor was chosen. Its irradiation position is located outside the graphite reflector of the reactor core and has a 90% thermalized neutron flux of approximately $1\cdot10^{11}$ cm^{-2} s^{-1} at full reactor power. All experiments were conducted within 1 h in order to provide the utmost comparable neutron flux conditions to all samples.

First test series

For the comparison of materials based on dissolved gold compounds, three substances were tested: aqueous solutions of tetrachloroauric(III) acid $H[AuCl_4]$; aqueous solutions of tetraamminegold(III) nitrate, $[Au(NH_3)_4](NO_3)_3$ (TAGN); and, in order to simulate the dissolution of metallic gold in *aqua regia*, $H[AuCl_4]$ in 30% hydrochloric acid (HCl, Merck™, suprapure). TAGN was synthesized according to literature (Eq. 2) [4, 5].

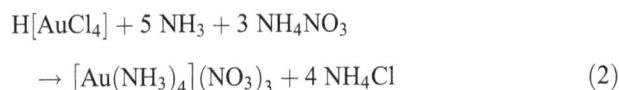

$$H[AuCl_4] + 5\ NH_3 + 3\ NH_4NO_3$$
$$\rightarrow [Au(NH_3)_4](NO_3)_3 + 4\ NH_4Cl \qquad (2)$$

In particular, 250 mg of $H[AuCl_4]\cdot3\ H_2O$ (Sigma Aldrich™, 99.9+%) were dissolved in 5 ml of a solution that has been saturated with NH_4NO_3 at room temperature. In the next step, $NH_{3\ (g)}$ steam from an ammonia solution bottle was bubbled through the solution with a syringe until white crystals started precipitating or the yellow solution lost its color almost abruptly. Precipitation was completed overnight and yielded a significant amount of crystalline and very pure TAGN (as confirmed by elemental CHN analysis). The purity of the crystals with respect to the residual chlorine was checked with INAA providing a negative result.

Stem solutions of TAGN and $H[AuCl_4]\cdot3\ H_2O$ were prepared by dissolving these solids in triply distilled H_2O. In both cases, the resulting formal gold concentration was 1.00 mg/ml. The Au^{3+} concentration in both stem solutions was analyzed by total reflection X-ray fluorescence (TXRF, Atomika 8030 C) [6]. TXRF was chosen as an independent method to confirm the formal concentrations of gold (1,023±35 µg/ml for TAGN, 1,088±71 µg/ml for H[AuCl$_4$]).

Five replicates were prepared from each of the three liquid flux monitor materials. Aliquots corresponding to 50 µg Au were pipetted and weighed into cylindrical polyethylene (PE) irradiation vials (diameter, 12.5 mm; fluid level, 42 mm). In order to provide the utmost accurate sample geometry, each vial was filled with 5 ml of triply distilled H$_2$O (in the case of TAGN and H[AuCl$_4$]) or 5 ml 30% HCl (for the experiment "H[AuCl$_4$] in HCl", samples S1-S5), respectively.

After a preliminary experiment, we observed signs of radiolysis in some of the solutions. In the course of radiolysis, Au^{3+}-containing solutions generally tend to change their color towards red, pink, or brown. Sometimes even precipitation of brownish flakes occurs. Hence, potassium cyanide (KCN) was added to the TAGN solutions in an ion ratio Au^{3+}/CN$^-$ = 1:8 for stabilization [7] (see Fig. 1). Addition of KCN could not be done with the acidic solutions based on tetrachloroauric(III) acid because of the possible formation of very poisonous hydrocyanic acid (HCN) gas. It should be noted at this point that aqueous solutions of H[AuCl$_4$] should not be neutralized with ammonia because of the possible formation of fulminating gold (Knallgold) [4].

A foamed PE plastic lid was placed on the surface of the liquid, which allowed any expansion of the sample (due to temperature and the formation of radiolytic gases). Then the vial was sealed with paraffinic wax. All monitors of the first test series were irradiated sequentially in the mentioned position for 2 min each in order to cancel out flux variations.

The liquid monitors could be measured immediately (TAGN) after irradiation or after 24 h (for the Cl-containing monitors), respectively. The activity was determined by measurement with a 226-cm^3 HPGe detector (Canberra™, detector model GC5020; 2.0 keV resolution at 1,332 keV ^{60}Co peak; 52.8% relative efficiency), connected to a PC-based multichannel analyzer with preloaded filter. The detector efficiency for the 411-keV peak of ^{198}Au was calibrated with a QCY48 (Amersham® Ltd.) solution. The necessary measurement times were usually quite short (until the uncertainty due to counting statistics was <1%).

Second test series

For the second test series, the TAGN monitors (Z1-Z4) were recycled after sufficient cooling time. The second test series comprised the experiments shown in Table 1.

The gold wire (Alfa Aesar; 99.998%) had a thickness of 0.1 mm and a length of 42 mm (equal to the filling level); the gold foil had a diameter of 5 mm and a thickness of 0.05 mm. The foil was placed on the half height of the filling level of the liquid samples. In order to grant a stable position of the wires and foils in the water-filled samples, water was supplied in the form of aqueous cornstarch pudding (96% water), which had a sufficient viscosity to prevent any unwanted movement of the monitor. The cadmium shielding (the cartridge as well as the "sandwich") had a thickness of 1 mm. For the measurement, the gold foils and wires, respectively, were dissolved in the minimum amount of *aqua regia* and filled to a level of 5 ml, equivalent to the geometry of the liquid monitors.

In order to yield sufficient activity with the cadmium-shielded samples, the irradiation time was increased to 5 min in the second test series. The cadmium-shielded samples as well as the TAGN vials were measured immediately after irradiation. The unshielded metallic flux monitors required a cooling time of 3 weeks.

Fig. 1 Changes in color due to radiolysis in the neutron flux monitor based on H[AuCl$_4$] in H$_2$O (*right*), compared to the TAGN-based monitor (*left*)

Table 1 Neutron flux monitor samples irradiated in the second test series

Sample code	Gold sample description
Z1, Z2	TAGN in water
Z3, Z4	TAGN in water, vial irradiated in cadmium cartridge
F9	Foil in air
F1	Foil in water
F3	Foil in water, vial irradiated in cadmium cartridge
F5	Foil in water, foil placed inside a cadmium "sandwich"
F4	Foil in aqueous liquid with chlorine content of 29.2 wt.%
D1	Wire in air
D2	Wire in water
D3	Wire in water, vial irradiated in cadmium cartridge

Results and discussion

The results of both test series are tabulated in Tables 2 and 3. Although the irradiation conditions were comparable in all experiments of the first series, the specific activities obtained show three sharply distinguished groups. Obviously, the chlorine content in the sample affects the performance of the respective neutron flux monitor. Compared to the TAGN-based monitor (Z samples), activation of Au is suppressed in $H[AuCl_4]$ in water (W samples) and, to a much greater extent, in $H[AuCl_4]$ in concentrated hydrochloric acid (S samples).

The specific activities of the gold foils have been corrected with a factor G according to the self-shielding correction established by de Corte [3] (Eq. 3).

$$\Phi_{korr} = \frac{\Phi}{G} \tag{3}$$

The correction factor G is calculated as shown in Eqs. 4, 5, 6, where t is the foil thickness and N is the number of atoms per unit volume. For a foil thickness of 0.05 mm, Eq. 6 applies. In our case, G is 0.968.

$$G = \frac{3}{4y^3} \left[y^2 - \frac{1}{2} + \left(\frac{1}{2} + y \right) \cdot e^{-2y} \right] \tag{4}$$

$$\xi = N \cdot t \cdot \sigma \tag{5}$$

$$y = \frac{3}{2} \xi \tag{6}$$

Table 2 Specific activities of the neutron monitor samples after irradiation (first test series); irradiation time of 120 s

Sample code	Sample	Specific activity [Bq/g]
Z1	TAGN in water	5.02E+07
Z2		5.10E+07
Z3		5.30E+07
Z4		5.23E+07
Z5		5.11E+07
W1	$H[AuCl_4]$ in water	4.56E+07
W2		4.54E+07
W3		4.53E+07
W4		4.49E+07
W5		4.46E+07
S1	$H[AuCl_4]$ in HCl with chlorine content of 29.2 wt.%	3.73E+07
S2		3.53E+07
S3		3.74E+07
S4		3.67E+07
S5		3.71E+07

Analytical uncertainty due to counting statistics <2%

Table 3 Specific activities of the neutron monitor samples after irradiation (second test series); irradiation time of 300 s

Sample	Specific activity [Bq/g]
Z1	1.23E+08
Z2	1.33E+08
Z3	5.02E+07
Z4	4.92E+07
F9	9.59E+07
F1	9.48E+07
F3	2.75E+07
F5	2.10E+07
F4	8.85E+07
D1	9.21E+07
D2	1.17E+08
D3	7.76E+06

Analytical uncertainty due to counting statistics <2%

Figure 2 shows the specific activities of the neutron flux monitors made of solid gold foils and wires, once fixed in air in the irradiation vial (F9, D1), once immersed in water (F1, D2). Whereas no significant effect can be observed for the gold foil when irradiated in air or in water; the gold wire shows a significantly higher specific activity when immersed in an aqueous medium (+26%).

This effect must be due to neutron scattering in the water-filled vial, causing an increasing probability for neutron capture. Interestingly, this effect has not been observed with the gold foil and only for the wire. We plan to elucidate these contradictory results with gold foils and wires in future studies.

In Fig. 3, the influence of chlorine onto the activation of the gold monitor is shown, in particular the comparison between solid and liquid materials. The graph on the left-hand side shows the specific activity of the liquid monitors TAGN in water (Z1–Z5), $H[AuCl_4]$ in water (W1–W5), and

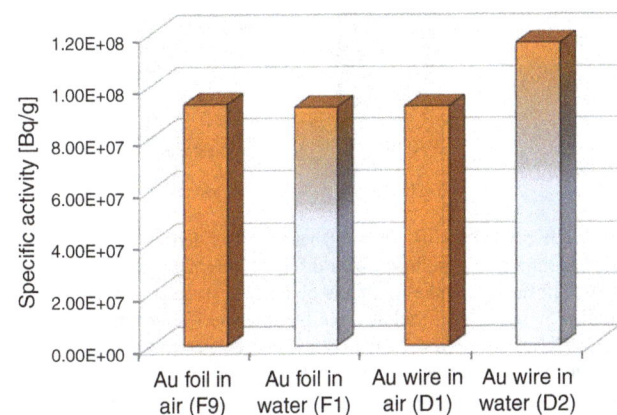

Fig. 2 Specific activities of [198]Au obtained by neutron irradiation of gold foils and wires irradiated in air and in water, respectively

H[AuCl₄] in concentrated hydrochloric acid (S1–S5) (calculated for 300 s activation time based on the 120-s experiments). In comparison, the graph on the right-hand side illustrates the specific activity of the gold foil immersed in water and in chloride solution with identical chlorine content like the hydrochloric acid above.

The negative influence of chlorine on the activation of gold is confirmed by this result; this effect is much enhanced for the liquid samples containing dissolved gold ions compared with the sample containing a solid gold foil. Even the small chlorine content in the samples containing H[AuCl₄] dissolved in water (W1–W5) causes a significant decrease of the activation of gold by approximately 12.3%. The experiment shows that the neutron scattering in the dissolved (aqueous) monitor materials causes not only an overall higher specific activity of ^{198}Au but also a reinforced relative attenuation of its activity in the presence of neutron absorbing impurities or matrix elements (compared with solid gold monitors surrounded by neutron absorbers).

The cross section for neutron capture is strongly neutron energy dependent, hence some insight can be gathered from shielding the thermal neutron fraction with cadmium. Taking into consideration that the fast neutron flux in the used irradiation position is 10% of the total flux, as a first approach, one would expect a resulting ^{198}Au activity of significantly less than 10% for cadmium shielded monitors because the cross section for ^{197}Au$(n,\gamma)^{198}$Au is roughly two orders of magnitude lower for fast neutrons. Since the energy spectrum of neutrons passing the cadmium shielding are already in the epithermal energy range, the resonance

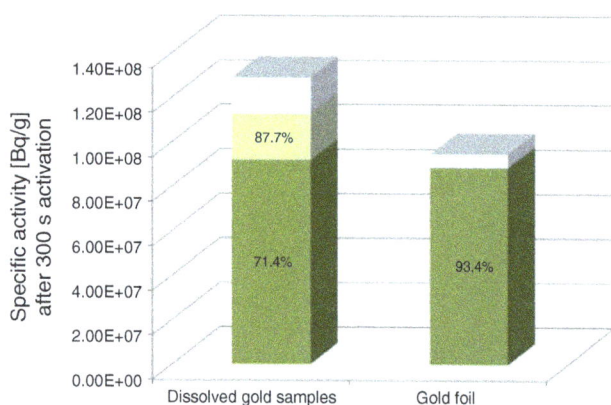

Fig. 4 Specific activities of ^{198}Au of Cd shielded (*gray*) and unshielded (*yellow*) neutron flux monitors. Shielding was performed in a Cd cartridge. TAGN in water (mean values of Z1 + Z2 vs. Z3 + Z4) is shown on the left-hand side; gold wire in water (D2 vs. D3) in the center; gold foil in water (F1 vs. F3) on the right-hand side

integral for the epithermal neutron capture by ^{197}Au, however, is in a position to cause an activation of the monitor of more than only 10% of the unshielded monitors.

We observed that scattering effects and the previously mentioned resonance integral yield higher activities in the aqueous samples (Fig. 4). Again, this effect is the greatest for the TAGN-based neutron flux monitors. From this result, the scattering inside the sample appears as approved; fast neutrons passing the cadmium shield are moderated inside the monitor to epithermal energies, thus coinciding with the relatively narrow resonance integral. Also, the scattering of neutrons in general inside the vial causes neutrons to stay longer inside the vial. This causes an over proportional activation of ^{197}Au. The same behavior is observed in the comparison of the Cd-shielded and unshielded gold foils immersed in water. In this context, the gold wire immersed in water and shielded with cadmium shows an unexpected low activation of approximately 7%, which will require more in-depth investigation, also with respect to the sample geometry, which may have an influence.

In another experiment, scattering effects and thus the influence of the resonance integral were minimized, by immersing a cadmium–gold–cadmium "sandwich" (F5) into a water-filled vial. In that case, any moderation towards the resonance integral in the space between the cadmium shielding and the gold foil was de facto excluded since practically no water could enter the space in-between. Hence, the activity of the "sandwich" was only 76% of the gold foil immersed in water and shielded with cadmium cartridge on the outside of the vial.

Conclusions

Neutron absorbers such as ^{35}Cl partly suppress the activation of ^{197}Au; this effect, however, depends not only on the

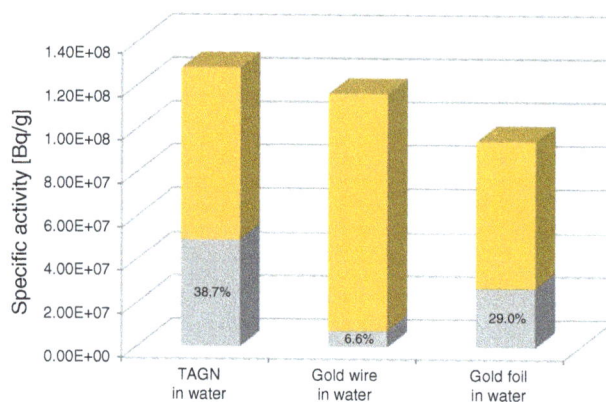

Fig. 3 Specific activities of ^{198}Au of neutron flux density monitors irradiated in an aqueous solution with different chlorine content. The *bar* on the left-hand side shows the dissolved gold samples: TAGN in water (Z1–Z5; *light blue*) yielding the highest specific activity, followed by H[AuCl₄] in water (W1–W5; *light green*) and finally H[AuCl₄] in hydrochloric acid with a chlorine content of 29.2 wt.% (S1–S5; *dark green*). The *bar* on the right-hand side shows the specific activities obtained by the gold foil in water (F1; *light blue*) and the gold foil immersed in aqueous NH₄Cl solution (F4; *dark green*) with a chlorine content of 29.2 wt.% as well

concentration of the absorber, but also on the state of the neutron flux density monitor. In solid samples (such as gold foils), the effect shows to be humble; in liquid samples of dissolved gold compounds, the effect is significantly amplified.

The results evidence that scattering occurs in hydrogen-rich sample matrices which is shown by the fact that cadmium-shielded aqueous samples show an over proportional activation. Hence, fast neutrons must be moderated to the epithermal energies covered by the resonance integral.

The TAGN-based neutron flux monitor appears as a very promising material for the manufacturing of neutron flux density monitors in liquid form, in case the neutron flux in an aqueous sample (or any other hydrogen-rich matrix) needs to be determined [8]. It is produced in a straightforward synthesis and from relatively cheap raw materials and can be recycled after a couple of weeks cooling.

The insight of this study with respect to neutron scattering in hydrogen-rich matrices must be taken into account also for neutron activation analysis (NAA); sample and standard must have a similar matrix with respect to its neutron scattering properties, otherwise the effect of increased activation as well as of enhanced self-shielding are underestimated. Also, for absolute NAA methods using single comparators, the assumption that a metal foil can supply sufficient information for a highly scattering matrix, is perhaps a simplification which may lead to relevant deviations in the subsequent calculations. For example, the determination of trace elements in crude oil, as recently shown in [9] may be affected by the effects discussed above. Perhaps a TAGN-based neutron flux monitor would be helpful in such analytical protocols. Under this point of view, the use of artificial synthetic standard materials in NAA should only be regarded as an approximation because they do not necessarily reflect the influence on the neutron flux caused by the various possible matrices in samples of various types.

Acknowledgments We gratefully acknowledge the financial support by the Vienna University of Technology in the course of patenting and the help by Angelika Valenta in the patenting process. Further, we thank both anonymous reviewers who helped us find the right track for this study.

References

1. Myerscough L (1973) The nuclear properties of gold. Gold Bull 6:62–68
2. Vermaercke P, Sneyers L, Bruggeman M, Wispelaere A, De Corte F (2008) Neutron spectrum calibration using the Cd-ratio for multi-monitor method with a synthetic multi-element standard. J Radioanal Nucl Chem 278:631–636
3. De Corte F (1969) Activeringsanalyse van sporenelementen in silicium. Dissertation, Ghent University
4. Steinhauser G, Evers J, Jakob S, Klapötke TM, Oehlinger G (2008) A review on fulminating gold (Knallgold). Gold Bull 41:305–317
5. Weishaupt M, Straehle J (1976) Crystal structure and vibrational spectrum of tetraamminegold(III) nitrate. Z Naturforsch B: Anorg Chem, Org Chem 31B:554–557
6. Wobrauschek P (2007) Total reflection x-ray fluorescence analysis—a review. X-Ray Spectrum 36:289–300
7. Steinhauser G (2008) Quantification of the abrasive wear of a gold wedding ring. Gold Bull 41:51–57
8. Steinhauser G, Merz S, Villa M (2010) Neutronenflussmonitor. VUT, Austrian patent 509490; application no. 706/2010
9. El-Khayatt AM (2010) Elemental analysis of Egyptian crude oils by INAA using rabbit irradiation system at ETRR-2 reactor. Appl Radiat Isot 68:2438–2442

17

Spinodal decomposition related to age-hardening and cuboidal structures in a dental low-carat gold alloy with relatively high Cu/Ag content ratio

Ji-In Jeong · Hyung-Il Kim · Gwang-Young Lee · Yong Hoon Kwon · Hyo-Joung Seol

Abstract A dental Au–Ag–Cu–Pd alloy with a relatively low Au content and a high Cu/Ag content ratio was examined to determine the correlation between the microstructural changes by the spinodal decomposition and age-hardening behaviour using a hardness test, X-ray diffraction study, field emission scanning electron microscopy and energy-dispersive X-ray spectrometry. Separation of the parent α_0 phase occurred by spinodal decomposition during aging at 350 °C after the solution treatment at 750 °C, and not by a nucleation and growth mechanism, resulting in the formation of the stable Ag-rich α_1 and AuCu I phases through a metastable state. Hardening resulted from the coherency lattice strain which occurred along the a-axis between the metastable Ag-rich α_1' and AuCu I' phases. In addition, lattice distortion occurred along the c-axis between the stable Ag-rich α_1 and AuCu I phases due to the tetragonality of the AuCu I ordered phase. The transformation of the stable Ag-rich α_1 and AuCu I phases from the metastable state introduced the formation of the fine and uniform cuboidal structures, which compensated for the increased gap in the lattice parameters through the phase transformation. Replacement of the fine cuboidal structures with the coarser lamellar structures occurred without a phase transformation, and resulted in softening by reducing the interfaces between the stable Ag-rich α_1 and AuCu I phases.

Keywords Spinodal decomposition · Cuboidal structures · Age-hardening · X-ray diffraction · Metastable AuCu I' phase

J.-I. Jeong · H.-I. Kim · G.-Y. Lee · Y. H. Kwon · H.-J. Seol (✉)
Department of Dental Materials, Institute of Translational Dental Science, School of Dentistry, Pusan National University, Beomeo-Ri, Mulgeum-Eup, Yangsan-Si, Gyeongsangnam-Do 626-814, South Korea
e-mail: seolhyojoung@daum.net

Introduction

Dental prostheses made with dental alloys must have sufficient mechanical properties to bear the stress from mastication. For that reason, dental alloys are subjected to proper heat treatment to be age-hardened. The hardening mechanism shows a range of aspects depending on the chemical composition of alloys [1]. In Au–Ag–Cu alloys, hardening occurs through the combined effects of ordering by the Au–Cu system and phase decomposition by the Ag–Cu system. In addition, such a hardening process can occur by spinodal decomposition in Au–Ag–Cu alloys under a specific composition and temperature range. Hamasaki et al. [2] reported that the precipitation of the metastable Cu_3Au ordered phase and spinodal decomposition were represented by two types of solution treatment temperatures, which resulted in different aging behaviours in the alloy composed of 45 Au–24.5 Ag–24.5 Cu–5 Pd–1 Pt (wt.%). Nakagawa and Yasuda [3] reported three types of age-hardening behaviours depending on aging temperatures in the alloy composed of 27.4 Au–17.4 Ag–55.2 Cu (at.%): (1) a dual mechanism of spinodal decomposition and Cu_3Au ordering, (2) a single mechanism of spinodal decomposition, and (3) a single mechanism of nucleation and growth of silver-rich precipitates.

The alloy that was age-hardened by spinodal decomposition showed finer and uniform structures than those induced by a nucleation and growth mechanism [4, 5]. Such fine microstructures will be beneficial to the tarnish and corrosion resistance as well as the biocompatibility for the use of these alloys in dental practice. In addition, a softening mechanism, such as a grain boundary reaction, can be more delayed than that by a nucleation and growth mechanism due to the formation of the microstructures that are similar to the single-phased structures [6]. Therefore, heat treatment conditions for hardening are less influenced by the aging time due to a delay of the grain boundary reaction. On the other hand, the correlation between the microstructural changes by spinodal decomposition and the age-hardening behaviours of the alloys has not been elucidated completely. Therefore, it is not easy to control both the microstructures and hardness for the optimal conditions in dental practice.

This study examined the age-hardening behaviour of an Au–Ag–Cu–Pd alloy with a relatively low Au content and a high Cu/Ag content ratio, which has the possibility of age-hardening by spinodal decomposition. The purpose of this study was to elucidate the spinodal decomposition related to age-hardening and cuboidal structures in a dental low-carat gold alloy composed of 31.88 Au–39.58 Ag–23.71 Cu–3.54 Pd (at.%) with minor ingredients using a hardness test, X-ray diffraction study, field emission scanning electron microscopy and energy-dispersive X-ray spectrometry.

Materials and methods

Specimen alloy

The specimen used in this study was a commercial type III dental low-carat gold alloy (AURIUM 50Y, Aurium® Research U.S.A., San Diego, CA, USA) according to the ISO classification (ISO 22674: 2006 (E)). This type of alloy is based on the ternary system of Au, Ag and Cu for the fabrication of single units and long bridges. The condition of the hardening heat treatment indicated by the manufacturer was 350 °C for 15 min. Table 1 lists composition of the specimen alloy in weight percent, which was supplied by the manufacturer. The sum of the Zn and Ir content was 1 wt.%. Therefore, the atomic ratio for these elements was converted from the values that were set to 0.5 wt.%, respectively.

Heat treatment

The specimens were subjected to a solution treatment to obtain a supersaturated solid solution of a single phase at 750 °C for 15 min in a vertical furnace under an argon atmosphere to inhibit oxidation, and then quenched in ice brine to prevent the formation of equilibrium phases. Subsequently, as recommended by the manufacturer, the solution-treated specimens were subjected to an isothermal aging treatment at 350 °C for various times in a molten salt

bath (25 KNO_3+30 KNO_2+25 $NaNO_3$+20 $NaNO_2$, wt.%), and then quenched in ice brine.

Hardness test

To investigate the hardening and softening behaviour during the aging process, the hardness of the heat-treated plate specimens was measured using a Vickers micro-hardness tester (MVK-H1, Akashi Co., Akashi, Hyogo, Japan) with a 300-gf load and a holding time of 10 s. The result values are the average of five measurements.

XRD study

XRD study was performed to examine the phase transformation during the aging process. Powder specimens with particle sizes below 45 μm were filed using a diamond disc and passed through a 330-mesh screen. The specimens were then mixed with alumina powders with particle sizes of 1 μm to prevent sintering agglomeration during heat treatment. Subsequently, the powder specimens were subjected to vacuum sealing in silica tubes and heat treatment, and the alumina powders were filtered from the heat-treated specimens. XRD profile was recorded by X-ray diffractometer (XPERT-PRO, Philips, Eindhoven, Netherlands) at 30 kV and 40 mA using Ni-filtered Cu kα radiation as incident beam.

FE-SEM observation

FE-SEM micrographs of the heat-treated plate specimens were taken to examine the microstructural changes during the aging process. The heat-treated specimens were treated with a polisher and etched with an aqueous solution containing KCN of 10 % (potassium cyanide) and $(NH_4)_2S_2O_8$ (ammonium persulfate) of 10 %. The surfaces of the heat-treated specimens were observed by FE-SEM (JSM-6700 F, Jeol, Akishima-shi, Tokyo, Japan) at 15 kV.

EDS analysis

The EDS (INCA X-Sight, Oxford Instruments Ltd., Oxford, UK) profile of the heat-treated plate specimen was recorded at 15 kV to examine the changes in the elemental distribution during the aging process.

Results and discussion

Age-hardening behaviour

Figure 1 shows the isothermal age-hardening curve of the plate specimen solution-treated at 750 °C for 15 min and aged at 350 °C for various times (~20,000 min). The hardness

Table 1 Composition of the specimen alloy

Composition	Au	Ag	Cu	Pd	Zn	Ir
wt.%	50.0	34.0	12.0	3.0	<1	<1
at.%	31.88	39.58	23.71	3.54	0.96	0.33

Fig. 1 Isothermal age-hardening curve of the plate specimen solution-treated at 750 °C for 15 min and then aged at 350 °C for various times until 20,000 min

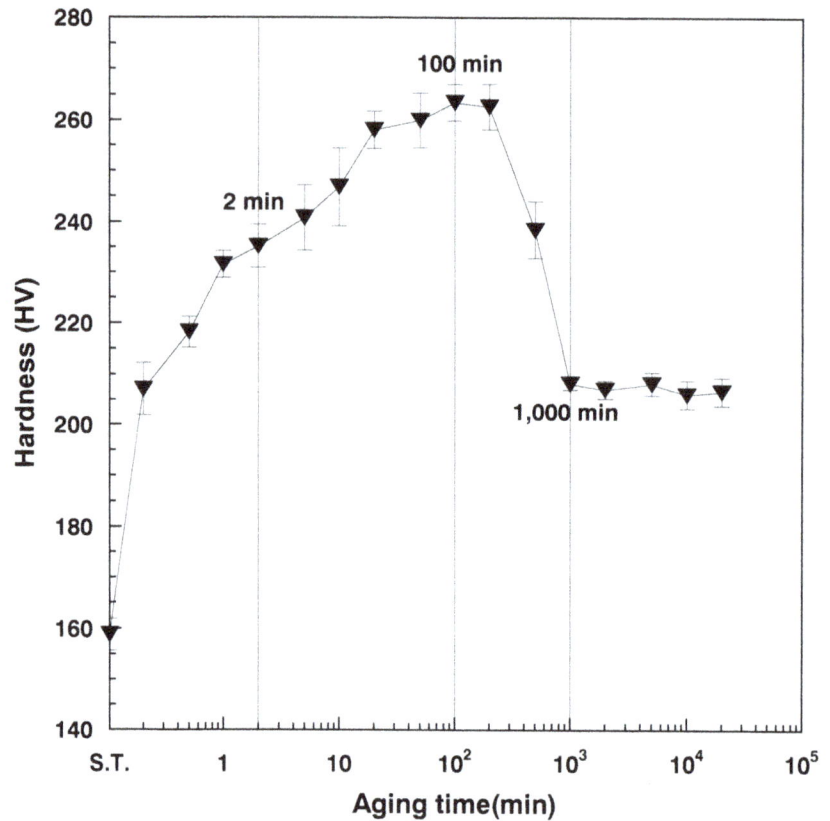

increased rapidly within only 12 s without an incubation period. At an aging time of 2 min, the hardness increased to 235 HV, which was approximately 1.5 times higher than that of the solution-treated specimen (159 HV). Subsequently, the increasing rate of the hardness decreased gradually and finally, the maximum value (263 HV) was obtained at an aging time of 100 min. After maintaining the maximum value for 200 min, the hardness decreased rapidly to 208 HV until 1,000 min, after which, there was no further decrease in hardness until an aging time of 20,000 min.

Phase transformation

Figure 2 presents the XRD profile of the powder specimens solution-treated at 750 °C for 15 min and aged at 350 °C for various times (~20,000 min). In the XRD patterns of the specimen solution-treated at 750 °C for 15 min, a single α_0 phase with a face-centered cubic (f.c.c.) structure and a lattice constant of $a_{200}=3.97$ Å was observed. By aging the specimen isothermally at 350 °C for 20,000 min, the α_0 phase was separated into an f.c.c. Ag-rich α_1 phase with a lattice constant of $a_{200}=4.06$ Å, and an AuCu I phase with a face-centered tetragonal (f.c.t.) structure and lattice constants of $a_{200}=3.89$ Å and $c_{001}=3.67$ Å.

In the XRD patterns of the specimen aged at 350 °C for 12 s, both sides of the (111) α_0 diffraction peak broadened as

the phase decomposition was initiated. Compared to the isothermal age-hardening curve, this corresponds to the rapid increase in hardness that occurred within 12 s. In the XRD patterns at an aging time of 2 min, when the hardness increased apparently to 235 HV, the diffraction peaks for the stable Ag-rich α_1 and AuCu I phases were observed clearly at lower and higher angles of the (111) α_0 diffraction peak, respectively. In addition, the diffraction peaks of the metastable Ag-rich α_1' and AuCu I' phases with higher intensity were observed closer at both sides of the (111) α_0 diffraction peak. Furthermore, side bands appeared much closer at both sides of the (111) α_0 diffraction peak, and moved closer with increasing aging time, as observed from the vertical solid lines in Fig. 2 [2, 3, 6, 7]. The appearance of the side bands and rapid increase in hardness suggest that separation of the parent α_0 phase occurred by spinodal decomposition, not by a nucleation and growth mechanism at 350 °C, and the apparent increase in hardness until 2 min was caused mainly by the formation of the metastable Ag-rich α_1' and AuCu I' phases through this process.

In the XRD patterns at an aging time of 10 min, when the hardness increased to 250 HV, the superlattice (110) AuCu I diffraction peak ($2\theta=32.6°$) appeared as atomic ordering progressed. In the diffraction peak of the Ag-rich phase, the (111) peak intensity of the stable phase (α_1) was stronger than that of the metastable phase (α_1'), but the (200) peak intensity

Fig. 2 XRD profile of the powder specimens solution-treated at 750 °C for 15 min and aged at 350 °C for various times until 20,000 min. (prime symbol (′): metastable phase, I: AuCu I)

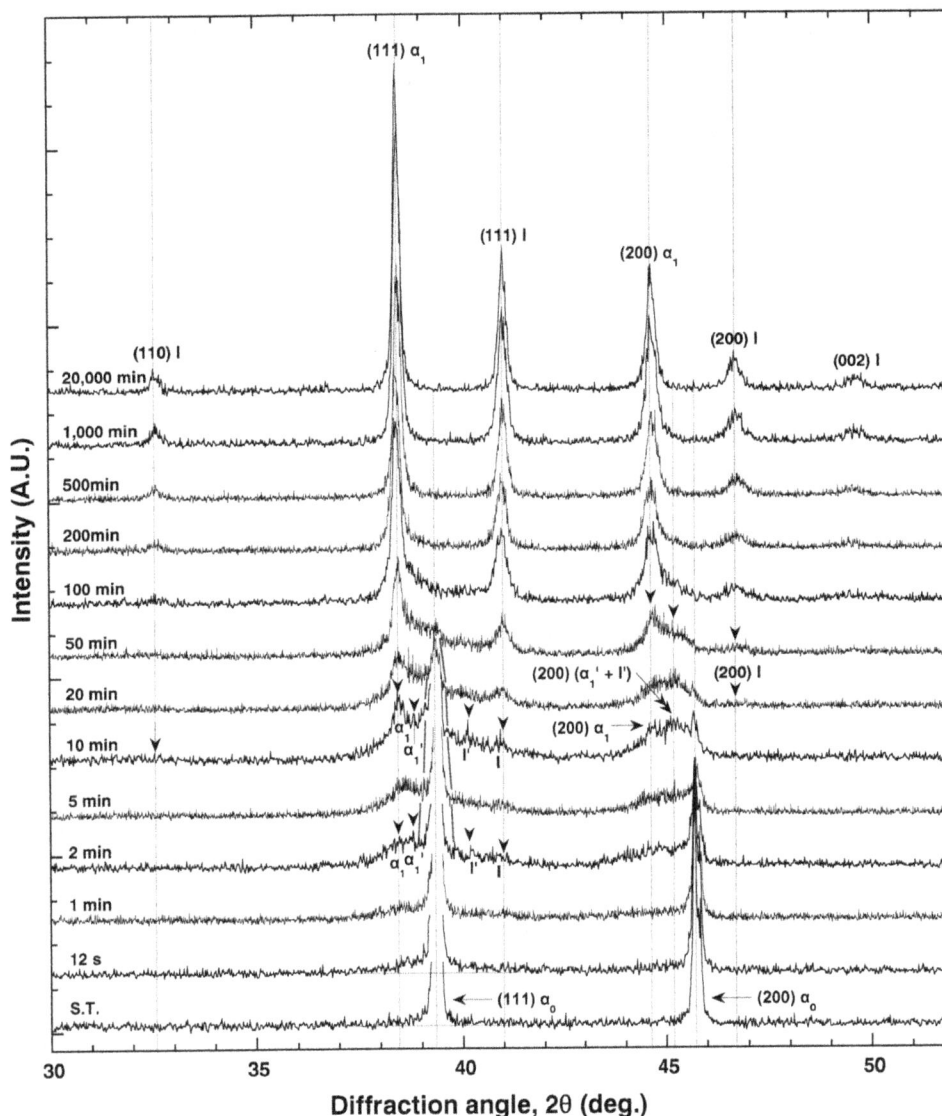

of the stable phase (α_1) was weaker than that of the metastable phase (α_1'). In addition, the (200) AuCu I′ diffraction peak was not observed, even though the intensity of the (111) AuCu I′ diffraction peak was relatively strong. From such a fact, the (200) diffraction peaks of the metastable Ag-rich α_1' and AuCu I′ phases appeared to be superimposed, as marked by a double arrow. Thus, these phases must have had coherent interfaces at the (200) plane [8]. Therefore, the hardening resulted from the coherency lattice strain that occurred along the a-axis between the metastable Ag-rich α_1' and AuCu I′ phases, in addition to the lattice distortion that occurred along the c-axis between the stable Ag-rich α_1 and AuCu I phases by the tetragonality of the AuCu I ordered phase.

In the XRD patterns at an aging time of 20 min, when the hardness value slightly increased to 258 HV, the (200) AuCu I diffraction peak appeared at a higher diffraction angle ($2\theta = 46.7°$) than the common (200) (Ag-rich α_1'+AuCu I′) diffraction peak. At an aging time of 50–200 min, the intensity

of the (200) diffraction peaks for the stable Ag-rich α_1 and AuCu I phases became stronger at both sides of the common (200) (Ag-rich α_1'+AuCu I′) diffraction peak, as marked by the arrow. Therefore, the coherency along the (200) plane was lost as the metastable phases transformed to the stable Ag-rich α_1 and AuCu I phases.

Subsequently, the diffraction peaks of the metastable Ag-rich α_1' and AuCu I′ phases disappeared, and there were no further changes in the XRD patterns after 1,000 min. The hardness increased at a decreasing rate during the formation of the stable Ag-rich α_1 and AuCu I phases from the metastable state. Therefore, the hardening effect by the transformation of the stable Ag-rich α_1 and AuCu I phases from the metastable state was weaker than that by the spinodal decomposition of the parent α_0 phase to the metastable Ag-rich α_1' and AuCu I′ phases. This resulted from the microstructural changes related to the phase transformation, as is discussed in the "Microstructural changes" section.

Figure 3 presents the full width at half maximum (FWHM) graph of the powder specimens solution-treated at 750 °C for 15 min and aged at 350 °C for various times (~20,000 min). The FWHM of the parent α_0 phase increased as the diffraction peaks of the metastable Ag-rich α_1' · AuCu I' and stable Ag-rich α_1 · AuCu I phases appeared. Subsequently, the FWHM of the parent α_0 phase decreased as the α_0 phase disappeared. The FWHM of the stable Ag-rich α_1 and AuCu I phases reached the maximum value at an aging time of 20 and 100 min, respectively. The FWHM then declined to the minimum value at an aging time of 500 and 1,000 min, respectively. This suggests that the changing aspect of the FWHM was similar to that of the isothermal age-hardening curve. The increase in the FWHM and its subsequent decrease by a phase transformation indicate the formation of lattice strain and its subsequent release in the alloy [9–13]. Therefore, it is considered that both the stable Ag-rich α_1 and AuCu I phases were related to hardening and subsequent softening.

Microstructural changes

Figure 4 shows FE-SEM micrographs of the plate specimens solution-treated at 750 °C for 15 min (a) and aged at 350 °C for 10 min (b), 100 min (c), 500 min (d), 1,000 min (e) and 20,000 min (f) at magnifications

of ×2,000 (1), ×8,000 (2) and ×20,000 (3). In the specimen solution-treated at 750 °C for 15 min (a), an equiaxed structure of a single phase was observed. In the specimen aged at 350 °C for 10 min (b), parts of the grain boundaries were replaced with lamellar structures, whereas no apparent changes were observed in the grain interior. In the corresponding XRD patterns, five phases of the parent α_0, metastable Ag-rich α_1' · AuCu I' and stable Ag-rich α_1 · AuCu I were detected. From the fact that the diffraction peaks of the stable Ag-rich α_1 and AuCu I, which are the final product phases, appeared within a short period of aging time (12 s), it is thought that the lamellar structures in the grain boundaries were transformed directly from the parent α_0 phase to the stable Ag-rich α_1 and AuCu I phases without a metastable state considering that the atomic diffusion in the grain boundaries occurs relatively fast [14]. The peak intensity of the stable phases was relatively high in the XRD patterns at 10 min, whereas the ratio of the lamellar structures in the matrix was relatively low in the corresponding FE-SEM micrographs. Therefore, it is thought that parts of the grain interior were also composed of the stable Ag-rich α_1 and AuCu I phases. Hence, the grain interior was in the process of spinodal decomposition from the parent α_0 phase to the metastable Ag-rich α_1' and AuCu I' phases, which had coherency, or was in the process of a subsequent transformation from a metastable state to the stable Ag-rich α_1 and AuCu I phases.

Fig. 3 FWHM graph obtained from the (111) diffraction peaks of the α_0, stable Ag-rich α_1 and AuCu I phases with aging at 350 °C

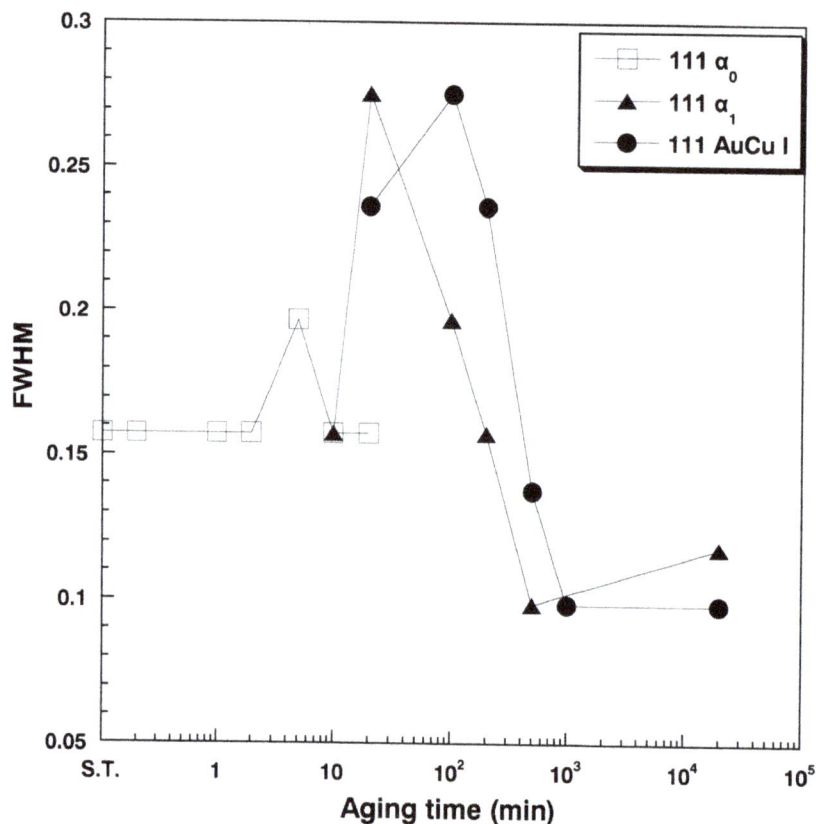

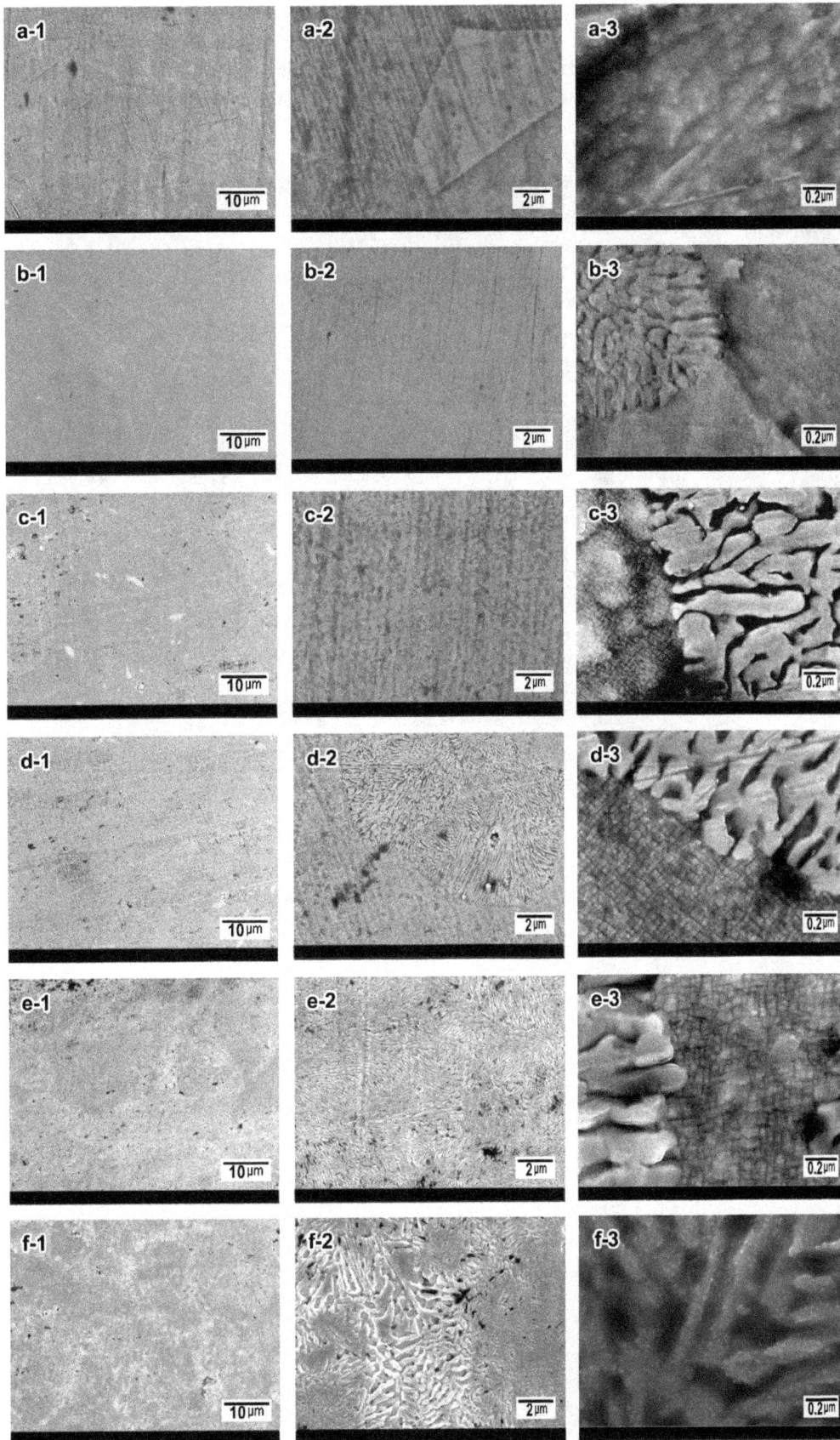

Fig. 4 FE-SEM micrographs of the plate specimens solution-treated at 750 °C for 15 min (**a**) and aged at 350 °C for 10 min (**b**), 100 min (**c**), 500 min (**d**), 1,000 min (**e**) and 20,000 min (**f**) at magnifications of×2,000 (*1*), ×8,000 (*2*) and×20,000 (*3*)

In the specimen aged at 350 °C for 100 min (c), the growth of lamellar structures was barely observed at the grain boundaries, and the grain interior was replaced with fine cuboidal structures surrounded by a solute-depleted matrix.

Such fine and uniform structures are characteristic that are obtained through spinodal decomposition [3, 5]. The XRD patterns at an aging time of 10–100 min showed that the transformation of the metastable Ag-rich α_1' and AuCu I'

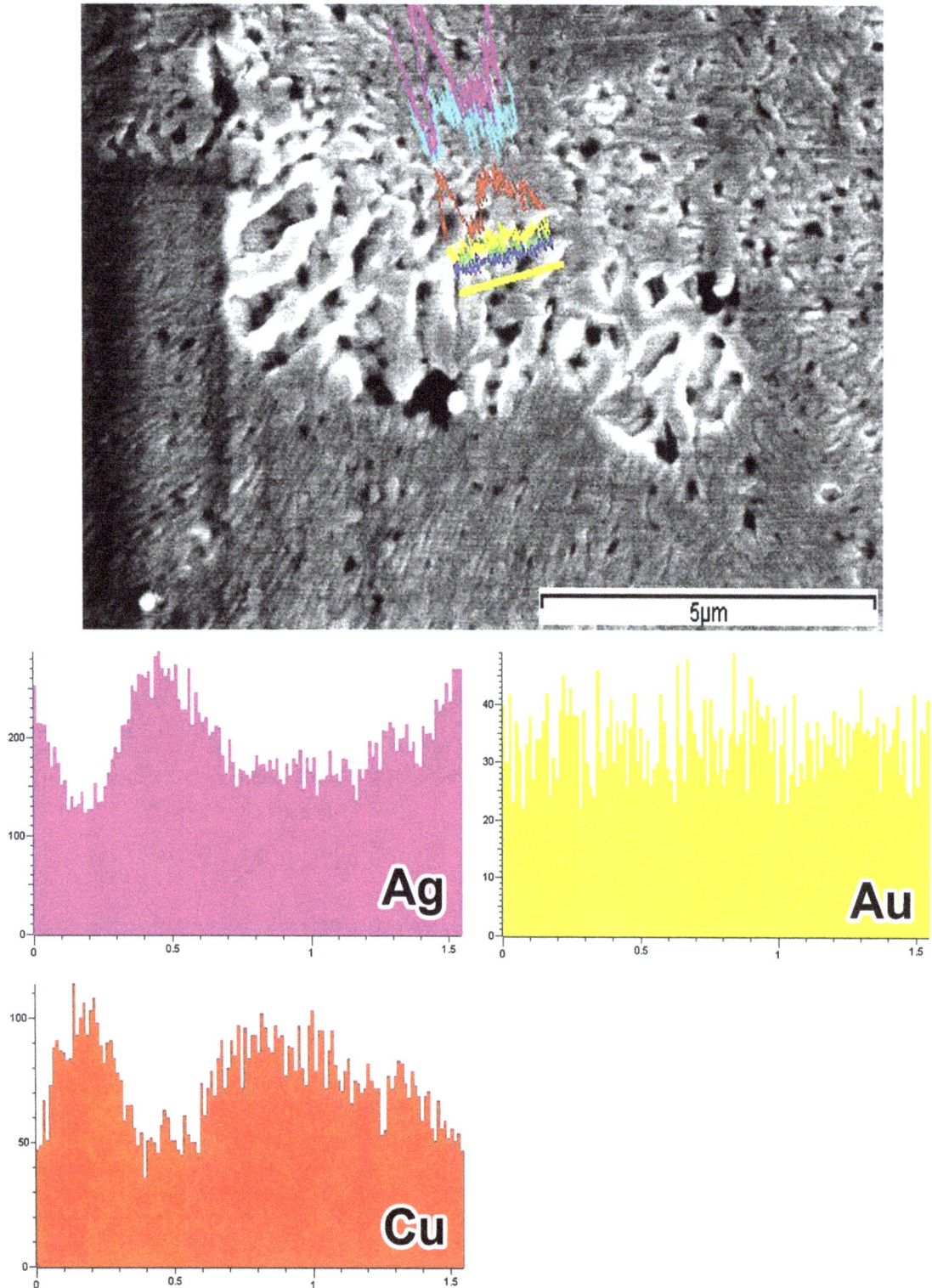

Fig. 5 FE-SEM micrograph and EDS line profile of the plate specimen aged at 350 °C for 20,000 min after the solution treatment at 750 °C for 15 min

phases to the stable state was almost complete. Therefore, it is considered that not only the lamellar structures at the grain boundaries but also the cuboidal structures and surrounding matrix were composed of the stable Ag-rich α_1 and AuCu I phases, and their interfaces lost coherency. The cuboidal structures with incoherent interfaces observed at 100 min were much coarser than the grain interior structures that were observed at 10 min. A comparison with the isothermal age-hardening curve showed that this period corresponded to the stage that the hardness increased at a decreasing rate. Therefore, it is considered that the formation of these cuboidal structures compensated for the increase in internal lattice strain resulted from the increased gap in lattice parameters through the phase transformation from the metastable to stable state.

In the specimen aged at 350 °C for 500 min (d), the cuboidal structures became slightly coarser, and approximately half of the matrix was replaced with the lamellar structures formed by the grain boundary reaction. A comparison with the isothermal age-hardening curve showed that this period corresponded to the stage where a rapid decrease in hardness was observed. Therefore, the replacement of the fine cuboidal structures with the coarser lamellar structures without a phase transformation resulted in the release of internal lattice strain by reducing the interfaces between the two phases of the stable Ag-rich α_1 and AuCu I. This can be supported by a study of a low-carat gold alloy, which showed that an extremely restricted grain boundary reaction apparently retarded the softening of the spinodally decomposed cuboidal structures [5].

In the specimen aged at 350 °C for 1,000 min (e), the lamellar structures did not coarsen further, but they grew and covered most of the matrix. The cuboidal structures were left in an extremely limited area. A comparison with the isothermal age-hardening curve showed that this period corresponded to the stage that the hardness declined to the minimum value. Therefore, the decreasing rate of hardness is thought to be in proportion to the amount of the lamellar structures. In the specimen aged at 350 °C for 20,000 min (f), a second grain boundary reaction was initiated to form coarser lamellar structures by consuming the preformed fine lamellar structures from the grain boundaries. Because it occurred after the phase transformation was complete, it is considered that a second grain boundary reaction occurred to reduce the interface energy, regardless the phase transformation. A second grain boundary reaction in dental alloys was reported to cause a decrease in hardness [15]. On the other hand, further softening was not observed in the present study due to the slow progress of the reaction.

Elemental distribution

Figure 5 presents FE-SEM micrograph of the plate specimen aged at 350 °C for 20,000 min after the solution treatment, and

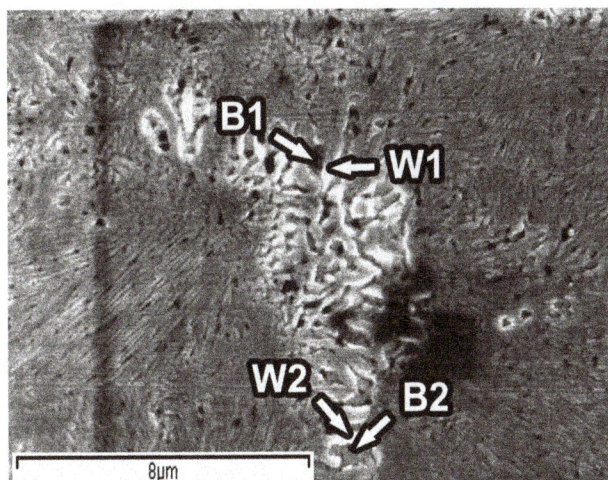

Fig. 6 FE-SEM micrograph of the plate specimen aged at 350 °C for 20,000 min after the solution treatment at 750 °C for 15 min

its EDS line profile obtained by crossing the coarsened lamellar structure. The result showed that the distribution of Ag and Cu was inversely proportional to each other whereas the Au content was relatively homogeneous.

The elemental distribution of the lamellar structure was detected more precisely by EDS point analysis (Fig. 6). Table 2 lists the EDS point analysis result for each neighbouring layer of the lamellar structures (black:B/white:W). The result showed that the coarse lamellar structure consisted of alternate Ag-rich and Cu-rich layers, and the Pd and Zn content was concentrated in the Cu-rich layer. These minor ingredients resulted in a dissimilarity in the lattice constants of the AuCu I phase with the values ($a=3.966$ Å, $c=3.673$ Å) reported by Villars and Calvert [16]. The contrary distribution of Ag and Cu resulted from the solubility limit for each other, whereas Au was soluble with Ag and Cu at all atomic ratios [17]. In the present study, at the point where the maximum hardness was obtained, the grain interior had fine and uniform structures that were similar to the single-phased structure due to spinodal decomposition. In addition, the grain boundary reaction was restrained until that point. Such aspects of the alloy will be beneficial to tarnish and corrosion resistance as well as biocompatibility for use in dental practice.

Table 2 Point analysis result by EDS of the specimen at the regions indicated in Fig. 6

Region (at.%)	Au	Ag	Cu	Pd	Zn	Ir
B1	33.0	54.2	12.8	0	0	0
B2	33.3	57.3	9.4	0	0	0
W1	30.6	6.1	55.0	4.8	3.5	0
W2	28.7	3.5	60.1	4.6	3.1	0

Conclusions

This study examined the spinodal decomposition related to age-hardening and cuboidal structures in a dental low-carat gold alloy with a relatively high Cu/Ag content ratio, and the following results were obtained.

(1) Separation of the parent α_0 phase occurred by spinodal decomposition during aging at 350 °C after the solution treatment at 750 °C, not by a nucleation and growth mechanism, resulting in the formation of the stable Ag-rich α_1 and AuCu I phases through a metastable state.

(2) Hardening resulted from the coherency lattice strain that occurred along the a-axis between the metastable Ag-rich α_1' and AuCu I′ phases, in addition to the lattice distortion that occurred along the c-axis between the stable Ag-rich α_1 and AuCu I phases due to the tetragonality of the AuCu I ordered phase.

(3) The transformation of the stable Ag-rich α_1 and AuCu I phases from the metastable state induced the formation of the fine and uniform cuboidal structures, which compensated for the increased gap in lattice parameters through a phase transformation.

(4) The replacement of fine cuboidal structures with coarser lamellar structures occurred without a phase transformation, and resulted in softening by the reducing interfaces between the stable Ag-rich α_1 and AuCu I phases.

Acknowledgements This research was supported by Basic Science Research Program through the National Research Foundation of Korea (NRF) funded by the Ministry of Education, Science and Technology (grant number: 2011–0010995).

References

1. Yasuda K (1987) Age-hardening and related phase transformations in dental gold alloys. Gold Bull 20:90–103
2. Hamasaki K, Hisatsune K, Udoh K, Tanaka Y, Iijima Y, Takagi O, Naruse S (1998) Ageing behaviour in a dental low-gold alloy with high copper content. J Mater Sci Mater Med 9:213–219
3. Nakagawa M, Yasuda K (1988) Age-hardening and the associated phase transformation in an Au–55.2at% Cu–17.4at% Ag ternary alloy. J Mater Sci 23:2975–2982
4. Hisatsune K, Udoh K, Sosrosoedirdjo BI, Tani T, Yasuda K (1991) Age-hardening characteristics in an AuCu–14at.% Ag alloy. J Alloys Compd 176:269–283
5. Park MG, Yu CH, Seol HJ, Kwon YH, Kim HI (2008) Age-hardening behaviour of a spinodally decomposed low-carat gold alloy. J Mater Sci 43:1539–1545
6. Udoh K, Fujiyama H, Hisatsune K, Hasaka M, Yasuda K (1992) Age-hardening associated with ordering and spinodal decomposition in a AgCu–40at% Au pseudobinary alloy. J Mater Sci 27:504–510
7. Yasuda K, Hisatsune K (1993) Microstructure and phase transformations in dental gold alloys. Gold Bull 26:50–66
8. Lee JH, Yi SJ, Seol HJ, Kwon YH, Lee JB, Kim HI (2006) Age-hardening by metastable phases in an experimental Au–Ag–Cu–Pd alloy. J Alloys Compd 425:210–215
9. Douglass DL (1969) Spinodal decomposition in Al/Zn alloys. J Mater Sci 4:130–137
10. Tanaka Y, Udoh K, Hisatsune K, Yasuda K (1998) Early stage of ordering in stoichiometric AuCu alloy. Mater Trans JIM 39:87–94
11. Kawashima I, Ohno H, Sarkar NK (2000) Effect of Pd or Au addition on age-hardening in AgMn-based alloys. Dent Mater 16:75–79
12. Shiraishi T, Ohta M (2002) Age-hardening behaviors and grain boundary discontinuous precipitation in a Pd-free gold alloy for porcelain bonding. J Mater Sci Mater Med 13:979–983
13. Suryanarayana C, Norton MG (2006) X-ray diffraction - a practical approach, 1st edn. Springer Science & Business Media, New York, p 89
14. Hisatsune K, Udoh K, Nakagawa M (1990) Aging behavior in a dental low carat gold alloy and its relation to CuAu II. J Less-Common Metals 160:247–258
15. Ohta M, Hisatsune K, Yamane M (1975) Study on age-hardenable silver alloy III on the ageing process of dental Ag Pd Cu Au alloy. J Jpn Soc Dent Appar Mater 16:87–92
16. Villars P, Calvert LD (1985) Pearson's handbook of crystallographic data for intermetallic phases, American Society for Metals, Metals Park, pp 416–417
17. Massalski TB (1990) Binary alloy phase diagrams, 2nd edn. ASM International, Materials Park, pp 12–13, 28–29, 358–362

18

Soldering of non-wettable Al electrode using Au-based solder

Fengqun Lang · Hiroshi Nakagawa · Hiroshi Yamaguchi

Abstract In manufacturing three-dimensional SiC power modules, the Al electrode of SiC power devices should be soldered to the substrate. However, the Al electrode is difficult to be bonded by a solder due to the naturally formed aluminum oxide on it. In this paper, we describe an effective approach for soldering the non-wettable Al electrode by fabricating a Au-stud bump in the Al electrode together with a Au-20 wt% Sn or Au-12 wt% Ge solder. The soldering initiated at the Au bump and spreaded on the Al electrode. The soldering featured as reactive wetting, realized by the reaction of liquid Au in the Au-base solder and the Al electrode. The activation energy of the Au-20 wt% Sn soldering the Al electrode was $Q=159$ kJ/mol. A continuous Au_4Al layer formed at the Au-20 wt% Sn bond interface. The shear strength exceeded 60 MPa, ~1 order magnitude higher than the required shear strength. For the bond with Au-12 wt% Ge solder on the Al electrode with a Au bump, the liquid Au-12 wt% Ge solder reacted with the solid Al electrode and formed a Au-Ge-Al solid solution after solidification. The shear strength of the Au-12 wt% Ge solder on the Al electrode with a Au bump was beyond 50 MPa. Little electrical characteristics of the SiC-SBD changed after the Al electrode was bonded to a circuit substrate using this technology.

Keywords Au-Sn alloy · Au-Ge alloy · Wetting · Soldering · Aluminum electrode · Metallization

F. Lang (✉) · H. Nakagawa · H. Yamaguchi
Advanced Power Electronics Research Center, National Institute of Advanced Industrial Science and Technology (AIST),
Tsukuba 305-8568, Japan
e-mail: fqlang03325@yahoo.co.jp

Present Address:
F. Lang
R & D Partnership for Future Power Electronics Technology, c/o, AIST, Tsukuba, Japan

Introduction

Development of the next-generation energy-saving and miniaturized power converters using the newly developed low-power conversion loss and high-temperature resistant SiC power devices is under way worldwide [1]. The SiC power devices can operate above 200 °C and can withstand reflow temperature of 500 °C [2]. The upside electrodes of almost all the SiC and Si power devices are metalized with Al or its alloys for its desirable properties such as ease of patternability, high conductivity, and good adherence [3, 4]. At present, two-dimensional (2-D) fabrication technology is the mainstream in power electronics manufacture with Si power devices. The Al electrodes are connected to the electrode in the substrate or other devices with Al or Au wires using ultrasonic energy and a pressure [3, 5, 6]. There are urgent requirements of miniaturized power modules. Three-dimensional (3-D) integration of power modules using SiC power devices makes it possible to miniaturize the volume of power modules by providing double-side heat dissipation and reducing electrical noise [7, 8]. In fabricating a 3-D power module, the upper and lower side electrodes of power devices should be soldered to the substrates. However, Al is very reactive with atmospheric oxygen, and a thin and chemically stable Al oxide film (~4 nm thickness) immediately forms on any exposed aluminum surface [9, 10]. This aluminum oxide film hinders the bonding of a solder to the Al electrode.

Although solderablity of the Al electrode can be realized by photolithography wafer process combined with electroplating, evaporation, zincate pretreatment, and Ni electroless deposition followed by Au electroless plating, single chip with Al electrode is unsuitable [11, 12]. Further, the process is complex and cost-ineffective. It is difficult to apply electroplating or electroless plating process or physical vapor deposition to deposit solderable metal pads on single chip. However, most of the devices are supplied as discrete chips. The

Sn3.5Ag4Ti(Ce,Ca) with a melting point (M.P) of ~220 °C has ever been reported to be able to solder Al alloys [13, 14]. The soldering temperature above 800 °C is so high that could damage the SiC power devices and exceeds the M.P of the Ag-Cu-Ti bond for the Cu pattern layer on the ceramic substrate. Further, the M.P of this solder is too low to endure the operation temperature above 200 °C. Sn-3.5Ag-xTi was used to bond the Al alloys at 250 °C. But a two-step soldering process is needed by brushing the bonded solder pad formed in the first step soldering. Sn-xZn solders (M.P=198.5 °C) have ever been reported to be able to bond the Al alloys [15–17]. The M.P of all these solders is too low to be used in high temperature electronics (>200 °C). Au-20 wt% Sn eutectic solder (M.P=278 °C) and Au-12 wt% Ge eutectic solder (M.P=356 °C) are extensively used in high-power electronics and optoelectronics packaging. These Au-based solders feature as good mechanical properties, high electrical and thermal conductivity along with excellent corrosion resistance [18, 19]. In this paper, we describe a new method for soldering an Al electrode using Au-based solders together with fabricating a tinny Au bump (~80 μm diameter) in the Al electrode. The reactive wetting between the Au-based solders and the Al electrode, the wetting rate, the active energy of the reactive wetting, and the bond strength were investigated.

Experimental procedures

SiC Schottky Barrier Diode (SBD) power device (reverse voltage 1,200 V; continuous current 20 A) chips with an Al upside electrode (anode) were used in this experiment. The Al electrode has an area of 2.13 L × 2.13 W (mm^2) and a thickness of 4 μm. Beneath the Al electrode is a thin tungsten (W) layer. A 25-μm-diameter Au-1 wt% Pd wire was used to form the Au bumps in the Al electrode using a Au stud bump bonder (model, SBB-1, Shinkawa Co., Ltd, Japan). The Au stud bumps were prepared with a forming voltage of 2,200 V and a current of 4 mA at 300 °C in air. The Au bump had a base diameter of 82 μm. Since the surficial status of the Al electrode has strong influence on the spreading behavior of a liquid alloy [20], the Al electrode without heavy scratches were used.

Solder pastes are commonly used in electronics manufacturing for their flexibility in bonding pads with different sizes and configurations, in addition to the surface cleaning effect of the mixed flux. The Au-20 wt% Sn (M.P, 278 °C, hereafter denoted as Au-20Sn) eutectic solder paste was used to bond the Au-bumped Al electrode in this research work. The Au-20Sn solder paste contains Au-20Sn particles with diameters <32 μm. The Au-20Sn paste was dispensed on the Au-bumped Al electrode of SiC-SBD. Then, the SiC-SBD device was placed in a reflow analyzer system (Model:VISTA

7, Okuhara Electric Cop., Japan; maximum reflow temperature, 350 °C). The reflow was performed in N$_2$ atmosphere (oxygen content <16 ppm). The ramp-up rate from soak temperature to reflow temperature was 1 °C/s. The cooling rate after reflow was ~3 °C/s. The reflow temperature was set to be 300, 320, 330, and 340 °C, respectively. Four milligrams solder paste was used for each reflow temperature. The images of the solder ball were recorded by a CCD camera during reflow. The area of the image of the solder ball was measured using Bersoft Image Measurement software. The radius of the image was calculated from the measured area. The bond strength of the solder to the Al electrode was measured using a die shear tester (Model 4000, Dage Holdings, Ltd., UK). The tests were carried out at room temperature with a displacement rate of 150 μm/s. To investigate the cross-sectional microstructure of the bond of the solder on the Al electrode, some of the soldered specimens were fractured. The microstructure of the specimens was also observed and analyzed with the scanning electron microscope (SEM, LEO/EEiSS GEMINI FE-SEM, Model Supra 35, Carl Zeiss SMT Ltd., Germany) equipped with an energy dispersed X-ray analyzer (EDX, Genisis Spectrum, Model LEO 35, EDXA Ltd., USA).

For comparison, the Pb-10 wt% Sn high temperature solder paste (solidus temperature, 275 °C; hereafter termed as Pb-10Sn) and Sn-3Ag-0.5Cu (wt%) eutectic solder paste (M.P, 217 °C; hereafter termed as Sn3Ag0.5Cu) were used to solder the *single Au-bumped Al electrode* (hereafter denoted as SA-BA), respectively. The reflow was carried out with the reflow analyzer system. The reflow peak temperature and time for Pb-10Sn and Sn-3Ag-0.5Cu were 350 °C for 200 s and 250 °C for 200 s, respectively.

The Au-bumped Al electrode was also bonded using a high temperature solder Au-12 wt% Ge eutectic solder paste (M.P, 356 °C; hereafter denoted as Au-12Ge). This paste contains Au-12Ge alloy particles with diameters <32 μm. The reflow process was carried out at 430 °C for 180 s with a high temperature vacuum reflow system (Model SRO-704, ATV Technology GmbH, Germany; maximum reflow temperature, 500 °C). The SiC-SBDs with the SABA were bonded to a Cu/Ni (P)-metalized Si$_3$N$_4$ substrate with the Au-12Ge solder paste in a N$_2$ atmosphere. The shear strength of the bond was tested at room temperature. The cross sections of the bond were observed and analyzed with SEM/EDX.

In order to investigate the alloying effect of the Au-based solders with the Al electrode, Au-20Sn and Au-12Ge solder sheets were used to bond the SABA by a reflow process without flux. The reflow was performed in the vacuum reflow system. The reflow was conducted at 350 °C for 120 s and 430 °C for 120 s for the Au-20Sn solder and Au-12Ge solder sheets, respectively. The surficial and fracture morphologies after reflow or after shear test were investigated with SEM/EDX.

To evaluate the electrical characteristics of the SiC-SBD after its SABA was bonded to a Cu(300 μm)/Ni (P)(5 μm)/Au (0.05 μm)-metalized Si_3N_4 (320 μm) ceramic substrate (the value in the parenthesis represents the thickness), the forward current–forward voltage (I_F–V_F) curves and reverse current–reverse voltage (I_R–V_R) curves were measured with a high-power curve tracer (model, 371A, Tektronix, Inc., USA). To protect the guard ring on the periphery of the anode of the SiC-SBD, the SABA was bonded to a Au-top-metalized Cu electrode spacer with a size of 1 L×1 W×0.4 T (in millimeter). The Cu electrode spacer was fabricated on the Cu/Ni (P)/Au-metalized Si_3N_4 ceramic substrate with Au-12Ge solder. Figure 1 shows the schematic of the experimental conditions and methods of the electrical test. The anode and the cathode of the electrical power were connected to the anode and cathode of the SiC-SBD, respectively (Fig. 1a). The voltage was applied and increased. The measured I–V curve was the I_F–V_F curve, which gives the on-state resistance (Ron) from the slope of the curve. Reversely, the cathode and anode of the electrical power were connected to the anode and cathode of the SiC-SBD, respectively (Fig. 1b). The measured I–V curve was the I_R–V_R curve, which represents the capability of withstanding the reverse voltage.

Results

Bonding with Au-20Sn paste

Figure 2 shows the morphology of the SABA after reflow with different solders. Panels a and b of Fig. 2

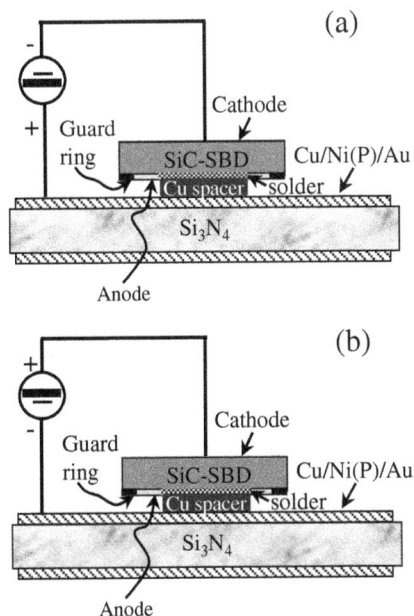

Fig. 1 Schematic of measuring the electrical properties of the bonded SiC-SBD with SABA

present the morphology of the SABA after reflow with Pb-10Sn paste at 350 °C for 200 s and Sn-3.5Ag-0.5 Cu paste after reflow at 250 °C for 200 s, respectively. Upon reaching the solidus temperature of the Pb-10Sn solder or the M.P of the Sn-3.5Ag-0.5Cu solder, the solder melted and formed a ball in the Au bump. Then, the solder ball moved freely on the Al electrode, without any bonding to the Al electrode. During reflow, the Au bump was dissolved into the solder, left a Au bump's root in the Al electrode. This means that Pb-10Sn and Sn-3Ag-0.5Cu solder pastes could not bond the SABA.

Panels c and d of Fig. 2 show the morphologies of the Al electrode without and with a Au bump after reflow with the Au-20Sn solder at 340 °C for 60 s, respectively. As shown in Fig. 2c, the Au-20Sn solder does not bond the Au bump-free Al electrode. During reflow, the molten solder formed a metallic ball and randomly moved on the surface of the Al electrode, without bonding to the Al electrode. However, for the Al electrode with single Au stud bump in its center, a Au-20Sn metallic ball formed upon the solder melted and was immediately trapped by the Au bump. The soldering initiated at the Au bump and then spreaded on the Al electrode.

Figure 3 illustrates the shear strength of the Au-20Sn solder on the Al electrode as a function of number of the Au bumps. The shear strength was calculated through dividing the shear force by the bonding area, which was measured by a digital microscope. Three samples were used for each number of Au bumps. The shear strength was not significantly affected by the number of Au stud bumps. This is due to, as will be shown in Figs. 4 and 5, the spreading rate of the Au-20Sn solder on the SABA at 340 °C was high and a Au bump was sufficient to complete the reaction. The shear strength exceeds 60 MPa, ~10 times higher than the required strength limit [21].

Figure 4a shows the cross-sectional fracture morphology of the Al electrode with a Au bump bonded with the Au-20Sn solder. The soldering was carried out at 340 °C for 60 s. The Au in the liquid solder reacted with the Al electrode forming a Au_4Al intermetallic compound (IMC) at the interface between the solder and the W layer. The Au_4Al layer exhibits a columnar structure. Reaction of Au with Al resulted in depletion of Au in the Au-20Sn solder. The AuSn (δ phase) in the solder is a hypereutectic phase, formed mainly due to the deviation from the eutectic composition.

Figure 4b presents the interfacial morphology of the solder and the SiC-SBD after shear test. The Au_4Al layer was observed at the Au-20Sn/W interface. Au_4Al IMC was also observed at the center of the interface (Fig. 4c). The spreading of the solder on the Al electrode is determined by the lateral

Fig. 2 Surficial morphology of
the Al electrode after reflow with
different solders: **a** Pb-10Sn on
the SABA after reflow at 350 °C
for 200 s; **b** Sn-3.5Ag-0.5 Cu
solder on the SABA after reflow
at 250 °C for 200 s; **c** Au-20Sn
solder on the Al electrode without
a Au bump after reflow at 340 °C
for 60 s; **d** Au-20Sn solder on the
SABA after reflow at 340 °C for
60 s

growth of Au$_4$Al layer. Clearly, the soldering of Au-20Sn solder on the Al electrode featured as reactive wetting.

Figure 5 depicts the interfacial morphology of the Au-20Sn solder and the SABA after shear test. The size of the Au$_4$Al grains was ~1 μm. The Au-Sn solder, which was identified by SEM/EDX, was observed in the Au$_4$Al grain boundaries.

Variations of the base radius of the Au-20Sn solder ball as a function of reflow time at various temperatures are shown in Fig. 6a. The reactive spreading rate generally exhibits a linear law [20, 22–24]. The solid lines plotted in the figure are the least-square fits of the data. The spreading rates are revealed in the slope of the lines. The spreading rate increases rapidly with increasing soldering temperature. This indicates that the

spreading of Au-20Sn solder on the SABA is a thermally activated process. The spreading rate for a reactive wetting is described as Eq. (1) [20, 24].

$$dR/dt = kexp(-Q/R*T) \tag{1}$$

Where dR/dt is the spreading rate, k is a constant, Q is the activation energy for the spreading, $R*$ is the universal gas constant, T is the temperature in Kelvin. The spreading rates were plotted with $1/T$, as shown in Fig. 6b. The soldering temperature has strong effect on the spreading rate. The spreading rate increased approximate 1 order magnitude when the soldering temperature was increased from 320 to 340 °C. At a lower reflow temperature as shown in our previous work [25], because of the low reaction rate of Au-20Sn solder with Al electrode, several hundreds of Au bumps were necessary to reach a high bond strength. This was cost-ineffective. At a higher reflow temperature as described in this work, a single Au bump was enough to reach a high bond strength due to the high reaction rate between the solder and the Al electrode.

The activation energy was calculated by fitting the data in Fig. 6b with Eq. (1). The activation energy is Q=159 kJ/mol. The spreading rate for Au-20Sn solder on the Al electrode can be expressed as Eq. (2) based the data in Fig. 6b, with a unit of meter/second.

$$dR/dt = 1.81E + 8exp(-19190/T) \tag{2}$$

Bonding with Au-12Ge paste

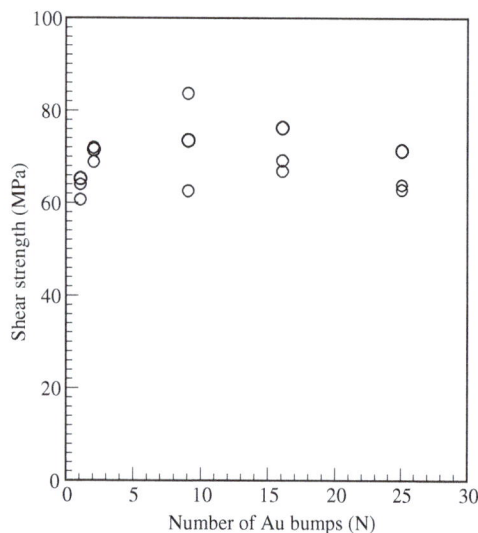

Fig. 3 Shear strength of the Au-20Sn solder on the SABA as a function of number of Au bumps. Reflow was conducted at 340 °C for 60 s

Figure 7 presents the SEM micrograph of a cross section perpendicular to the interface of the Au-12Ge solder on the

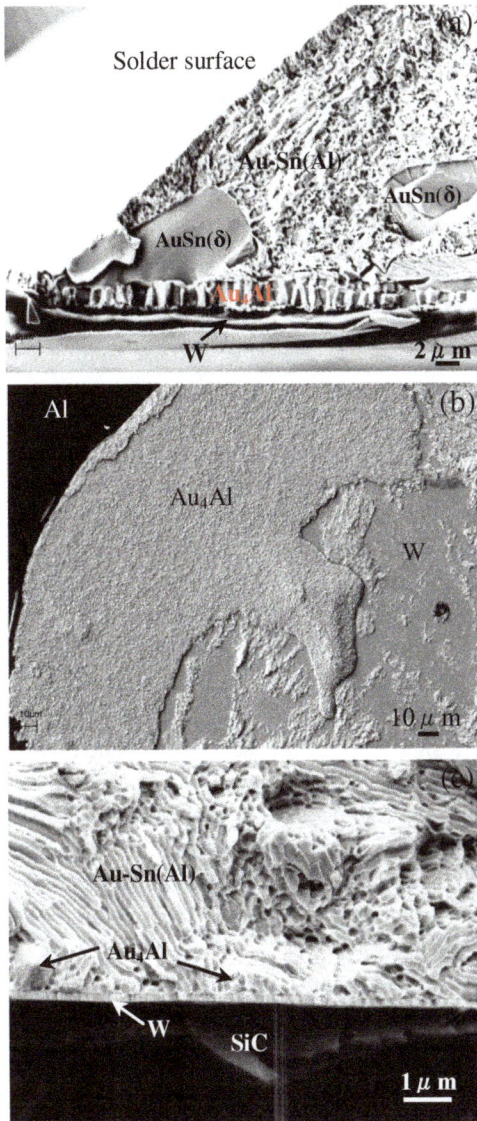

Fig. 4 SEM micrographs of the Au-20Sn solder on the SABA reflowed at 340 °C for 60 s. **a** Cross-sectional morphology with fracture of the front of the bond. **b** Interfacial morphology after shear test showing the formation of Au₄Al IMC at the interface of Au-20Sn/W layer. **c** Cross-sectional morphology with fracture in the center of the interface

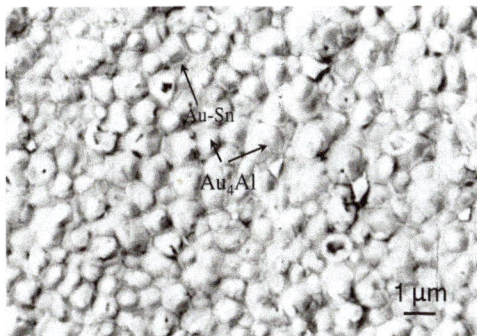

Fig. 5 Magnification of the interfacial morphology of the Au-20Sn solder and the Au₄Al IMC after shear test

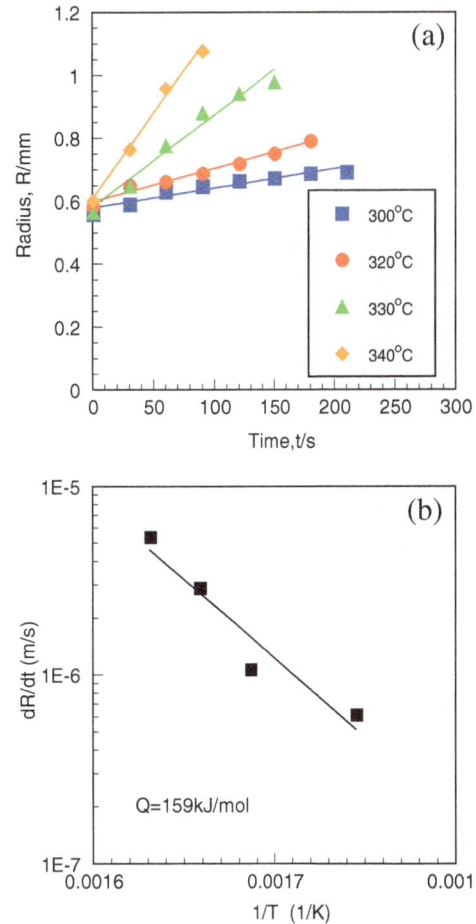

Fig. 6 **a** Spreading kinetics of the Au-20Sn solder on the SABA at different temperatures. Solder mass, 4 mg. **b** Plot of spreading rate versus $1/T$ for the isothermal spreading of Au-20Sn solder on the SABA

SABA cooled from 430 to 25 °C. The Au-12Ge alloy could well bond the Al electrode. The Au-12 Ge solder joined the W layer of the SiC-SBD after reflow. EDX analysis in the outlined area with dash lines in Fig. 7 reveals that the composition of the Au-Ge solder was 33Au-67Ge-5Al (at.%). This means that Al was dissolved into the Au-Ge solder after reflow.

Figure 8 illustrates the elemental mapping of Au, Ge, and Al in the fractured surface of SABA after shear test. Al was detected in the Au-Ge solder. This indicates that Al dissolved into the Au-Ge solder during reflow and dispersed in both the Au-rich phase and the Ge-rich phase after solidification.

Figure 9 shows the SEM micrograph of the interfacial morphology of the SABA after shear test. At the interface of the Au-12Ge solder and SABA, the specimen fractured in the Au-Ge (Al) solder near the W diffusion barrier.

Figure 10 depicts the shear strength of the SiC-SBD bonded on the Cu/Ni (P)-metalized Si₃N₄ substrate with Au-12Ge solder as a function of number of the Au bumps. The shear strength exhibits a value beyond ~50 MPa. The larger scatter of the shear strength may be due to the fact that some chips

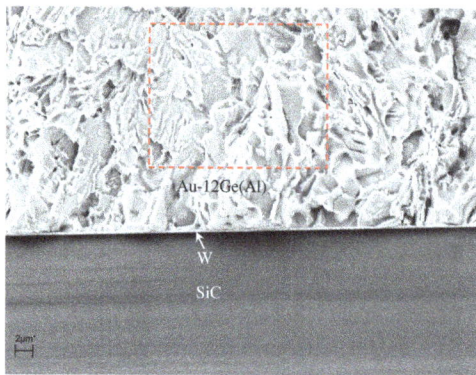

Fig. 7 SEM micrograph of a cross-sectional fractured morphology of the Au-12Ge solder on the SABA cooled from 430 to 25 °C

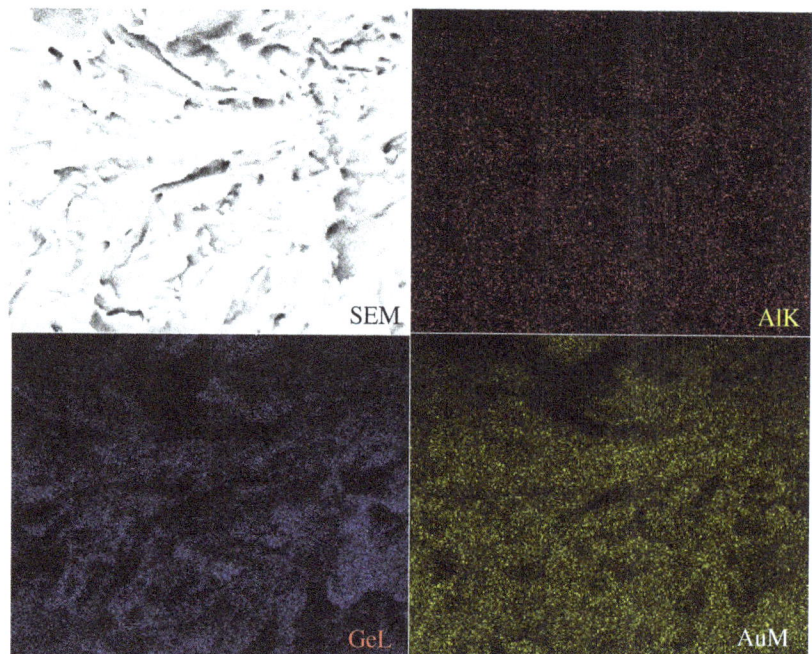

Fig. 9 SEM micrograph of the surficial morphology of the SABA after shear test

were destroyed during shear test which resulted in low shear strength. Similar to the bond with the Au-20Sn solder on the SABA, the bond strength of the Au-12Ge solder on the SABA was not significantly affected by the number of Au bumps. This is due to, as shown in Figs. 8 and 9, the reaction between the Au-12Ge solder and the Al electrode initiated at the Au bump during reflow, and a Au bump was sufficient to complete the reaction at a higher reflow temperature.

Bonding with flux-free solder Au-20Sn and Au-12Ge

To confirm the fact that the bond of the SABA with Au-based solders is realized by the reaction of the Au in the Au-based solder rather than the flux in the paste, the SABA was bonded with the Au-20Sn and Au-12Ge sheets in the vacuum reflow system, respectively. No flux was used. Figure 11a shows the surficial morphology of the wetting front of the Au-20Sn solder sheet after reflow at 350 °C in N_2 for 120 s. The solder

bonded the Al electrode well. EDX analysis in the surface of the solder reveals that the solder has a composition of Au-25Sn-9Al (at.%). Since the solder, as shown in Fig. 11b, is ~5 μm thick and the EDX analytical depth is ~2 μm, the detected Al was from the solder. This indicates that Al dissolved into the solder. Figure 11b shows the cross-sectional morphology of the Au-bumped Al electrode reflowed with the Au-20Sn solder at 350 °C for 120 s. The cross-sectional morphology is similar to that of the SABA bonded with the Au-20Sn paste (Fig. 4a). Al dissolved into the Au-20Sn solder to form a Au-Sn (Al) solder. Au_4Al was detected with SEM/EDX at the interface between the solder and the W diffusion barrier. This indicates that the bonding of the solder to the Al electrode was attributed to the alloying effect of the Au-20Sn solder rather than the flux.

Figure 12a shows the fracture morphology of the wetting front of the Au-12Ge sheet bonded on a SABA at 430 °C in N_2 for 120 s. No flux was used. Al dissolved into the solder.

Fig. 8 Elemental mappings of Au, Ge, and Al in the fractured surface of SABA after shear test. The SABA was bonded with the Au-12Ge solder at 430 °C for 120 s

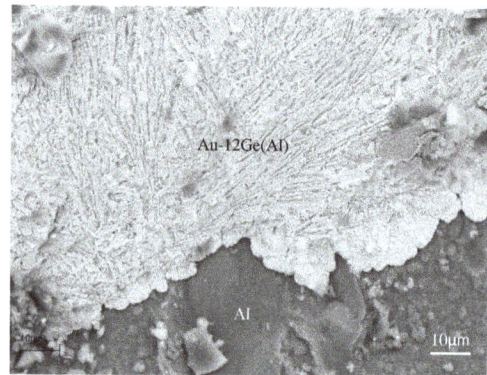

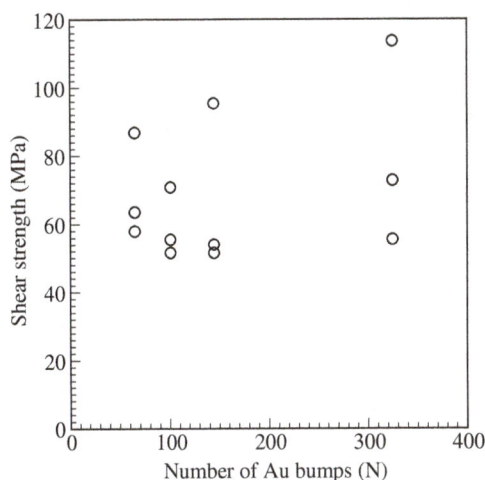

Fig. 10 Shear strength of the SiC-SBD bonded on the Cu/Ni (P)-metalized Si_3N_4 substrate with Au-12Ge solder paste as a function of number of the Au bumps

The solder bonded the W diffusion barrier well. EDX analysis in the fractures surface of the solder reveals that the solder has a composition of 23Au-63Ge-14Al (at.%). Figure 12b shows the EDX spectra of the outlined area with dash lines in Fig. 12a. An Al spectrum was clearly detected in addition to Au and Ge spectra.

These results reveal that the reaction of Au in the Au-based solder with the Au-bumped Al is responsible for the wettability.

Electrical characteristics

The SiC-SBDs handle the power energy by having the current flow from the anode to the cathode while blocking it from the cathode to the anode. The electrical characteristics of SiC-SBD are characterized by I_F-V_F and I_R-V_R curves. A guard ring on the periphery of the anode of the SiC-SBD is needed for improving the blocking voltage [26]. Hence, to protect the guard ring of the device, space between the guard ring and substrate should be prepared. The Au-top-metalized Cu electrode spacer on the metallic layer of the substrate could meet this requirement. Figure 13 shows the optical morphology of the substrate with a Cu electrode spacer on the Cu/Ni (P)/Au-metalized Si_3N_4 ceramic substrate for bonding the SABA. The Cu electrode was bonded to the metallic layer of the substrate with Au-12Ge solder. A space of ~0.4 mm between the guard ring of the SiC-SBD and the metallic layer was formed after reflow with the Au-20Sn solder.

Panels a and b of Fig. 14 show the I_F-V_F and I_R-V_R curves of the SiC-SBD with its Al electrode bonded by the Au-20Sn solder on a Cu/Ni(P)/Au-metalized Si_3N_4 substrate, respectively. The space between the guard ring and the substrate was filled with silicon epoxy underfill. Very little change was observed in the I_F-V_F curves before and after soldering the SABA (Fig. 14a). The leakage current was very low, <9 µA

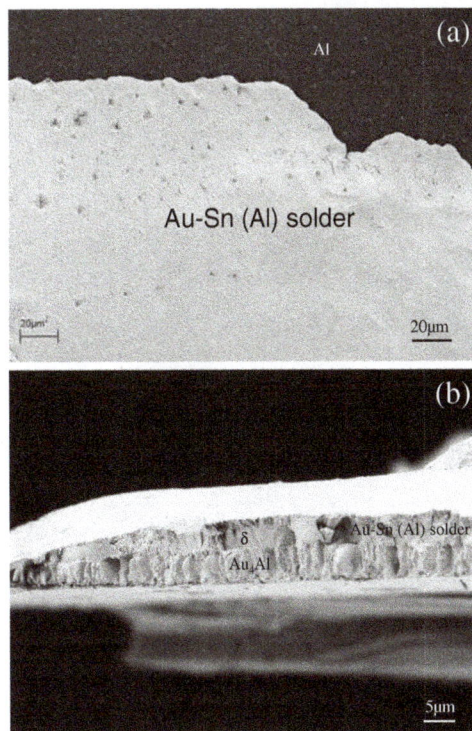

Fig. 11 a Surficial morphology of the wetting front of the Au-20Sn solder sheet and b cross-sectional morphology of the Au-bumped Al electrode reflowed with the Au-20Sn solder sheet at 350 °C for 120 s

even at 1,200 V (Fig. 14b). This means that the SiC-SBD had high blocking voltage after its Al electrode was soldered to a

Fig. 12 a Fracture morphology of the Au-12Ge solder sheet bonded on the Au-bumped Al electrode without a flux. b X-ray spectra of the fractured solder

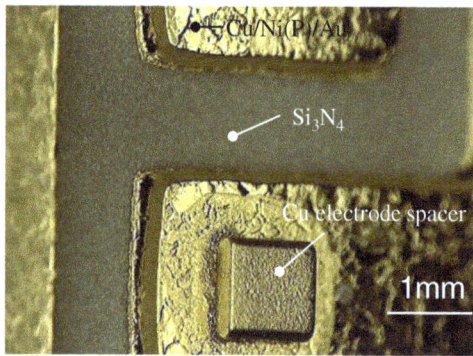

Fig. 13 Optical morphology of the substrate with a Cu electrode spacer on the Cu/Ni (P)/Au-metalized Si_3N_4 ceramic substrate for bonding the SABA. The image was taken by tilting the sample 27° from the horizontal level to show the three-dimensional shape of the Cu electrode spacer

substrate with a Cu electrode spacer. These results reveal that little electrical characteristics of the SiC-SBD changed after its Al electrode was bonded to a substrate using this technology.

The SABA was also bonded to the Cu electrode spacer which had already been bonded to the Cu/Ni(P)/Au-metalized

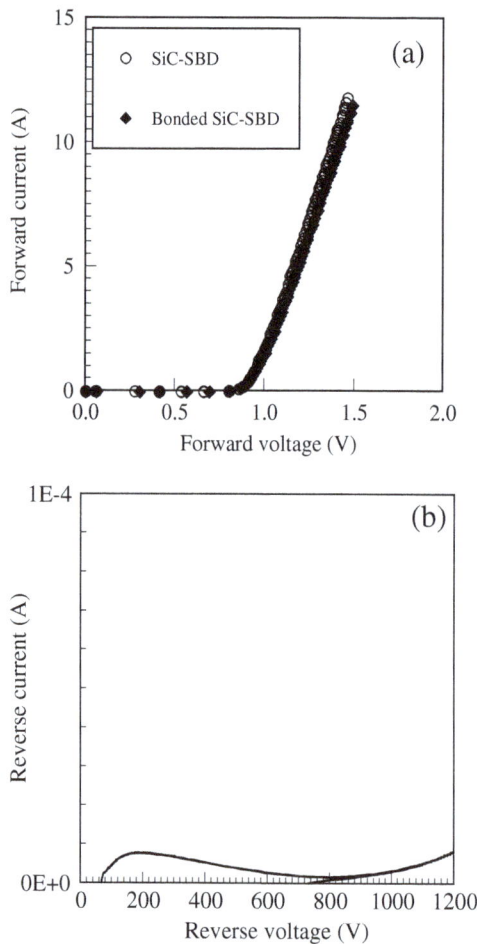

Fig. 14 Electrical characteristics of the SiC-SBD before and after soldering with the Au-20Sn solder. **a** Forward current–forward voltage curves before and after soldering the Al electrode. **b** Reverse current–reverse voltage curves after reflow

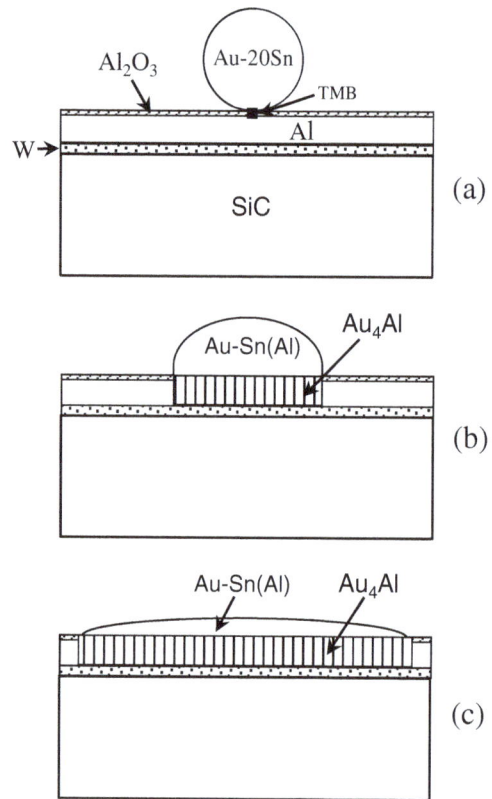

Fig. 15 Schematic of the reactive wetting of the Au-based solder on the Al electrode. **a** Capture of the solder ball by a TMB. **b** Reaction between the solder and the Al electrode initiate at the TMB. **c** Reactive wetting proceed

Si_3N_4 substrate with the Au-12Ge solder. The space between the gard ring and the substrate was also filled with silicon epoxy filler. As the case of Au-20Sn bonded SABA, very little change in the I_F–V_F curve was observed. The Ron changed from the 50.2 to 50.5 mΩ. The leakage current measured from the I_R–V_R curve at 1,200 V was very low, ~3 μA. Hence, it can be established that very little change occurred after the SABA of the SiC-SBD was bonded to a substrate with a Cu electrode spacer with the Au-12Ge solder, as the case of the Au-20Sn bonded SABA.

Discussion

The Pb-based solder was unable to bond the SABA. This is due to the fact that this solder has less reactivity with Al. The eutectic Al-Pb-Sn only contains 0.08 wt% Al [27]. That is to say, in view of thermodynamics, very little amount of Al is dissolved in the Pb-Sn solder. Hence, the Pb-Sn solder almost did not react with the Al electrode at the Au bump. As a result, the Pb-10 wt% Sn solder did not bond the Al electrode. For the Sn-based solder bonding on the SABA, as neither Sn-Al IMC nor Sn-Al solubility exists at room temperature in the Al-Sn binary phase diagram [28], the Sn-based solder is difficult to react with the Al electrode. Although Al can react with Ag, the activity of Ag in the solder was not enough to form a Ag-Al intermetallic or to form a Ag solid solution under the reflow conditions. Hence, the Sn-based solder is hard to be used to bond the SABA.

Unlike the Pb-based and Sn-based solders, Au in the Au-based solders can react with Al to form the solid solution and the Au-Al IMCs. Figure 15 shows the schematic of the reactive wetting of the Au-20Sn solder on an Al electrode with a tinny metal block (TMB) such as a Au bump. During reflow, a solder ball forms and immediately fixed by the wettable TMB (Fig. 15a). Since the aluminum oxide film is disrupted at the TMB, reaction between the liquid Au in the molten solder and solid-state Al becomes possible. The reaction initiates at the TMB. The thin aluminum oxide film is destroyed with the progress of the reaction. A continuous Au_4Al IMC layer forms at the interface between the solder and the W layer (Fig. 15b). With reactive spreading of the Au-based solder on the Al electrode, the device is bonded by the Au-based solder (Fig. 15c). The wetting behavior of the Au-20Sn solder on the SABA is similar to that of spreading of Cu-40 at.% Si alloy on vitreous carbon [20]. In this system, wetting is prompted by the formation of a continuous Au_4Al layer with an active energy of 159 kJ/mol.

During reflow of Au-12Ge solder on the SABA, the reaction of liquid Au-Ge solder with Al initiated at the Au bump. With increasing reflow time, the Al dissolved into the Au-Ge solder forming a ternary alloy of Au-Ge-Al. The Au-Ge-Al solder wetted the W diffusion barrier of the SiC-SBD. After solidification, the solder bonded the W diffusion barrier, resulting in bond of the SiC-SBD by the Au-Ge-Al solder.

To insure the high blocking voltage of the SiC-SBD, a space between the guard ring of the device and the substrate are needed. Since wetting of the solder on the SABA initiate at the TMB and the solder spreads with a spreading rate as described in Eq. (2), the soldering area can be controlled by the reflow temperature and time. Hence, a reflow temperature and time can be set to allow the solder unable to reach the guard ring. On the other hand, a Cu electrode spacer on the substrate with a size smaller than the SABA can make sure a space between the SABA and the substrate.

Conclusions

The non-solderable Al electrode can be soldered by the Au-based solders along with fabricating a TMB in the Al electrode. Soldering initiated at the TMB (e.g., a Au bump) and a reactive spreading pervaded on the Al electrode. A continuous Au_4Al intermetallic compound layer formed at the bonding interface by reaction of the liquid Au from the Au-20Sn solder with the solid Al electrode. The activation energy of the Au-

20Sn soldering the single Au-bumped Al electrode is $Q = 159$ kJ/mol. The shear strength exceeded 60 MPa, ~10 times higher than the required strength limit. Similarly, the Au-12Ge solder reacted with the Al electrode resulting in a bonding of the Au-12Ge with the SABA. The shear strength of the Al electrode by the Au-12Ge solder was beyond 50 MPa. The reaction of Au in the Au-based solder with the Au-bumped Al is responsible for the wettability. Little electrical characteristics of the SiC-SBD changed after bonding the Al electrode using this technology.

Acknowledgments This work was financially supported by the NEDO (New Energy and Industrial Technology Development Organization, Japan) project, Development of Inverter Systems for Power Electronics, and partially supported by the project of Development of the Next-Generation Power Electronics Technology. This research work was mainly conducted in Energy Semiconductor Electronics Research Laboratory of AIST. The authors thank Dr. H. Ohashi, Dr. M. Aoyagi, and Dr. Y. Hayashi for their discussions.

References

1. Chinthavali M, Ozpineci B, Tolbert LM, Zhang H (2009) Summary of SiC research for transportation applications at ORNL. Mater Sci Forum 600–603:1239–1242
2. Virshup A, Liu F, Buchholt F, Spetz AL, Porter LM (2011) Improved thermal stability observed in Ni-based ohmic contacts to n-type SiC for high-temperature applications. J Electron Mater 40:400–405
3. Harman GG (2010) Wire bonding in microelectronics, 3rd edn. McGraw-Hill, New York
4. Li DM, Li Q, Pan F (2008) Improvement in power durability of Al electrode films used in SAW devices by Zr additive and Ti underlayer. J Electron Mater 37:180–184
5. Gan CL, Francis C, Chan BL, Hashim U (2013) Extended reliability of gold and copper ball bonds in microelectronic packaging. Gold Bull 46:103–115
6. Xu H, Qin I, Clauberg H, Chylak B, Acoff VL (2013) Behavior of palladium and its impact on intermetallic growth in palladium-coated Cu wire bonding. Acta Mater 61:79–88
7. Yoon SW, Glover MD, Mantooth HA, Shiozaki K (2013) Reliable and repeatable bonding technology for high temperature automotive power modules for electrified vehicles. J Micromech Microengn 23(1):015017
8. Kit C, Schirmer KC, Rowden B, Mantooth HA, Simon S, Ang SS, Balda JC (2011) Packaging and modeling of SiC power modules. ECS Trans 41:183–188
9. Mozetic M, Zalar A, Cvelbar U, Babic D (2004) AES characterization of thin oxide films growing on Al foil during oxygen plasma treatment. Interface Anal 36:986–988
10. Campbell T, Kalia PK, Nakano A, Vashishta P, Ogata S, Rodgers S (1999) Dynamics of oxidation of aluminum nanoclusters using variable charge molecular-dynamics simulations on parallel computers. Phys Rev Lett 82:4866–4869

11. Azumi K, Yugiri T, Seo M, Fujimoto S (2001) Double zincate pretreatment of sputter-deposited Al films. J Electrochem Soc 148: C433–C438.

12. Lee SK, Lee JH, Kim YH (2007) Nucleation and growth of zinc particles on an aluminum substrate in a zincate process. J Electron Mater 36:1442–1447

13. Koleňák R, Šebo P, Provazník M, Koleňáková M, Ulrich K (2011) Shear strength and wettability of active Sn3.5Ag4Ti(Ce, Ga) solder on Al₂O₃ ceramics. Mater Des 32:3997–4003

14. Wang WL, Tsai YC (2012) Microstructural characterization and mechanical property of active soldering anodized 6061 Al alloy using Sn–3.5Ag–xTi active solders. Mater Charact 68:42–48

15. Wang Z, Wang HY, Liu LM (2012) Study on low temperature brazing of magnesium alloy to aluminum alloy using Sn–xZn solders. Mater Des 39:14–19

16. Huang ML, Huang YZ, Ma HT, Zhao J (2011) Mechanical properties and electrochemical corrosion behavior of Al/Sn-9Zn-xAg/Cu joints. J Electron Mater 40:315–323

17. Movahedi M, Kokabi AH, Hosseini HR (2009) An investigation on the soldering of Al3003/Zn sheets. Mater Charact 60:441–446

18. Tan QB, Deng C, Mao Y, He G (2011) Evolution of primary phases and high-temperature compressive behaviors of as-cast AuSn20 alloys prepared by different solidification pathways. Gold Bull 44:27–35

19. Liu YC, Teo JWR, Tung SK, Lam KH (2008) High-temperature creep and hardness of eutectic 80Au/20Sn solder. J Alloys Compd 448:340–343

20. Eustathopoulos N, Nicholas MG, Drevet B (1999) Wettability at high temperatures, 1st edn. Pergamon, Oxford

21. International Standard IEC 60749-19

22. Landry K, Eustathopoulos N (1996) Dynamics of wetting in reactive metal/ceramic systems: linear spreading. Acta Mater 44:3923–3932

23. Eustathopoulos N (1998) Dynamics of wetting in reactive metal ceramic systems. Acta Mater 46:2319–2327

24. Peebles DE, Peebles HC, Ohlhausen JA (1998) Kinetics of the isothermal spreading of tin on the air-passivated copper surface in the absence of a fluxing agent. Colloids Surf A 144:89–114

25. Lang FQ, Hayashi Y, Nakagawa H, Aoyagi M, Ohashi H (2009) A novel three-dimensional packaging method for Al-metalized SiC power devices. IEEE Trans Adv Packgn 32:773–779

26. Kim SJ, Oh DJ, Yu SJ (2006) Breakdown voltage characteristics of SiC Schottky Barrier Diode with Aluminum deposition edge termination structure. J Korean Phys Soc 49:S768–S773

27. Kartzmark R (1956) The systems aluminum–lead–tin and aluminum–tin. Thesis of MS degree, Univ Manitoba

28. Okamoto H (2000) Desk handbook phase diagrams for binary alloys. ASM International, Ohio

19

Lamellar-forming grain boundary reaction related to age-hardening mechanism in an Au-Pt-Pd-In metal-ceramic alloy

Sung-Min Kim · Hyung-Il Kim · Byung-Wook Jeon · Yong Hoon Kwon · Hyo-Joung Seol

Abstract This study examined the relationship between the lamellar-forming grain boundary reaction and the change in hardness during the aging process of an Au-Pt-based metal-ceramic alloy composed of 76.6 Au–9.9 Pt–9.3 Pd–1.7 In–1.2 Ag–0.56 Sn (wt%) with minor ingredients. The phase decomposition of the parent Au-rich α phase occurred within a very short time (30 s) by aging at 550 °C after a solution-treatment at 950 °C due to the solubility limits of Au and Pt in each other, which initiated a lamellar-forming grain boundary reaction. The observed hardening was attributed to both the grain interior and grain boundary precipitate. On the other hand, the grain boundary precipitate comprising the fine lamellar structure caused more powerful hardening from the early stage of the aging process. The alternate layer of the grain boundary lamellar structure was composed of Pt-, In- and Sn-depleted Au-rich α_1 phase as well as precipitated Pt-, In- and Sn-concentrated β_1 phase. The extremely fine nature of the grain boundary lamellar structure supplied large amounts of inter-phase boundaries, which contained lattice strain by the difference in the lattice parameter between the α_1 and β_1 phases, resulting in hardening.

Keywords Lamellar-forming grain boundary reaction · Age-hardening · Au-Pt-Pd-In metal-ceramic alloy · Precipitation

S.-M. Kim · H.-I. Kim · B.-W. Jeon · Y. H. Kwon · H.-J. Seol (✉)
Department of Dental Materials, Institute of Translational Dental Science, School of Dentistry, Pusan National University, Beomeo-Ri, Mulgeum-Eup, Yangsan-Si, Gyeongsangnam-Do 626-814, South Korea
e-mail: seolhyojoung@daum.net

Introduction

Metal-ceramic restorations are still the most reliable methods in dental prosthetics, and a range of metal-ceramic alloys is used in dentistry [1]. Among the others, dental gold alloys for metal-ceramic restorations consist basically of Au and Pd as the principle ingredients, as well as Pt, Ag, Sn, In and Fe as minor ingredients. Metal-ceramic alloys based on Au are hardened substantially by an appropriate heat treatment to resist the occlusal force in the oral environment [2, 3]. Minor ingredients, such as Pt, Sn, In and Fe, play important roles in precipitation hardening [4, 5].

In dental alloys hardened by precipitation, lamellar structures are normally observed in the grain boundaries [6–11]. These lamellar-forming grain boundary reactions were reported to be the mechanism for softening in most cases and sometimes for hardening depending on the alloy composition [6–11]. In a study with a dental casting alloy composed of 48.78 Ag–28 Pd–12.04 Au–9.12 Cu (wt%), the lamellar structure that formed by the first grain boundary reaction resulted in hardening, but the second grain boundary reaction produced a very coarse lamellar structure that resulted in softening [6]. In a study with a dental casting alloy composed of 47.5 Au–36 Ag–10.6 Cu–4 Pd–1 In–0.7 Zn–0.2 Ir (wt%), the formation of a grain boundary lamellar structure resulted in softening [11]. In the case of Pt-added-Au-based alloy for metal-ceramic restorations, precipitation hardening can occur by an appropriate heat treatment due to the solubility limits of Au and Pt in each other [12–14]. Therefore, the hardness of the alloy can be changed by lamellar formation during the firing process for porcelain. To use the Pt-added-Au-based metal-ceramic restoration under the best mechanical condition, it is important to determine the relationship between the lamellar-forming grain boundary reaction and the change in hardness. On the other hand, it is unclear if the lamellar-forming grain boundary reaction is the mechanism for hardening or softening in the Pt-added-Au-based metal-ceramic alloy [15]. This study examined the relationship between the lamellar-forming grain boundary reaction and the change in hardness during the aging process of a Pt-added-Au-based metal-ceramic alloy composed of 76.6 Au–9.9 Pt–9.3 Pd–1.7 In (wt%) with minor ingredients by hardness testing (HV), X-ray diffraction (XRD), field emission-scanning electron microscopy (FE-SEM) and energy-dispersive spectrometry (EDS).

Materials and methods

Specimen alloy

The specimen used in this study was a dental high-carat gold alloy (AURIUM 3, Aurium® Research USA, San Diego, CA, USA) of type III–IV according to the ISO classification (ISO 22674:2006 (E)). This type of alloy is based on the ternary system of Au, Pt and Pd for the fabrication of metal-ceramic prostheses. Table 1 lists the chemical composition of the specimen alloy supplied by the manufacturer, where only the minor ingredients that comprised less than 1 wt% were examined by X-ray fluorescence analysis (XRF) to determine the precise content.

Heat treatment

The specimens were subjected to a solution treatment at 950 °C for 10 min in a vertical furnace in an argon atmosphere to obtain a supersaturated solid of a single phase. The specimens were then quenched in ice brine to prevent the formation of an equilibrium phase. Subsequently, the solution-treated specimens were subjected to isothermal aging at 550 °C for various times in a molten salt bath (25 KNO_3+30 KNO_2+25 $NaNO_3$+20 $NaNO_2$, wt%) and quenched in ice brine.

Hardness test

To examine the hardening and subsequent softening mechanism during the aging process, the hardness of the heat-treated plate specimens was measured using a Vickers microhardness tester (MVK-H1, Akashi Co., Akashi, Hyogo Japan) using a 25- or 300-gf load and a holding time of 10 s. The values reported are the mean of five measurements.

XRD study

XRD (X'PERT PRO, PRO, Philips, Eindhoven, Netherlands) was performed to examine the phase transformation during the aging process. Powder specimens with a particle size below 45 µm were filed using a diamond disc and passed through a 330-mesh screen. The specimens were then mixed with alumina powders with a particle size of 1 µm to prevent sintering agglomeration during heat treatment. Subsequently, the powder specimens were subjected to vacuum sealing in silica tubes and heat treatment. The alumina powders were then filtered from the heat-treated specimens. The XRD profile was recorded at 30 kV and 40 mA using Ni-filtered Cu Kα radiation as the incident beam.

FE-SEM observation

FE-SEM image of the heat-treated plate specimens was taken to examine the microstructural changes during the aging process. The heat-treated specimens were treated with a polisher and etched in an aqueous solution containing 10 % KCN (potassium cyanide) and 10 % $(NH_4)_2S_2O_8$ (ammonium persulfate). The surfaces of the heat-treated specimens were observed by FE-SEM (JSM-6700 F, Jeol, Akishima-shi, Tokyo, Japan) at 15 kV.

EDS analysis

The EDS (INCA X-Sight, Oxford instruments Ltd., Oxford, UK) profiles of the heat-treated plate specimens were recorded at 15 kV to determine the changes in the elemental distribution during the aging process. The specimens were prepared in the same manner as used for the FE-SEM observations.

Result and discussion

Age-hardening behaviour

To determine the most effective aging temperature, which is related to age-hardening behaviour, the plate-like specimens, which were solution-treated at 950 °C for 10 min, were aged isochronally over the temperature range of 300–630 °C for 20 min. The microhardness was measured using a 300-gf load. Figure 1 shows the isochronal age-hardening curves of the specimen aged over the temperature range of 300 to 630 °C for 20 min. The specimen showed apparent age-hardenability

Table 1 Chemical composition of the specimen alloy	Composition	Au	Pt	Pd	In	Ag	Sn	Cu	Ir	Fe
	wt%	76.6	9.9	9.3	1.7	1.2	0.56	0.3	0.27	0.17
	at.%	68.61	8.95	15.42	2.61	1.96	0.83	0.83	0.25	0.54

Fig. 1 Isochronal age-hardening curves of the specimen aged at temperature ranging from 300 to 630 °C for 20 min

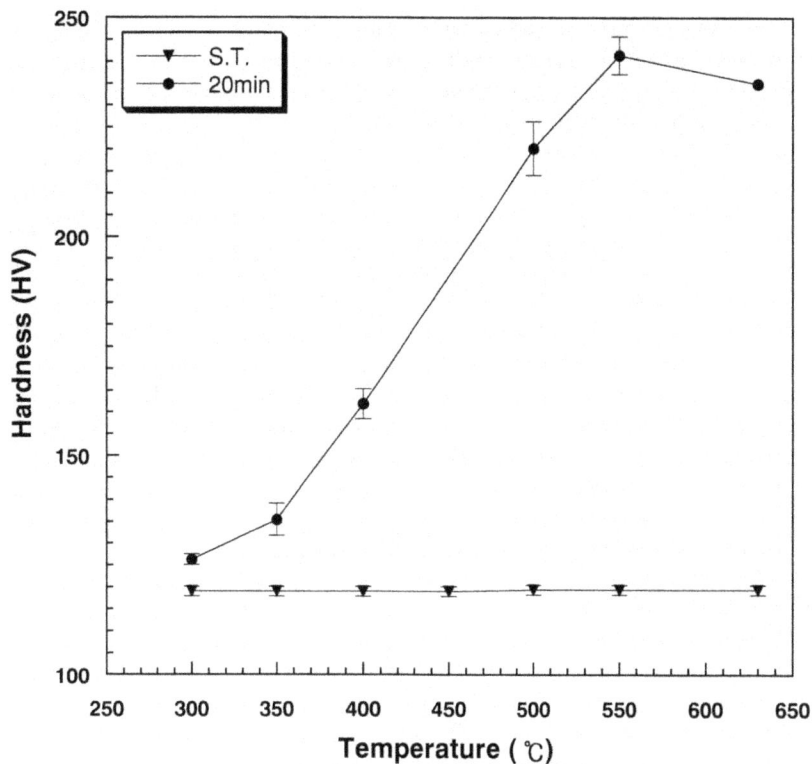

at an aging temperature of 550 °C. Therefore, the isothermal age-hardening behaviour was observed at 550 °C to evaluate the age-hardenability of the specimen.

Figure 2 shows the isothermal age-hardening curve of the specimen solution-treated at 950 °C for 10 min and then aged

at 550 °C for various times until 20,000 min. The microhardness was measured using a 300-gf load without dividing the grain interior and grain boundary. The hardness obtained at 2 min was approximately two times higher than that of the solution-treated condition. Thereafter, the hardness increased

Fig. 2 Isothermal age-hardening curve of the specimen aged at 550 °C

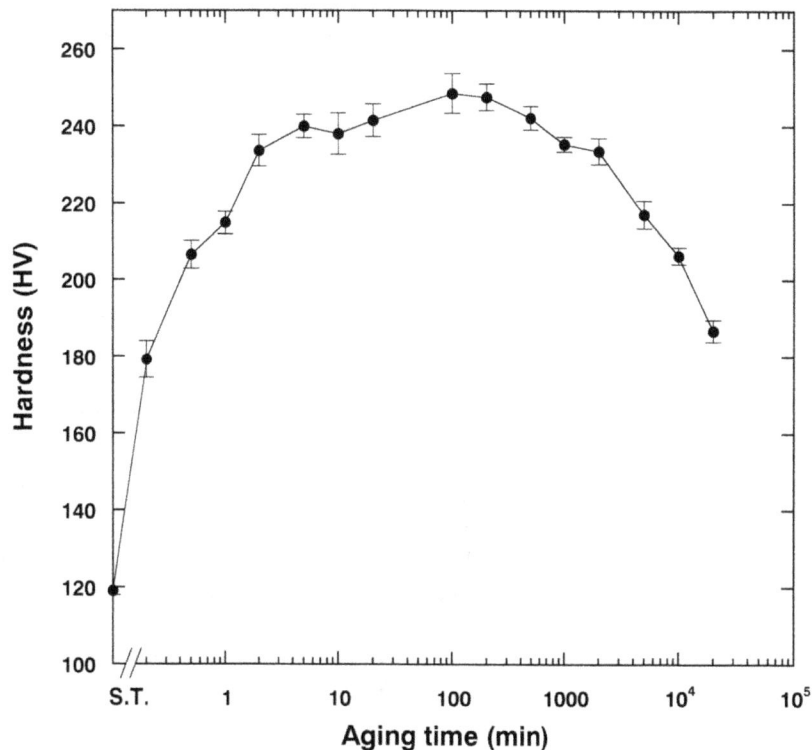

gradually with aging time and reached a maximum at 100 min. After maintaining the maximum value until 200 min, the hardness decreased gradually until 2,000 min and then decreased apparently until 20,000 min.

Phase transformation

The variations of the XRD patterns during isothermal aging were examined to clarify the relationship of the phase transformation and hardness change in the early and later stage of the aging process. Figure 3 presents the variations of the XRD patterns of the specimens solution-treated at 950 °C for 10 min and then aged at 550 °C for various times until 20,000 min. The XRD pattern of the solution-treated (S.T.) specimen showed a single Au-rich α phase with a face-centred cubic (f.c.c) structure and a lattice parameter of a_{200}=4.0222 Å. This value was slightly smaller than the reported lattice parameter of Au (a=4.0786 Å) [16]. In the XRD pattern of the specimen aged at 550 °C for 30 s, the (111, 200) α diffraction peaks broadened asymmetrically to the lower diffraction angle side. And, weak diffraction peak in the higher diffraction angle side of the main peak was observed. Therefore, the phase decomposition of the parent α phase was initiated within 30 s, resulting in an apparent increase in hardness by 100 HV, as shown in Fig. 2.

By prolonged aging until 20,000 min, the XRD patterns did not show further changes. The asymmetrical shape of the main peaks resulted mainly from peak overlap of the solute-

Fig. 3 Variation of the XRD patterns of the specimens during isothermal aging at 550 °C with aging time

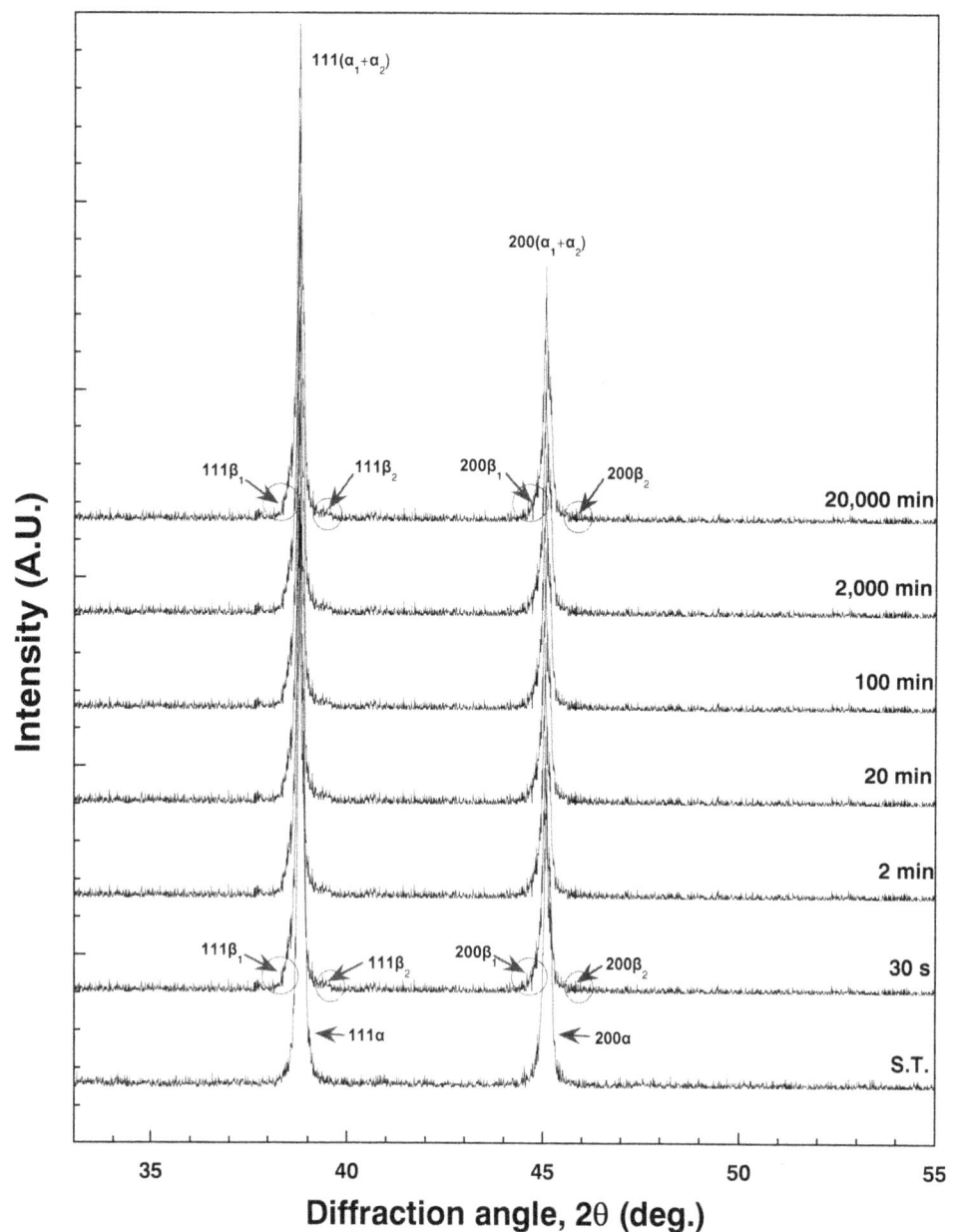

depleted parent phase and precipitated phase (β_1) due to the small gap in the lattice parameters. The weak diffraction peak in the higher diffraction angle side of the main peak was from the f.c.c β_2 phase which had the lattice parameter, $a_{200}=$ 3.986 Å. This value was very similar to the lattice parameter of the Pt$_3$In phase, $a=3.992$ Å [17]. The lattice parameter of the parent Au-rich α phase showed little change from $a_{200}=$ 4.0222 Å to $a_{200}=4.0231$ Å by a transformation into the solute-depleted Au-rich phases ($\alpha_1+\alpha_2$). This will be further mentioned in elemental distribution part.

The specimen used in this study was a gold alloy containing approximately 9 wt% Pd and Pt, respectively. The alloy composed of only Au and Pd shows no precipitation phenomena, because it forms a complete solid solution in all atomic ratios [14]. On the other hand, precipitation can occur in the Au-Pt alloy due to the differences in solubility in each other at the temperatures for aging (550 °C) and solution-treatment (950 °C) [14]. From the alloy composition, the precipitated phases (β_1 and β_2) were believed to be Pt-concentrated phases, even though the lattice information was unclear due to XRD peak overlap. To confirm this, the specimen was analysed by FE-SEM and EDS.

Microstructural changes

Figure 4 shows FE-SEM images of $\times 300$ (1), $\times 8,000$ (2) and $\times 60,000$ (3) magnifications for the specimens solution-treated at 950 °C for 10 min (Fig. 4a), and aged at 550 °C for 30 s (Fig. 4b), 2 min (Fig. 4c), 100 min (Fig. 4d), 2,000 min (Fig. 4e) and 20,000 min (Fig. 4f). In the specimen solution-treated at 950 °C for 10 min (Fig. 4a), an equiaxed structure of a single phase was observed. The lamellar-forming grain boundary reaction was progressed slightly. Such a grain boundary reaction must have occurred instantly during quenching after a solution treatment, considering the XRD results that revealed peak broadening within very short aging time (30 s).

In the specimens aged for 30 s (Fig. 4b), the grain boundary nano-sized precipitate grew towards the grain interior, and after further aging for 2 min (Fig. 4c), they grew slightly more, as marked by the arrows. By aging the specimen for 100 min (Fig. 4d), at which time the maximum hardness was obtained, lamellar structures composed of a precipitate and solute-depleted matrix coarsened apparently and replaced approximately half of the matrix, whereas no apparent change was observed in the grain interior. In the specimens aged for 2,000 min (Fig. 4e), the lamellar structure became much coarser, and the lamellar-forming grain boundary reaction was stopped without initiating a second grain boundary reaction which produces a much coarser lamellar structure. In addition, the fine grain interior precipitate was observed throughout the entire grain interior by microstructural coarsening. These changes decreased the hardness slightly

(-15 HV), from the age-hardening curve shown in Fig. 2. After further aging for 20,000 min (Fig. 4f), the precipitate in the grain boundary and grain interior coarsened slightly more. Such a change resulted in an apparent decrease in hardness (-50 HV).

The above result confirmed hardening by precipitation and the lamellar-forming grain boundary reaction. To characterise the relationship of the lamellar-forming grain boundary reaction with the change in hardness, the microhardness was measured again using a 25-gf load to obtain data separately in the grain interior (G.I) and grain boundary (G.B) region of the specimens aged for 2 min (Fig. 4c), 100 min (Fig. 4d) and 2,000 min (Fig. 4e). The results are listed in Table 2. In the specimen aged for 2 min (Fig. 4c), as shown in Table 2, each hardness value in the G.I and G.B precipitate was much higher than that of the single-phased solution-treated specimen (119 HV). Moreover, the hardness in the G.B precipitate was much higher than that in the G.I. Therefore, both the grain interior and grain boundary precipitate attributed to hardening, but the grain boundary precipitate comprising the fine lamellar structure caused more powerful hardening in the early stage of the aging process.

In the specimen further aged for 100 min (Fig. 4d), at which time the maximum hardness was obtained by measuring using a 300-gf load, the precipitate at the grain boundary was coarsened slightly. This change, however, did not decrease the hardness in the grain boundary according to the data measured using the 25-gf load. On the other hand, the hardness in the G.I increased to become similar to that of the G.B precipitate. Therefore, both the grain boundary and grain interior precipitate contributed to the maximum hardness in the intermediate stage of the aging process.

In the specimen aged for 2,000 min (Fig. 4e), at which time the hardness measured using a 300-gf load decreased slightly (-15 HV), only the hardness in the G.B precipitate decreased slightly (-20 HV) according to the data measured using a 25-gf load. This was expected from the apparent coarsening of the grain boundary precipitate. The coarsening of the microstructure also progressed in the grain interior slightly, but a high hardness was maintained. Therefore, in the later stage of aging process, microstructural coarsening and resulting softening occurred faster in the grain boundaries than in the grain interior.

Ohta et al. reported that the first and second grain boundary reaction produced a lamellar structure with different widths, resulting in hardening and softening, respectively, in the alloy composed of 48.78 Ag–28 Pd–12.04 Au–9.12 Cu (wt%) [6]. In the present study, the formation of a lamellar structure at the grain boundaries occurred immediately as the hardening mechanism, and the subsequent slow increase in width of the lamellar structure resulted in softening without a second grain boundary reaction. Therefore, it is believed that the

Fig. 4 FE-SEM images of ×300 (*1*), ×8,000 (*2*) and ×60,000 (*3*) magnifications for the specimens solution-treated at 950 °C for 10 min (**a**), and aged at 550 °C for 30 s (**b**), 2 min (**c**), 100 min (**d**), 2,000 min (**e**) and 20,000 min (**f**)

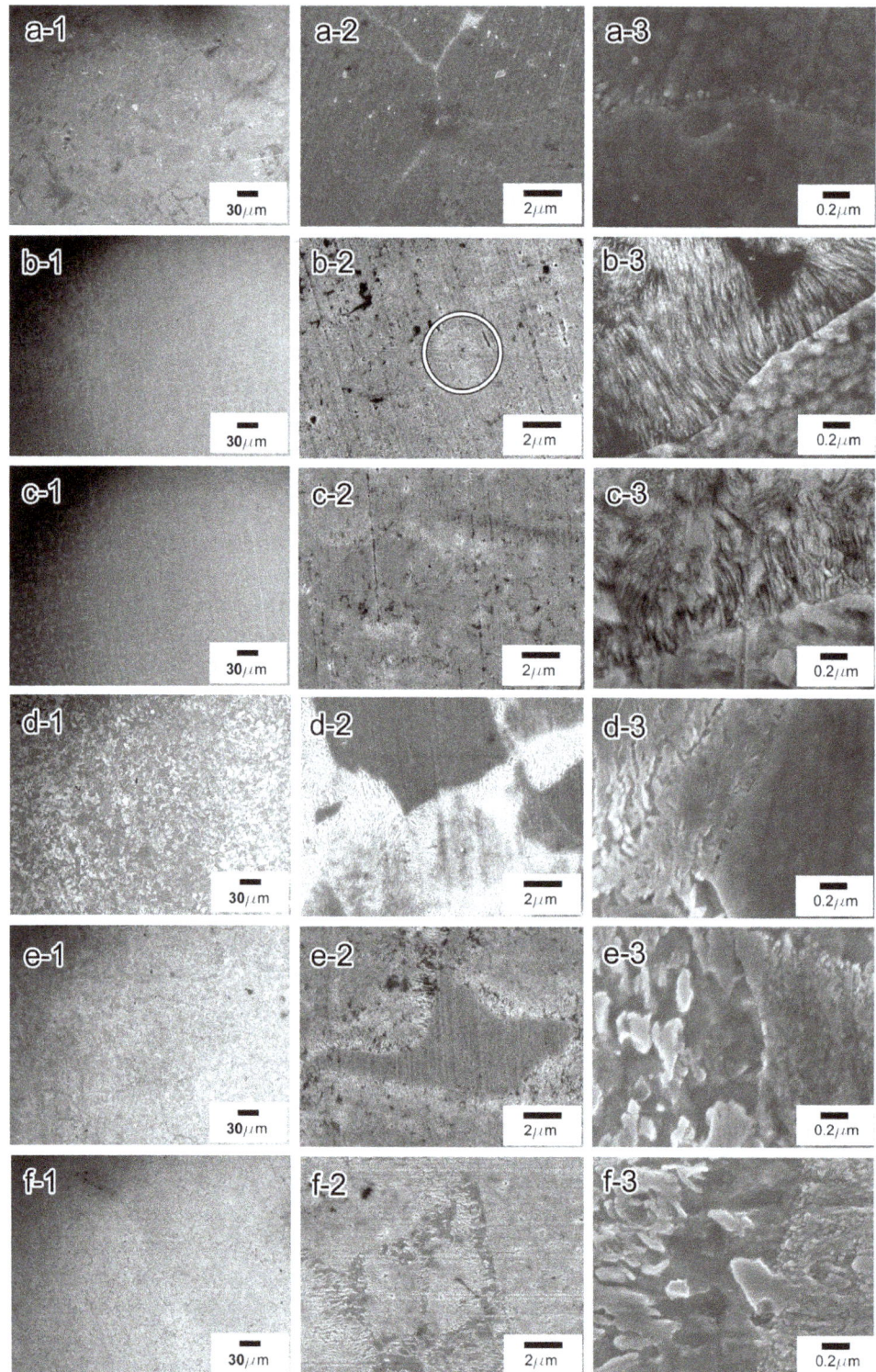

extremely fine nature of the lamellar structure supplied large amounts of inter-phase boundaries, which contain lattice strain by difference in lattice parameter between the solute-depleted Au-rich layer and the precipitated layer of the lamellar structure. Therefore, the subsequent microstructural coarsening that reduces the inter-phase boundaries results in softening, as has also been observed in various dental age-hardenable alloys [18–21]. Considering the immediate lamellar formation, the restorations made by the Pt-added-Au-based metal-ceramic alloys can be strengthened easily during the firing process for porcelain by the introduction of a fine lamellar structure.

Table 2 Vickers microhardness numbers (HV) measured with a 25-gf load in the grain interior (G.I) and grain boundary (G.B) precipitate of the specimens aged at 550 °C for various times

Aging time Region	2 min	100 min	2,000 min
G.I	188.4 (±4.7)	244.4 (±5.2)	242.9 (±10.9)
G.B	246.1 (±15.1)	247.2 (±7.7)	227.6 (±4.1)

Elemental distribution

The elemental distribution in the grain interior and grain boundary lamellar structure was observed by EDS for the specimens solution-treated and aged at 550 °C for 20,000 min. Figure 5 presents FE-SEM images of the specimens solution-treated at 950 °C for 10 min (Fig. 5a, ×2,000 magnification) and aged at 550 °C for 20,000 min (Fig. 5b, ×8,000 magnification) for EDS point analysis. Table 3 lists the EDS results of the matrix and matrix layer (M) and coarsened precipitate (P) layer of the grain boundary lamellar structure, and the G.I region. In Table 3, the elemental distribution in the matrix of the solution-treated specimen (a-M) was similar to the alloy composition listed in Table 1. The minor ingredients (Sn, Cu, Ir, Fe) were not detected by EDS.

In the solute-depleted matrix layer (b-M) of the grain boundary lamellar structure for the specimen aged at 550 °C

for 20,000 min, the Au content increased compared to that of the solution-treated specimen (a-M). The Pt content decreased, and there was no apparent change in the Pd content. Minor ingredients (In, Ag, Sn, Cu, Ir, Fe) were not detected. In the coarsened precipitate layer (b-P) of the grain boundary lamellar structure for the specimen aged at 550 °C for 20,000 min, the Au content decreased to almost half of the value in the matrix (b-M), and the Pt content increased approximately ten times the value in the matrix (b-M), as can be expected from the binary phase diagram of the Au-Pt system which showed the solubility limit of Au and Pt in each other [14]. The minor ingredients, In and Sn, were concentrated in the precipitate region (b-P). Pd is completely soluble with Au at all atomic ratios but has a solubility limit with Pt [14]. On the other hand, in the present study, the Pd content was slightly higher in the Pt-concentrated region (b-P) than in the Pt-depleted region (b-M). This appears to be because Pd normally tends to form a stable phase with elements having a relatively low melting temperature, such as In and Sn [21]. In the G.I region for the specimen aged at 550 °C for 20,000 min, the elemental distribution was similar to that of the single-phased matrix of the solution-treated specimen (a-M) due to the fine nature of the grain interior precipitates and solute-depleted matrix in the grain interior.

The above results were also revealed by EDS line analysis. Figure 6 shows the EDS line profile and FE-SEM image (×8,500 magnification) of the specimen aged at 550 °C for 20,000 min after a solution-treatment. The elemental distribution of Au was opposite to that of Pt, In and Sn in the coarsened precipitate layer and solute-depleted matrix layer of the grain boundary lamellar structure. The amount of Pd increased slightly in the precipitate region, whereas Ag, Fe, Cu and Ir were distributed relatively evenly in the solute-depleted matrix and precipitate region. Therefore, the Au-rich α_1 phase comprising the solute-depleted matrix layer of the grain

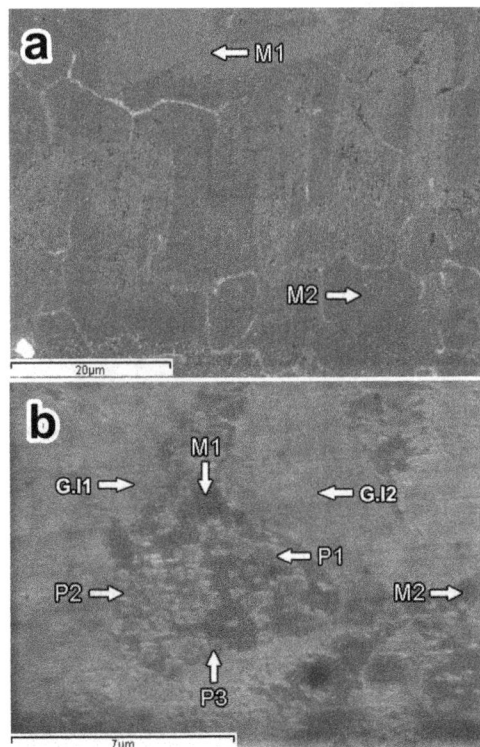

Fig. 5 FE-SEM images at ×2,000 and ×8,000 magnifications for EDS point analysis of the specimen solution-treated at 950 °C for 10 min (**a**) and aged at 550 °C for 20,000 min (**b**), respectively

Table 3 EDS analysis of the regions marked in Fig. 6 of the specimens solution-treated at 950 °C (a) and aged at 550 °C for 20,000 min (b)

Region (at.%)		Au	Pt	Pd	In	Ag	Sn
a	M1	73.88	8.19	15.67	2.26	0	0
	M2	70.96	8.64	15.64	2.66	2.10	0
b	M1	81.50	2.59	15.91	0	0	0
	M2	82.15	2.43	15.42	0	0	0
	P1	48.59	23.08	16.73	6.93	1.92	2.75
	P2	48.11	21.02	17.72	9.17	0	3.98
	P3	46.84	23.56	17.09	9.57	0	2.94
	G.I1	72.34	9.24	15.78	2.64	0	0
	G.I2	73.85	9.55	14.04	2.56	0	0

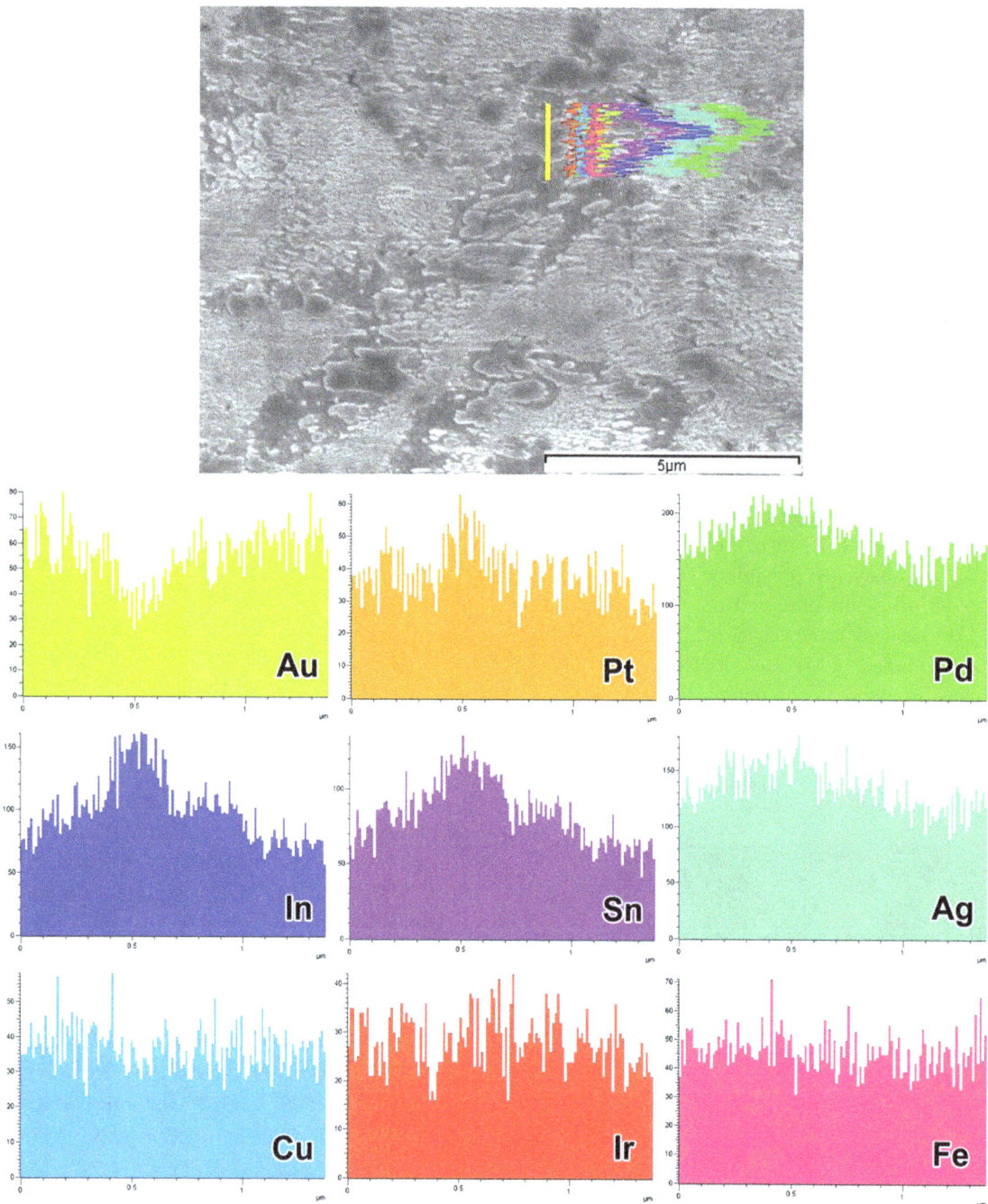

Fig. 6 FE-SEM images and EDS line profile of ×8,500 magnification for the specimen aged at 550 °C for 20,000 min after the solution-treatment at 950 °C for 10 min

boundary lamellar structure is a Pt-, In- and Sn-depleted phase compared to the parent Au-rich α phase, and the precipitated layer (β_1) of the grain boundary lamellar structure is an In- and Sn-concentrated Au-Pt-rich phase. The precipitated phase in the grain interior was not clear by EDS. However, considering the XRD results which showed the existence of the β_2 phase having lattice parameter close to that of the Pt_3In phase, the grain interior was possibly decomposed into the Pt_3In-based β_2

precipitates and Pt-, In-depleted Au-rich α_2 matrix, resulting in apparent hardening in the grain interior.

XRD revealed that the lattice parameter from the superimposed diffraction peaks of the solute-depleted Au-rich ($\alpha_1 + \alpha_2$) phases ($a_{200} = 4.0231$ Å) was similar to that of the parent Au-rich α phase ($a_{200} = 4.0222$ Å). This is because the precipitated β_1 and β_2 phases were formed by the co-precipitation of Pt and In, which has a smaller and larger atomic size than Au, respectively (distance of the closest approach, $D=2.7747$ Å for Pt, $D=3.2515$ Å for In and $D=2.880$ Å for Au) [16].

Conclusions

This study examined the relationship of the lamellar-forming grain boundary reaction with the change in hardness during the aging process of the Au-Pt-based metal-ceramic alloy composed of 76.6 Au–9.9 Pt–9.3 Pd–1.7 In–1.2 Ag–0.56 Sn (wt%) with minor ingredients.

1. The phase decomposition of the parent Au-rich α phase occurred within a very short time (30 s) by aging at 550 °C after a solution-treatment at 950 °C, which initiated the lamellar-forming grain boundary reaction.
2. Both the grain interior and grain boundary precipitate attributed to hardening, but the grain boundary precipitate comprising the fine lamellar structure caused more powerful hardening from the early stage of the aging process.
3. The alternate layer of the lamellar structure was composed of a Pt-, In- and Sn-depleted Au-rich α_1 phase and a precipitated Pt-, In- and Sn-concentrated β_1 phase.
4. The extremely fine nature of the grain boundary lamellar structure supplied large numbers of inter-phase boundaries that contained lattice strain due to difference in lattice parameter between the α_1 and β_1 phases, resulting in hardening.

Acknowledgments This research was supported by Basic Science Research Program through the National Research Foundation of Korea (NRF) funded by the Ministry of Education, Science and Technology (grant number: 2011-0010995).

References

1. Anusavice KJ (2006) Phillips' science of dental materials, 11th edn. Saunders WB, Philadelphia, pp 621–654
2. German RM (1980) Hardening reactions in a high-gold content ceramo-metal alloy. J Dent Res 59:1960–1965
3. O'Brien WJ, Kring JE, Ryge G (1964) Heat treatment of alloys to be used for the fused porcelain technique. J Prosthet Dent 14:955–960
4. Leinfelder KF, O'Brien WJ, Ryge G, Fairhurst CW (1966) Hardening of high-fusing gold alloys. J Dent Res 45:392–396
5. Craig RG (1989) Restorative dental materials. The CV Mosby Co., St Louis, p 499
6. Ohta M, Hisatsune K, Yamane M (1975) Study on the age-hardenable sillver alloy (3rd Report) III on the ageing process of dental Ag-Pd-Cu-Au alloy. J Jpn Soc Dent Appar Mater 16:87–92
7. Yasuda K, Ohta M (1980) Age-hardening characteristics of a commercial dental gold alloy. J Less-Common Metals 70:75–87
8. Yasuda K, Udoh K, Hisatune K, Otha M (1983) Structural change induced by ageing in commercial dental gold alloys containing palladium. Dent Mater 2:48–58
9. Hisatsune K, Hasaka M, Sosrosoedirdjo BI, Udoh K (1990) Age-hardening behaviour in a palladium-based dental porcelain-fused alloy. Mater Charact 25:177–184
10. Yasuda K, Hisatsune K (1993) Microstructure and phase transformations in dental gold alloys. Gold Bull 26:50–66
11. Jeon GH, Kwon YH, Seol HJ, Kim HI (2008) Hardening and mechanisms in an Au-Ag-Cu-Pd alloy with In additions. Gold Bull 41:257–263
12. Wise EM, Crowell W, Eash JT (1932) The role of the platinum metals in dental alloys II. Trans Met Soc AIME 99:363–412
13. Wise EM, Eash JT (1933) The role of the platinum metals in dental alloys III. Trans Met Soc AIME 104:276–307
14. Massalski TB (1990) Binary alloy phase diagrams, 2nd ed. ASM International, Materials park, pp 409-410 (Au-Pd), pp 414-416 (Au-Pt), pp 3033-3034 (Pd-Pt)
15. Hisatsune K, Tanaka T, Udoh K, Yasuda K (1977) Ageing reactions in a high carat gold alloy for dental porcelain bonding. J Mater Sci 8:277–282
16. Culity BD (1978) Elements of X-ray diffraction, 2nd edn. Addison-Wesley publishing Co., Inc., Massachusetts, pp 506–507
17. Villars P, Calvert LD (1985) Pearson's handbook of crystallographic data for intermetallic phases. American Society for Metals, Metals Park, p 2562
18. Hisatsune K, Otha M, Shiraishi T, Yamane M (1982) Age hardening in a dental white gold alloy. J Less-Common Metals 83:243–253
19. Udoh K, Hisatsune K, Yasuda K, Otha M (1984) Isothermal age-hardening behavior in commercial dental gold alloys containing palladium. Dent Mater 3:253–261
20. Hisatsune K, Tanaka Y, Udoh K, Yasuda K (1997) Ageing reactions in a high carat gold alloy for dental porcelain bonding. J Mater Sci Mater Med 8:277–282
21. Kim HI, Park YH, Lee HK, Seol HJ, Shiraishi T, Hisatsune K (2002) Precipitation hardening in a dental low-gold alloy. Dent Mater 22:10–20

20

Nanostructured and nanopatterned gold surfaces: application to the surface-enhanced Raman spectroscopy

A. Bouvrée · A. D'Orlando · T. Makiabadi · S. Martin ·
G. Louarn · J. Y. Mevellec · B. Humbert

Abstract Surface-enhanced Raman spectroscopy (SERS) has enormous potential for a range of applications where high sensitivity needs to be combined with good discrimination between molecular targets. However, the SERS technique has trouble finding its industrial development, as was the case with the surface plasmon resonance technology. The main reason is the difficulty to produce stable, reproducible, and highly efficient substrates for quantitative measurements. In this paper, we report a method to obtain two-dimensional regular nanopatterns of gold nanoparticles (AuNPs). The resulting patterns were evaluated by SERS. Our bottom-up strategy was divided into two steps: (a) nanopatterning of the substrate by e-beam lithography and (b) electrostatic adsorption of AuNPs on functionalized substrates. This approach enabled us to highlight the optimal conditions to obtain monolayer, rows, or ring of AuNPs, with homogeneous distribution and high density (800 AuNPs/μm^2). The nanostructure distributions on the substrates were displayed by scanning electron microscopy and atomic force microscopy images. Optical properties of our nanostructures were characterized by visible extinction spectra and by the measured enhancements of Raman scattering. Finally, we tried to demonstrate experimentally that, to observe a significant enhancement of SERS, the gold diffusers must be extremely closer. If electron beam lithography is a very attractive technique to perform

reproducible SERS substrates, the realization of pattern needs a very high resolution, with distances between nanostructures probably of less than 20 nm.

Keywords Surface-enhanced Raman scattering · SERS · Gold nanoparticles · Nanolithography

Introduction

Assembling nanoparticles into one-, two-, and three-dimensional (1D, 2D, and 3D) arrays attract considerable interest due to their capabilities to generate strong enhancement of electromagnetic field in their near surroundings. This effect is also successfully exploited in Raman spectroscopy, spectroscopy based on the inelastic scattering of a monochromatic light after its interaction with a molecule for instance. Thus, this improvement of the scattering, so called "surface-enhanced Raman scattering," is used for the characterization of materials or more recently in the design of chemical and biochemical sensors [1–3].

Use gold nanoparticles (AuNPs) in aqueous solutions as a way for enhancing Raman scattering was first demonstrated by Creighton et al. [4]. This observed enhancement was due to the interaction of light with metallic nanoparticles. This interaction can be characterized by strong optical resonances due to the excitation of localized surface plasmons [5]. One of the main consequences of plasmon excitation is the high electromagnetic field created at the surface of the nanoparticle. This field is then involved in enhancing the efficiency of Raman process. It should be stressed here that the intensity of the electromagnetic field depends strongly on the nanoparticle size, shape, intercenter distance, and interparticle separation [3].

A. Bouvrée · A. D'Orlando · T. Makiabadi · S. Martin ·
G. Louarn (✉) · J. Y. Mevellec · B. Humbert
Institut des Matériaux Jean Rouxel (IMN), CNRS, Université de Nantes, 2 rue de la Houssinière, BP 32229, 44322 Nantes cedex 3, France
e-mail: guy.louarn@cnrs-imn.fr

As a consequence, numerous papers in the literature have developed smart strategies to obtain Raman active substrates, with controlled uniform roughness, or regular nanostructure distributions, with good reproducibility [1–3, 6–8]. In parallel, many approaches were proposed to optimize and control the surface-enhanced Raman scattering efficiency of colloidal AuNPs assembled on substrates into 2D nanostructures [9–19].

Here, we report an alternative procedure for preparing patterned AuNP arrays by combining electron beam lithography which allows control of lateral dimensions down to 20 nm, with self-assembly of silanes on silicon oxide in the micrometer and submicrometer range to direct the AuNPs assembly. A thin polymer layer on a substrate was patterned. The substrate consisted of a polymer relief pattern with exposed substrate regions in between, where (3-aminopropyl)-dimethylethoxysilane (APDMES) was grafted. This substrate could be converted to chemically patterned substrates by removing the polymer template. Then, substrates were used to attach AuNPs by immersion. In this manner, close packing was found on micrometer-sized features, and typical confined particle geometries were observed on submicrometer features. The self-assembly of monolayers of AuNPs on organosilane-coated substrates yields macroscopic surfaces that are highly active for surface-enhanced Raman spectroscopy (SERS) [1, 19]. The strategy delineated herein was the utilization of gold particles to form a well-ordered close-packed 2D nanostructure. Then, thanks to e-beam lithography, given a regular and uniform distribution of lines and circles, we were able to perform a Raman study as a function of the density of nanoparticles adsorbed on the surface and the distance between them.

Materials and methods

Materials

HAuCl$_4$ for AuNPs preparation was bought from Fluka. Other chemical materials, such as sodium citrate (99 %), methanol (99.8 %), APDMES (97 %), and crystal violet (95 %), were purchased from Sigma-Aldrich. Deionized water with a specific resistance of 18 MΩ was used for all preparations.

Instrumentation

Raman spectra were obtained and recorded on a spectrometer multichannel Jobin-Yvon T64000 (HORIBA) spectrometer using the 676.4 nm line of krypton ion lasers. All the experiments were carried out under a confocal microscope to focus the excitation line on the nanopatterned area. A holographic NOTCH filter or a double subtractive stage with gratings with 1,800 grooves mm^{-1} may be used to remove the Rayleigh scattering. A single dispersion stage with 1,800 grooves mm^{-1} grating is placed in front of the detector. The spectral resolution, with excitation at 676.4 nm is 2 cm^{-1}. Raman spectrometer optically conjugated with an Olympus Model BX51 upright microscope was equipped with a motorized XY stage with a step of 80 nm. This setup allowed us to obtain a better lateral spatial resolution than 2 μm [20].

Topographic and morphologic measurements were recorded using scanning electron microscopy (SEM—JEOL JSM 6400 F1) and atomic force microscopy (AFM VEECO Nanoscope IIIa), respectively. The UV-visible absorbance spectra between 200 and 800 nm were obtained with a Varian Cary 2300 spectrometer. The particle size distribution of suspensions was determined by dynamic light scattering using a Zetasizer Nano ZS instrument equipped with a 633-nm He–Ne laser (Malvern Instruments Ltd., Malvern, Worcestershire, UK), which measured the scattered light at an angle of 173° and at a temperature of 25 °C.

Methods

Gold colloids were prepared by modifying the method of citrate thermal reduction reported in [21, 22]. Typically, in the process of thermal reduction, gold colloids were prepared by adding 1 ml of 1 wt% HAuCl$_4$ aqueous solution and 1.5 ml of 38.8 mM sodium citrate aqueous solution drop at drop into 90 ml boiling water under stirring. Then, after 10 min, the solution becomes ruby red signature of the action of citrate ions as both reducer and stabilizer. Indeed, as shown in Fig. 2a, citrate ions are adsorbed on the AuNP, and they modify the charge of AuNPs surface (Zeta potential about −25 mV at pH 3). Finally, the solution was cooled naturally to room temperature with continued stirring. Then, the AuNPs aqueous solution has been concentrated by centrifugation for 10 min (7,800 rpm–10,000g). After this centrifugation step, a relatively high concentration of AuNPs has been estimated about 0.96±0.12 nmol/L from optical absorption spectra [23].

Preparation of SERS substrates

First step The electron beam nanopatterned substrates were produced from the technological platforms of the IMCN Institute, UCL, Belgium using the method described in reference [24]. Sketches of the protocol of silanization of patterned lines and rings are presented in Fig. 1. After deposition of PMMA by "spin-coating" on a silicon substrate, different patterns have been formed by electron beam lithography. From this process, the patterns (grooves) in the PMMA provides a very high resolution (about 10 nm) [24, 25] and the film is very resistant to the silanization step. Various geometrical structures were processed, such as concentric rings and line networks. The geometric characteristics of the different samples studied in this work were outlined in the electronic

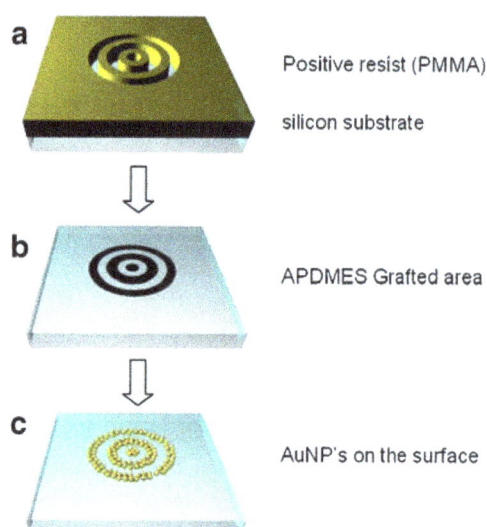

Fig. 1 Sketches of the AuNPs immobilization process on e-beam-patterned substrates (Coll. Alain Jonas, UCL); **a** substrate after e-beam writing step; **b** APDMES-coated substrate step; **c** AuNPs immobilization on APTMS-coated glass or silicon substrates

supplementary material (Fig. S1). Acetone was used to dissolve the PMMA mask at this step.

Second step Immediately after a piranha treatment, the substrate was placed on the sample holder in a Schlenk tube which was heated at 80 °C. The Schlenk tube was closed hermetically with a rubber septum. This tube was then purged several times before to be filled with N_2. APDMES (0.1 mL) was added through the rubber septum in the Schlenk tube, and the system was maintained overnight. At the end, the substrates were placed in a beaker containing dry toluene.

Third step The prepared substrate was then dried for about 30 min at about 100 °C and then rinsed again three times with methanol. Then, the silanized substrate was immersed for different times (also called "soaking times" in the following) in an AuNPs solution which had been concentrated (0.96± 0.12 nmol/L) by centrifugation for 10 min (10,000g). Finally, the substrate was rinsed with distilled water.

Raman experiments

SERS spectra of 1 μM crystal violet adsorbed on SERS substrates were carried out after dipping the substrates in a 1-μM dye aqueous solution. In this work, the data were collected immediately after immersion in the dye solution (3 min) without further rinsing, and the main experimental conditions can be summarized as: λ_{exc}=676.4 nm, power of the Raman laser P=0.1 mW, 2 μm diameter spot, and the typical exposure times were 30 s per point. We focused our attention on the C–C in-plane stretching vibration pointed at 1,620 cm^{-1}. The acquisitions were repeated (five to ten

Raman spectra for each experimental condition in our present study) at different locations in order to give a full statistical understanding of the enhancement factor.

Results and discussion

The surface morphology and the size of AuNPs were observed by transmission electron microscopy (TEM) and atomic force microscopy (AFM) (height image). In Fig. 2b, the TEM image shows nearly spherical and smooth AuNPs. From the TEM images, the average diameter of particles has been evaluated about 22 nm in complete agreement with the dynamic light scattering results (see supplementary materials, Fig. S2).

Then, these gold particles have been adsorbed on APDMES-coated glass slides thanks to electrostatic interactions between the positively charged amino-terminated groups of APDMES and the citrate anions available on the nanoparticle surface (Fig. 2a) [26]. It should be pointed here that the citrate ions are only adsorbed on the AuNPs; hence, the NPs are electrostatically adsorbed via citrates on the amine groups of silanes (Fig. 2c) [10]. In this way, the electrostatic adsorption implies that the NPs can be desorbed upon changes of ionic strength and pH. In order to prevent the desorption of AuNPs and according to the reference [19], significant attention has been paid to the drying after silanization (30 min, 100 °C) and after nanoparticles immobilization (10 min, 90 °C). This protocol is a needful step to prepare stable SERS active surfaces and to increase the reproducibility and stability of the surface. This drying probably reduced hydration of citrate anions adsorbed on the gold surface, and it allows a closer interaction of neighboring nanoparticles in the monolayer structure [19]. However, the stability and the regularity of the NP layers at different media and in presence of different chromophores will have to be considered carefully before each experiment.

Figure 2d, e shows the SEM and AFM (height) images of the AuNPs adsorbed onto modified solid substrates after soaking in AuNPs aqueous solution at 40 °C for 30 min. AuNPs in rather well-defined 2D structures were successfully constructed on glass or silicon slides. In fact, different temperatures have been tested from 20 to 50 °C. As presented in Fig. S3 (in the supplementary materials) at 50 °C and higher, a homogeneous distribution of AuNPs on the surface was not obtained. Indeed, a higher temperature increases the mobility of the nanoparticles on the surface, and the interparticle repulsion was screened. In the same way, for low soaking times, the particles are isolated but too far to induce high local electric field. After 30 min at 40 °C, the film demonstrated a close-packed colloid monolayer, while aggregation started afterwards (after 60 min and more). Thus, 40 °C and 30 min were fixed in our soaking process. In these conditions, the particles were closely spaced, and the film of AuNPs was relatively

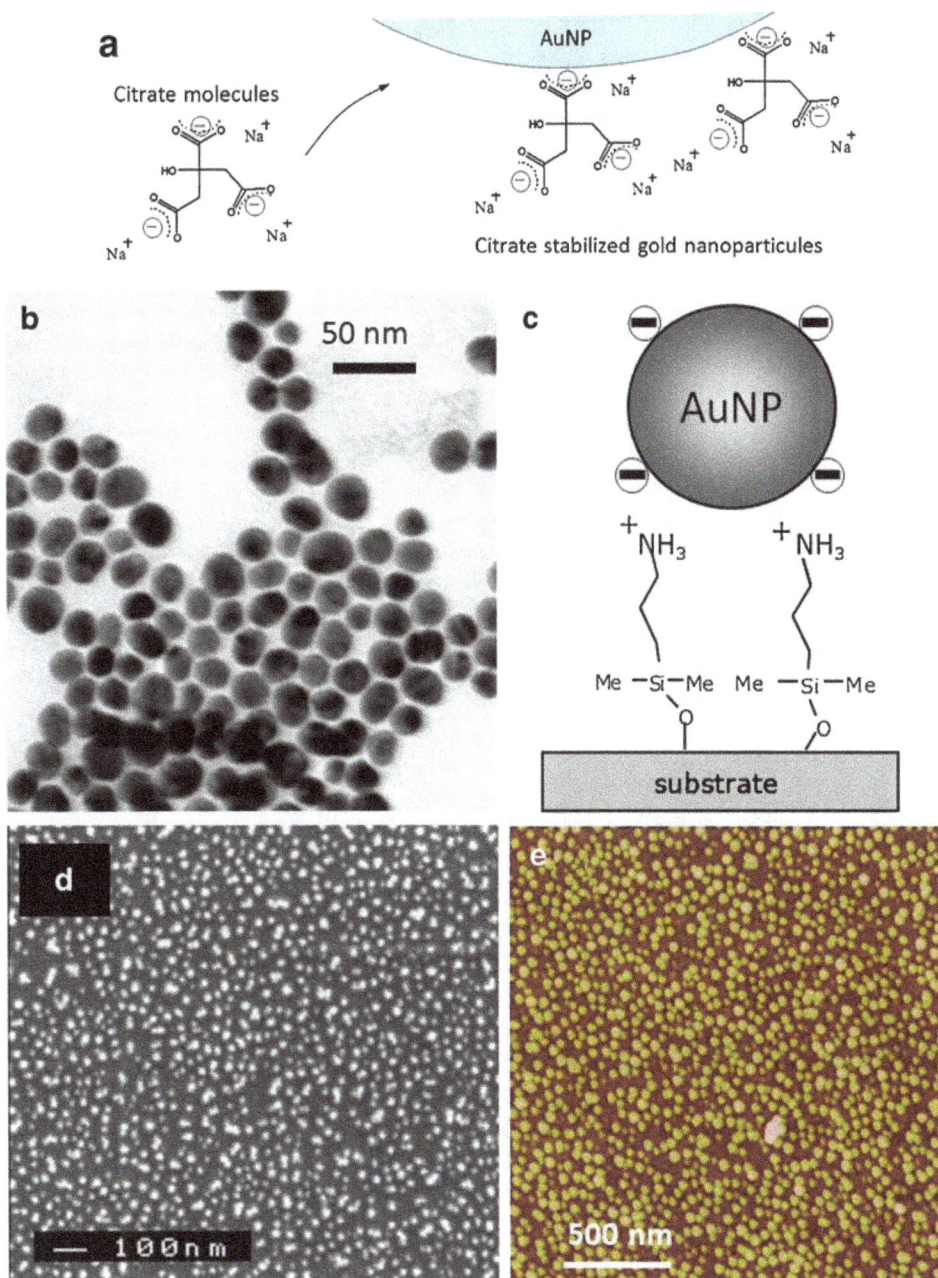

Fig. 2 a Schematic representation of the electrostatic adhesion process of citrate occurring during the synthesis of AuNPs; **b** TEM image of the AuNPs obtained by reducing HAuCl4 with Na-citrate; **c** schematic representation of AuNPS immobilization on APDMES-coated glass substrates; **d** SEM image and **e** tapping mode AFM image of immobilized AuNPs on APDMES-coated glass prepared by soaking the substrate in an AuNPs aqueous solution (0.96± 0.12 nmol/L) for 30 min and at 40 °C

uniform with a rather good regularity over wide areas. These observations confirmed previous studies [27–29].

UV-visible adsorption spectra of colloidal suspension and substrates

The absorbance of the gold nanoparticle samples was measured by optical spectroscopy. A typical UV-visible adsorption spectrum of the gold colloidal solution is shown in Fig. 3a(a). The adsorption spectra show a single maximum at 520 nm in aqueous medium. This absorption band results when the incident photon frequency is resonant with the collective oscillation of the conduction band electrons and is known as the SPR [30, 31].

Figure 3a(b–d) shows the extinction spectra of substrates after the immobilization of nanoparticles for different soaking times (3, 30, and 60 min) in the colloidal solution of AuNPs. The extinction spectrum contains two bands. The first one was pointed at 545 nm and corresponds to isolated AuNPs in the vicinity of the glass substrate. The observed shift from 520 to 545 nm was due to the modification of the surrounding medium. Concerning the second bands (about 660 nm) indicates the interaction of close nanoparticles and results from dipole–dipole interactions. As presented in Fig. 4a, this coupling between nanoparticles likely results from the increase of

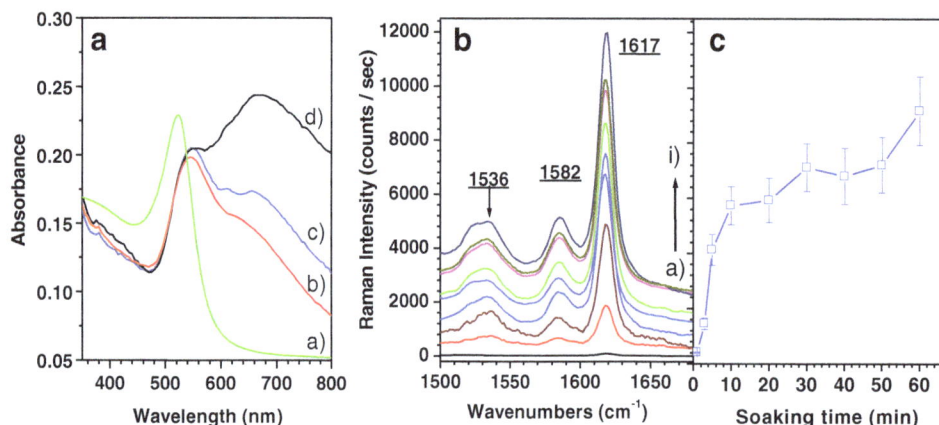

Fig. 3 **a** Extinction spectra of AuNPs (mean diameter of about 22 nm) in water (*a*) and electrostatically adsorbed on APDMES-coated glass as a function of the soaking time of the substrate in a solution of AuNPs [(*b*) 3 min, (*c*) 30 min, (*d*) 60 min]. **b** SERS spectra recorded on a substrate dipped in a 1-μM crystal violet aqueous solution (*a*–*i*): 0, 3, 5, 10, 20, 30, 40, 50, and 60 min, respectively; **c** Raman intensity of the band at

1,620 cm^{-1} of crystal violet as a function of the soaking time of the substrate in the colloidal solution. The acquisitions of data were repeated (five to ten Raman spectra for each experimental condition in our present study) at different locations in order to give a full statistical understanding of the enhancement factor. The intensity was deducted from the baseline in **b**

the density of nanoparticles on the surface (from 300 to 800 AuNPs/μm^2). In consequence, from a rough approximation, we estimated that the mean inter-AuNPs of distance decreased from 35 to 14 nm, as illustrated in Fig. 4b. It should be pointed here that these numerical values were estimated from SEM images (see Fig. S3 in complementary materials). The calculations of density were performed five times for each experimental condition and at different locations.

Raman enhancement measurements

SERS spectra of crystal violet adsorbed from the dye 1 μM aqueous solution on the Au colloidal film were registered with a 676.4-nm excitation in the 1,500 to 1,700 cm^{-1} range (Fig. 3b). The SERS intensity varied appreciably with the soaking time, the density, and the distance between AuNPs immobilized on the substrates. Here, we focused our attention on the C–C in-plane stretching mode observed at 1,620 cm^{-1} [32].

Figure 3c shows the enhancement intensity as a function of the density of AuNPs adsorbed on the substrates (immersion time in the colloidal solution of AuNPs, see Fig. 4). The Raman intensity increases strongly at first and slower after 10 min of soaking time. Thus, as expected, gold surfaces with periodic nanoscale features enhance the Raman scattering. Indeed, the increase of the density of particles on the surface and the decrease of the distance between them enhances the local electromagnetic field and Raman scattering [18, 33]. Accordingly, we assume that the number of locations of "hot spots" grows, which leads to observe strong Raman scattering signals. Likely, the increased number of adsorbed AuNPs is not the only/main parameter leading to higher Raman signal. As we can see in Fig. 3a, the increased number of adsorbed AuNPs induced a new absorption band around 660 nm. It

could be stressed here that this absorption band was close to the excitation wavelength used in the Raman experiments (λex=676.4 nm). The correlation between the SERS intensity and the LSPR band at ca. 660 nm confirms once more the importance to optimize the experimental condition in this type of spectroscopy in order to simultaneously benefit of all enhancing parameters.

e-Beam nanopatterned substrates

In the first part of our work, we studied the impact of the interparticle distances of few nanometers which is the most important parameter to optimize for SERS efficiency (when the NP size is fixed). In order to obtain a best control on the regularity and on the reproducibility, thanks to the e-beam nanolithography, we have developed different subtracts with various patterns. We have chosen two kinds of patterns, lines and circles. The interline and inter-circle distances were set from 20 to 2,000 nm. Two line thickness values, 30 and 60 nm, had been achieved, corresponding to bands of two or three AuNP widths. Unfortunately, non-consistent results had been obtained on lines and circles with low interline distances (20 and 40 nm). So, we decided to suppress values concerning them. It should be pointed here that "line" patterns were sensitive to the light polarization, whereas "circle" patterns were nonsensitive.

Considering the preliminary study presented above, we chosen to immerse the nanopatterned samples for 30 min in the AuNP solution. In Fig. 5a–d, AFM and SEM images showing two examples of surface prepared by the method of controlled adsorption are presented. Observations show that the nanoparticles are deposited preferentially on the silanized patterns. The widths of the AuNP lines are the same order of magnitude as the patterns of the masks of PMMA. About three

Fig. 4 a Density of AuNPs on the APDMES-coated glass as a function of the soaking time. Values were estimated from FE-SEM images (see supplementary materials). The measures of density were performed five times for each experimental condition and at different locations (each measure on a 1-μm^2 area). **b** Mean interparticle distances estimated from FE-SEM image and rough calculation (based on a regular hexagonal arrangement of nanoparticles)

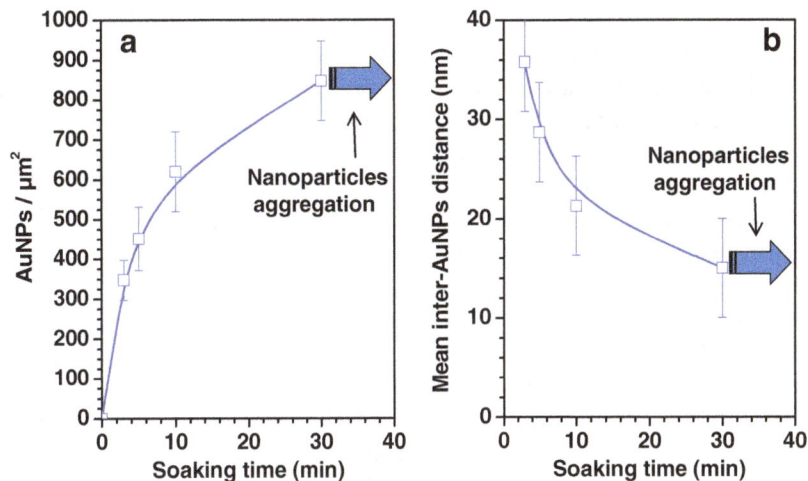

or four nanoparticles were arranged on the line width. In these regions, the height profiles are in the order of magnitude of diameters of nanoparticles in solution. This suggests that a monolayer of nanoparticles was adsorbed.

Figure 5e displays experimental Raman scattering results obtained on various patterns. The intensity (counts/s) of the band at 1,620 cm^{-1} (crystal violet) is reported as a function of the distance interlines. It should be noted that the intensity of Raman scattering varied significantly depending on the pattern. Signals recorded on concentric rings were lower than observed for lines in agreement with simple light polarization considerations.

We can highlight the effectiveness of patterns of lines separated by 60 nm. On these surfaces, the enhancement can be explained by tightening of the lines and the larger number of lines in the analysis surface. This observation is supported by the pattern where the lines are separated by 2 μm. In this case, the diameter of the incident laser beam is about 2 μm, and only one or two lines were involved in the Raman

experiments. The number of hot spots is limited (compared to other patterns), and therefore, we observe a low intensity of the Raman signal. However, no significant effect was demonstrated when spacing was beyond 40 nm. This observation confirms that the SERS experiments require that gold diffusers have to be very close and 40 nm is already a lot.

Conclusions

Immobilization of rather regular 2D nanostructures of AuNP was obtained on a APDMES-coated glass or silicon substrates with different soaking time in an AuNPs aqueous solution at 40 °C. Density and mean interparticle distances change as a function of soaking times were studied by SEM and optical spectroscopy (UV-vis). It was found that the red shift of the absorption is strongly dependent on the nanoparticles distribution on the surface. Then, the SERS activity observed from the gold surfaces was compared after dipping the substrates in

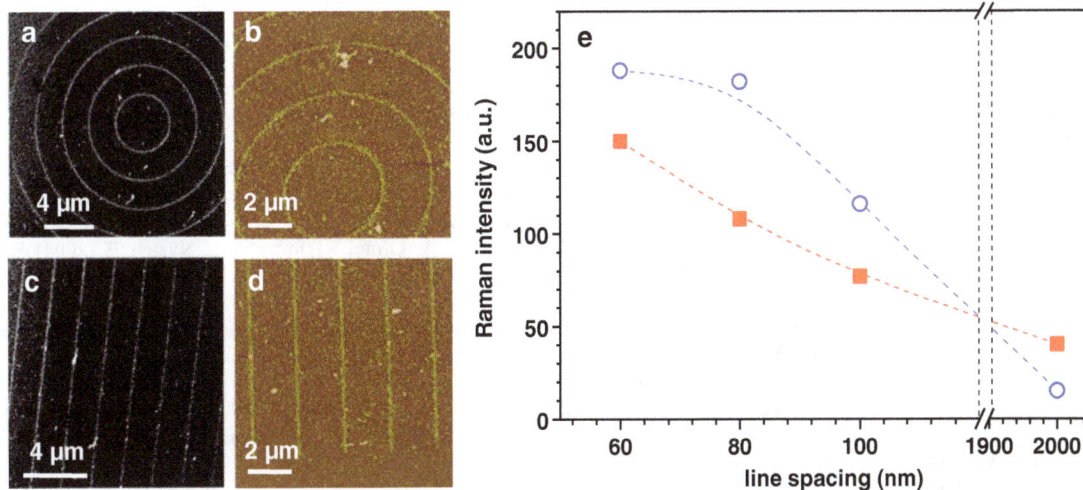

Fig. 5 SEM (**a** and **c**) images and tapping mode AFM images (**b** and **d**) of immobilized AuNPs on nanopatterned substrates (by immersing the substrate in a suspension of AuNPs). **e** Comparison of the Raman intensity measured for the band of crystal violet at 1,620 cm^{-1} as a function of the distance between AuNPs lines (*square* : 30 nm width of the line, *circle* : 60 nm width of the lines)

a 1-μM crystal violet aqueous solution. The SERS intensity varied significantly with the change in the nanoparticle density adsorbed on the substrates. The reasons for these changes were assigned to the distance inter-particles and the correlation between the experimental condition and the LSPR. Finally, controlled gold nanostructures on nanopatterned silicon substrates have been made. The followed method allowed the preparation for structured substrates with good resolution (patterns width inferior to 100 nm). The analysis of Raman spectra of crystal violet set down on these surfaces tends to demonstrate the influence of tightening the lines of each pattern. However, no effect had been put in evidence when interline was beyond 40 nm. From this work, we have tried to demonstrate experimentally that to observe a significant enhancement by SERS, the gold diffusers must be extremely closer. In this context, electron beam lithography will be a very competitive and attractive technique for the realization of SERS substrate if it allows to perform regular patterns separate of less than 20 nm.

Acknowledgments We thank Victor Le Nader at IMN for the helpful discussions.

References

1. Aroca R (2006) Surface-enhanced vibrational spectroscopy. Wiley, Chichester
2. Bell SEJ, Sirimuthu NMS (2008) Quantitative surface-enhanced Raman spectroscopy. Chem Soc Rev 37:012–1024.
3. Louis C, Pluchery O (2012) Gold nanoparticles for physics, chemistry and biology. Imperial College, London, p 395
4. Creighton JA, Blatchford CG, Albrecht MG (1979) Plasma resonance enhancement of Raman scattering by pyridine adsorbed on silver or gold sol particles of size comparable to the excitation wavelength. J Chem Soc Faraday Trans 75:790–798.
5. Novotny L, Hecht B (2006) Principles of nano-optics. Cambridge University, Cambridge
6. Schlegel VL, Cotton TM (1991) Silver-island films as substrates for enhanced Raman scattering: effect of deposition rate on intensity. Anal Chem 63:241–247.
7. Abu Hatab NA, Oran JM, Michael Sepaniak MJ (2008) Surface-enhanced Raman spectroscopy substrates created via electron beam lithography and nanotransfer printing. ACS Nano 2:377–385.
8. Zrimsek AB, Henry A-I, Van Duyne RP (2013) Single molecule surface-enhanced Raman spectroscopy without nanogaps. J Phys Chem Lett 4:3206–3210.
9. Zhu T, Yu HZ, Wang J, Wang YQ, Cai SM, Liu ZF (1997) Two-dimensional surface enhanced Raman mapping of differently prepared gold substrates with an azobenzene self-assembled monolayer. Chem Phys Lett 265:334–340
10. Zheng J, Zhu Z, Chen H, Liu Z (2000) Nanopatterned assembling of colloidal gold nanoparticles on silicon. Langmuir 16:4409–4412.
11. Baibarac M, Mihut L, Louarn G, Lefrant S, Baltog Y (2000) Doping and metallic-support effect evidenced on SERS spectra of polyaniline thin films. J Polym Sci Polym Phys 38:2599–2609.
12. Li X, Xu W, Zhang J, Jia H, Yang B, Zhao B, Ozaki Y (2004) Self-assembled metal colloid films: two approaches for preparing new SERS active substrates. Langmuir 20:1298–1304.
13. Maury P, Escalante M, Reinhoudt DN, Huskens J (2005) Directed assembly of nanoparticles onto polymer- imprinted or chemically patterned templates fabricated. Adv Mater 17:2718–2723.
14. Wang H, Levin CS, Halas NJ (2005) Nanosphere arrays with controlled sub-10-nm gaps as surface-enhanced Raman spectroscopy substrates. J Am Chem Soc 127:14992–14993.
15. Toderas F, Baia M, Baia L, Astilean S (2007) Controlling gold nanoparticle assemblies for efficient surface-enhanced Raman scattering and localized surface plasmon resonance sensors. Nanotechnology 18:255702.
16. Hossain MK, Shibamoto K, Ishioka K, Kitajima M, Mitani, Nakashima S (2007) 2D nanostructure of gold nanoparticles: an approach to SERS-active substrate. J Lumin 122:792–795.
17. Wang Y, Chen H, Wang E (2008) Facile fabrication of gold nanoparticle arrays for efficient surface-enhanced Raman scattering. Nanotechnology 318(1):82–7.
18. Konrad MP, Doherty AP, Bell SEJ (2013) Stable and uniform SERS signals from self-assembled two-dimensional interfacial arrays of optically coupled Ag nanoparticles. Anal Chem 85:6783–6789.
19. Hajduková N, Procházka M, Štěpánek J, Špírková M (2007) Chemically reduced and laser-ablated gold nanoparticles immobilized to silanized glass plates: preparation, characterization and SERS spectral testing. Colloids Surf A 301:264–270.
20. Grausem J, Humbert B, Spajer M, Courjon D, Burneau A, Oswalt J (1999) Near-field Raman spectroscopy. J Raman Spectrosc 30:833–840.
21. Turkevich J, Stevenson PC, Hillier J (1951) A study of the nucleation and growth processes in the synthesis of colloidal gold. Discuss Faraday Soc 11:55–75.
22. Frens G (1972) Controlled nucleation for the regulation of the particle size in monodisperse gold suspensions. Nature 241:20–22.
23. Haiss W, Thanh NTK, Aveyard J, Fernig DG (2007) Determination of size and concentration of gold nanoparticles from UV-vis spectra. Anal Chem 79:4215–4221.
24. Pallandre A, Glinel K, Jonas AM, Nysten B (2004) Binary nanopatterned surfaces prepared from silane monolayers. Nano Lett 4:365–371.
25. Baralia GG, Pallandre A, Nysten B, Jonas AM (2006) Nanopatterned self-assembled monolayers. Nanotechnology 17:1160.
26. Shipway AN, Katz E, Willner I (2000) Nanoparticle arrays on surfaces for electronic, optical, and sensor applications. Chem Phys Chem 1:18–52.
27. Grabar KC, Allison KJ, Baker BE (1996) Two-dimensional arrays of colloidal gold particles: a flexible approach to macroscopic metal surfaces. Langmuir 12:2353–2361.

28. Grabar KC, Freeman RG, Hommer MB, Natan MJ (1995) Preparation and characterization of Au colloid monolayers. Anal Chem 67:735–743.

29. Pazos-Perez N, Ni W, Schweikart A, Alvarez-Puebla RA, Fery A, Liz-Marzan LM (2010) Highly uniform SERS substrates formed by wrinkle-confined drying of gold colloids. Chem Sci 1:174–178.

30. Basu S, Kumar Ghosh S (2007) Biomolecule induced nanoparticle aggregation: effect of particle size on interparticle coupling. J Colloid Interface Sci 313:724–734.

31. Makiabadi T, Bouvrée A, Le Nader V, Terrisse H, Louarn G (2010) Preparation, optimization and characterization of SERS sensor substrates based on two dimensional structures of Au colloid. Plasmonics 5:21–29.

32. Canamares MV, Chenal C, Birke RL, Lombardi JR (2008) DFT, SERS, and single-molecule SERS of crystal violet. J Phys Chem C 112:20295–20300.

33. Campion A, Kambhampati P (1998) Surface-enhanced Raman scattering. Chem Soc Rev 27:241–250.

21

Phase diagram of Au-Al-Cu at 500 °C

Jyun Lin Li · Pei Jen Lo · Ming Chi Ho · Ker-Chang Hsieh

Abstract Diffusion couples and equilibrated alloys were used to construct the isothermal phase diagram of Au-Al-Cu at 500 °C. Electron microprobe analyses were performed to determine the phase compositions and phase relationships. Two ternary phases and 10 three-phase equilibriums were determined in this study. Four additional three-phase equilibriums were estimated to meet the criteria for phase relationships. The δ (Au_2Al) phase exhibited a wide range of solubility, and the lattice parameters were examined by X-ray. The solubility ranges of the binary intermetallic phases were also determined.

Keywords Phase diagram · Au-Al-Cu · Au-base alloy · Phase diagram · Diffusion couple

Introduction

Gold wire has been used as bonding wire in IC packages because of its ductility and anticorrosion characteristics. Recently, the increase in the I/O number of ICs from 200 to 300 has required diameter of gold wires to decrease from 25 μm to less than 20 μm to match the decrease in the area of the Al pad. Nevertheless, the stiffness of gold wire must be enhanced to avoid short circuits derived from wire sweep. Au-Cu alloy wire is one of the new alloy bonding wires that has been developed. The bonding wire reacts with the Al pad and forms intermetallic phases to maintain the electrical connection. A low-temperature Au-Al-Cu phase diagram could provide essential information to understand the reactions between the Au-Cu alloy wire and the Al pad. The stability of intermetallic phases is closely related to the reliability of wire bonding [1, 2].

A β ($Au_7Al_4Cu_5$) phase exhibiting martensite-like properties was developed [3–5]. The embossed and fine-layered surface of this material, which originates from a martensitic transformation, exhibits a characteristic shining color and is used in Spangold jewelry.

Binary systems in Au-Al-Cu systems have been studied in depth. Al-Au binary systems exhibit five intermediate phases at 500 °C, namely, $AuAl_2$, $AuAl$, Au_2Al, Au_8Al_3 (Au_5Al_2), and Au_4Al [6]. Al-Cu binary systems exhibit five intermediate phases at 500 °C, namely, θ (Al_2Cu), η_2 ($AlCu$), ζ_2 (Al_9Cu_{11}), δ (Al_2Cu_3), and γ_1 phases [7]. Au-Cu binary systems form solid solution at 500 °C [8].

In recent decades, several researchers have studied Au-Al-Cu ternary systems. The isothermal sections of Au-Al-Cu at 500 °C were reported by Levey et al. [9–11] and those at 750 °C by Bhatia et al. [12]. Levey et al. [13] also constructed isopleths for the 76 wt.% Au section. These previous phase-diagram studies were conducted by applying traditional bulk alloy equilibrium. Isothermal phase diagrams were constructed by preparing bulk alloys with various compositions and annealing them at a certain high temperature to reach phase equilibrium. This method is not suitable for constructing gold-based low-temperature phase diagrams because it is too expensive to prepare 30 to 50 samples without recovering the Au precious metal, and it would take much time to reach phase equilibrium because of the low temperature used. This study applied the diffusion couple method to construct the Au-Al-Cu ternary 500 °C phase diagram. This method uses little Au and easily attains the local phase equilibrium within 1 day. In this method, the interface and phases on both sides of the interface reach equilibrium, and the relationships among these phases are thereby elucidated. A detailed diffusion couple theory is available in [14].

J. L. Li · P. J. Lo · M. C. Ho · K.-C. Hsieh (✉)
Department of Materials and Optoelectronic Science, National Sun Yat-Sen University, 70, Lien-Hai Road, Kaohsiung 804, Taiwan
e-mail: khsieh@mail.nsysu.edu.tw

Table 1 Summary of phase equilibria in nine diffusion couples

Diffusion couple	Compositions (at.%) (Al/Au$_x$Cu$_{100-x}$)		Phase equilibria							
	Au	Cu								
D1	90	10	Al-AuAl$_2$	AuAl$_2$-AuAl	AuAl-δ (Au$_2$Al)	δ (Au$_2$Al)-α				
D2	80	20	Al-Al$_2$Cu-AuAl$_2$	AuAl$_2$-AuAl	AuAl-δ (Au$_2$Al)	δ (Au$_2$Al)-β				
D3	70	30	Al-Al$_2$Cu	Al$_2$Cu-AuAl$_2$	AuAl$_2$-AuAl	AuAl-δ (Au$_2$Al)-γ	β-α			
D4	60	40	Al-Al$_2$Cu	Al$_2$Cu-ε-AlCu	AlCu-AuAl$_2$	AuAl$_2$-γ	γ-β	β-α		
D5	50	50	Al-Al$_2$Cu	Al$_2$Cu-ε-AlCu	AlCu-Al$_9$Cu$_{11}$	Al$_9$Cu$_{11}$-Al$_2$Cu$_3$	Al$_2$Cu$_3$-AuAl$_2$	AuAl$_2$-γ	γ-β	β-α
D6	40	60	Al-Al$_2$Cu	Al$_2$Cu-ε-AlCu	AlCu-AuAl$_2$-Al$_9$Cu$_{11}$	Al$_9$Cu$_{11}$-Al$_2$Cu$_3$	Al$_2$Cu$_3$-AuAl$_2$	AuAl$_2$-γ	γ-β	β-α
D7	30	70	Al-Al$_2$Cu	Al$_2$Cu-ε-AlCu	AlCu-AuAl$_2$-Al$_9$Cu$_{11}$	Al$_2$Cu$_3$-AuAl$_2$	AuAl$_2$-γ	γ-β	β-α	
D8	20	80	Al-Al$_2$Cu	Al$_2$Cu-ε-AlCu	AlCu-Al$_9$Cu$_{11}$-AuAl$_2$	Al$_9$Cu$_{11}$-Al$_2$Cu$_3$-AuAl$_2$	Al$_2$Cu$_3$-AuAl$_2$-γ	γ-α		
D9	10	90	Al-Al$_2$Cu	Al$_2$Cu-ε-AlCu	AlCu-Al$_9$Cu$_{11}$-AuAl$_2$	Al$_9$Cu$_{11}$-Al$_2$Cu$_3$-AuAl$_2$	Al$_2$Cu$_3$-AuAl$_2$-γ	γ-α		

Experimental procedures

The alloys were prepared through mini arc melting of pure Au, Cu, and Al under a pure Ar atmosphere. Afterward, the samples were cast into a 2-mm diameter ingot bar. Nine Au-Cu binary alloys (D1–D9) and 12 Au-Al-Cu ternary alloys (A1–A12) were prepared for this phase diagram study. The alloy compositions are listed in Tables 1 and 2.

Figure 1 is the schematic diagram of the diffusion couple setup. The Au-Cu binary alloy ingots (D1–D9) were sliced into 1-mm-thick discs, and a pure aluminum bar with a 2-mm diameter was sliced into 2-mm-thick discs. The discs were joined to form a diffusion couple. The screw of the holder exerted pressure on the diffusion couple. The holder was sealed in a quartz tube to form a vacuum, maintained at 500 °C for 24 h, and then quenched in ice water. The ternary alloys (A1–A12) were sealed in the quartz tube vacuum. These alloy samples were maintained at 500 °C for 14–29 days.

An electron probe microanalyzer (EPMA) was used to determine the equilibrium phase compositions of the diffusion couple samples and ternary alloy samples. The isothermal phase diagram of the Au-Al-Cu annealed at 500 °C was constructed based on these results.

Results

The phase equilibria of the diffusion couples are summarized in Table 1. As shown in Figs. 2a, b, the D1 diffusion couple formed the equilibrium phase layers Al/AuAl$_2$/AuAl/δ (Au$_2$Al)/α (Au, Cu). The phase layer thickness of AuAl$_2$ and AuAl was approximately 10 µm each, and the δ (Au$_2$Al) phase was approximately 300 µm thick. The latter binary phase extends into ternary with large composition range in Cu and Au. The original interface was present within the δ (Au$_2$Al) phase (Fig. 2a), indicating that the diffusion rates of Cu and Au were faster than that of Al. As shown in Figs. 3a, b, c, the D5 diffusion couple comprised several phase layers: Al/Al$_2$Cu/AlCu and ε/Al$_9$Cu$_{11}$/Al$_2$Cu$_3$/AuAl$_2$/γ/β/α (Au, Cu). The line-scan imaging results of the D5 diffusion sample are shown in Fig. 3d. A two-phase layer (AlCu+ε) was in contact with the Al$_2$Cu phase layer; therefore, the Al$_2$Cu-ε-AlCu three-phase equilibrium was generated. The existence of this three-phase equilibrium was confirmed through the alloy sample A1 and the equilibrium microstructure (Fig. 4). The phase equilibria of the alloy equilibrium samples are summarized in Table 2. According to the results in Tables 1 and 2, the determined phase diagram is shown in Fig. 5. In the diffusion couple experiment, seven sets of three-phase equilibria, namely, Al-AuAl$_2$-Al$_2$Cu, Al$_2$Cu-ε-AlCu, γ-AuAl$_2$-Al$_2$Cu$_3$, AuAl-δ (Au$_2$Al) -γ, δ (Au$_2$Al)-α-β, AlCu-AuAl$_2$-Al$_9$Cu$_{11}$, and Al$_9$Cu$_{11}$-AuAl$_2$-Al$_2$Cu$_3$ were determined from the phase

Fig. 1 The sketch diagram of
diffusion couple setup

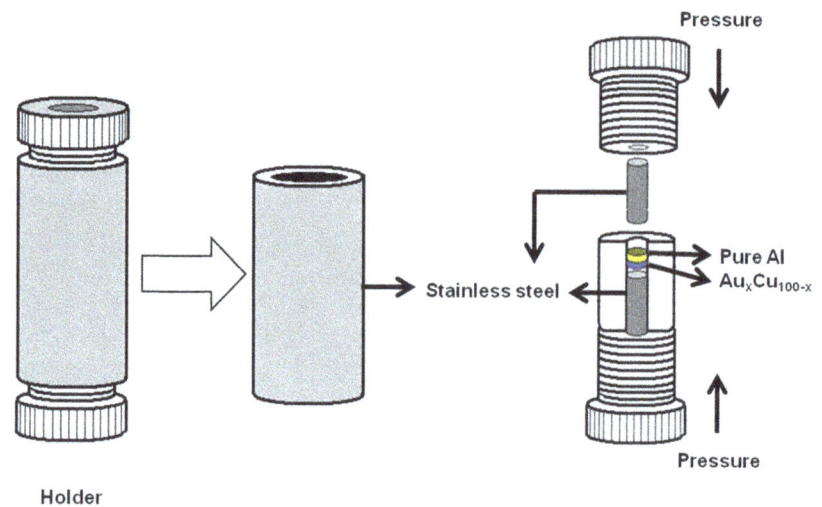

relationships, resembling the result of the Al_2Cu-ε-$AlCu$ three-phase equilibrium shown in Fig. 3c. The other three-

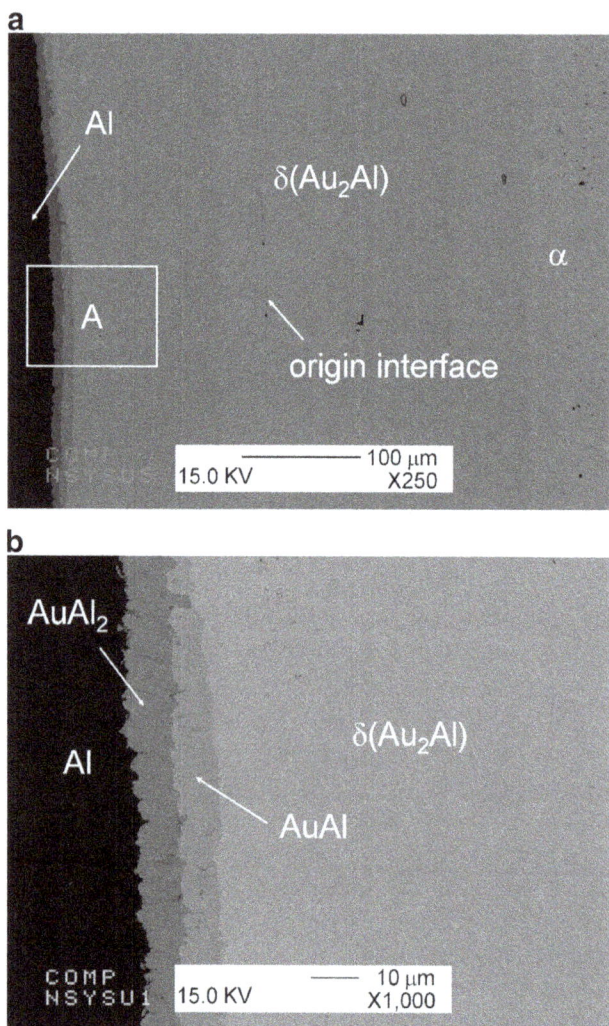

Fig. 2 a The whole view of D1 sample microstructure. **b** The detail phase layers of the A region in (**a**)

phase equilibria ($AuAl_2$-ε-$AlCu$, γ-$AuAl_2$-$AuAl$, and Au_8Al_3-δ (Au_2Al)-Au_4Al) were determined by applying the alloy equilibrium methods to alloy samples A2, A3, and A5 (Table 2). These 10 sets of three-phase equilibria are shown as solid lines in Fig. 5. In addition, the $AuAl_2$-ε-Al_2Cu three-phase equilibria were determined based on their two-phase equilibria. The other three sets of three-phase equilibria were estimated to meet the phase relationships shown as dashed lines in Fig. 5. The diffusion paths of D1 and D5 are plotted in Fig. 6, covering most of the binary and ternary phases. All of the binary intermediate phases in Al-Au [6] and Al-Cu [7] were identified in this study. For the Au-Cu [8] system, only the α (Au, Cu) phase appeared at 500 °C. The solubility of Al in the α (Au, Cu) phase was determined using an EPMA.

Figure 7 shows the Au-Al-Cu isothermal section obtained through bulk alloy equilibrium at 500 °C, reported by Levey et al. [9–11]. According to this study, the maximum Au solubility in the γ-$AlCu_2$ phase reached 48 at.%, agreeing with the result of Levey et al. [9–11]. The composition range and shape of the ternary β phase reported by these authors was confirmed in this study. In the D3–D7 samples, the β phase layer was present between the γ phase and α phase layers.

Ternary phase δ (Au_2Al) was also identified in this study. The homogeneity range of the δ (Au_2Al) phase was primarily determined from the composition range of the single-phase diffusion layer in the D1, D2, and D3 diffusion couples. The line-scan imaging results of the D1 diffusion sample are shown in Fig. 8. The peculiar concave shape was confirmed by analyzing the A4, A8, and A9 alloy samples, and the microstructure and compositions of tie lines between the δ (Au_2Al) and β phases are shown in Fig. 9. The crystal structure of δ (Au_2Al) phase was examined by X-ray in the A4, A8, A11, and A12 alloy samples. The lattice parameters are plotted and listed in Fig. 10.

The ε ternary phase identified in the D4–D9 diffusion couples comprised with AlCu phase as two-phase layer in

Fig. 3 a The whole view of D5 sample microstructure. **b** The detail phase layers of the A region in (**a**) as Al/Al$_2$Cu/AlCu, ε/Al$_9$Cu$_{11}$/Al$_2$Cu$_3$/AuAl$_2$/γ/β. **c** The detail phase layers of the A region in (**a**) as Al/Al$_2$Cu/AlCu, ε/Al$_9$Cu$_{11}$/Al$_2$Cu$_3$. **d** A composition profile of the D5 sample measured by EPMA

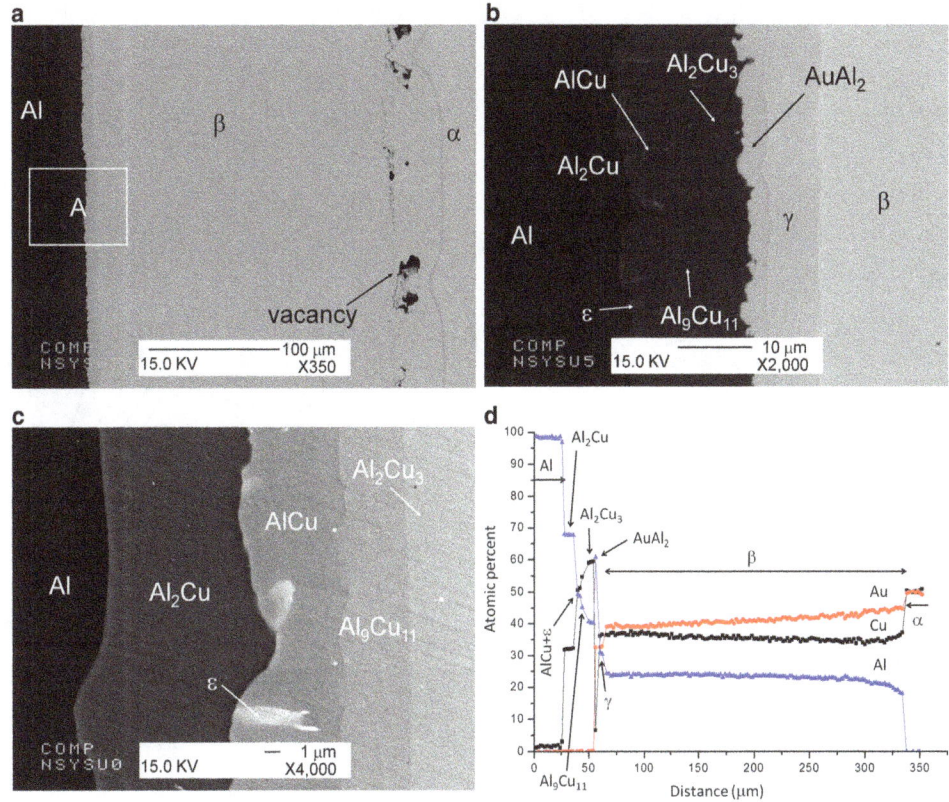

equilibrium with an AuAl$_2$ phase layer (Fig. 3c). The ε phase was not present as a single-phase layer in these diffusion couples. The composition range was determined using the three-phase equilibria of the Al$_2$Cu-ε-AlCu and AuAl$_2$-ε-

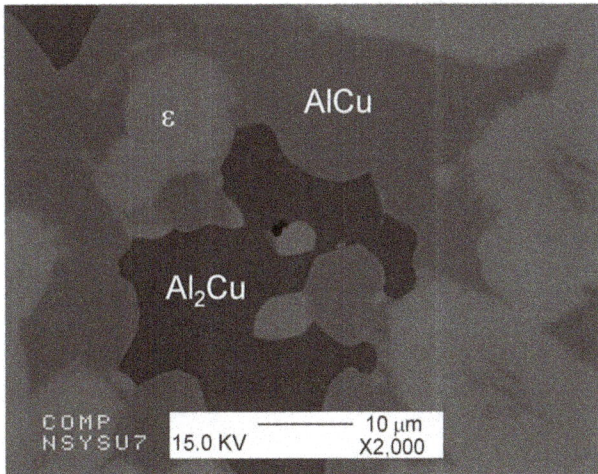

Fig. 4 The microstructure of A1 alloy sample including Al$_2$Cu, ε and AlCu phases

Color	Phase	Composition (at.%)		
		Au	Cu	Al
light gray	ε	4.21±0.30	33.39±0.57	62.40±0.58
dark gray	AlCu	none	49.48±1.12	50.49±0.13
black	Al$_2$Cu	none	32.31±1.81	67.67±1.81

Table 2 Summary of phase equilibria in 12 bulk alloy samples

Alloy	Alloy compositions (at.%) (Au$_x$Cu$_y$Al$_{100-x-y}$)			Phase equilibria
	Au	Cu	Al	
A1	2	39	59	Al$_2$Cu-ε-AlCu
A2	18	17	65	AuAl$_2$-ε-AlCu
A3	40	6	54	AuAl$_2$-AuAl-γ
A4	57.5	10	32.5	δ (Au$_2$Al)-β
A5	72	6	22	Au$_8$Al$_3$-δ (Au$_2$Al)-Au$_4$Al
A6	45	40	15	β-α
A7	47	39	14	β-α
A8	58	14	28	δ (Au$_2$Al)-β
A9	62	13	25	δ (Au$_2$Al)-β
A10	68	6	26	Au$_8$Al$_3$-δ (Au$_2$Al)
A11	68	11	21	δ (Au$_2$Al)
A12	62	6	32	δ (Au$_2$Al)

Fig. 5 The isotherm of Au-Al-Cu at 500 °C determined by this study

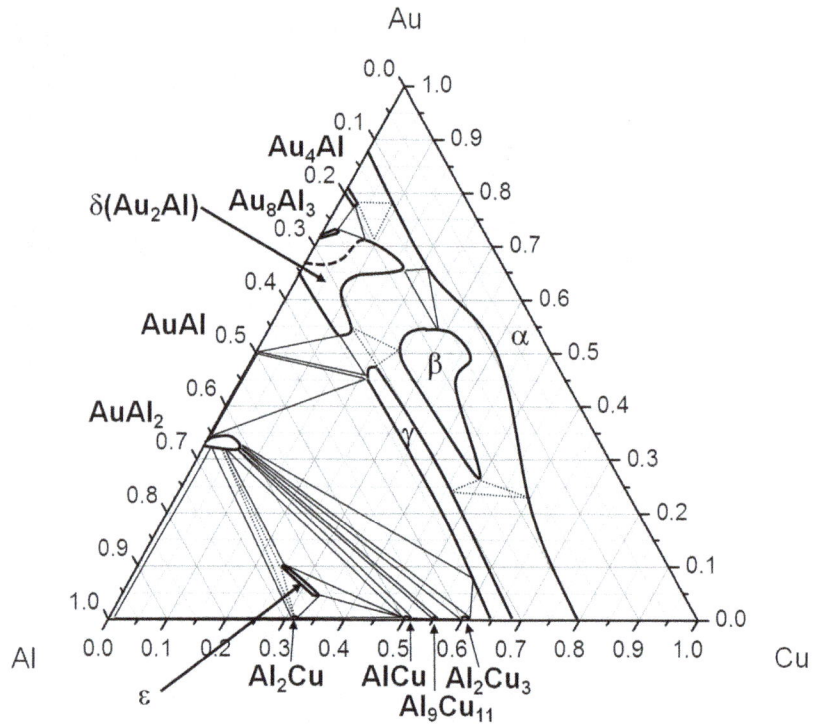

AlCu of the A1 and A2 alloy samples, respectively. The equilibrium microstructure and microprobe analysis results are shown in Figs. 4 and 11. The ε phase contained 25 to 33.4 at.% of Cu and 62.4 to 65 at.% of Al.

Discussion

In this study, a classical semi-infinite diffusion couple was applied, meaning that the end couples maintained their

Fig. 6 The diffusion paths of D1 and D5 samples

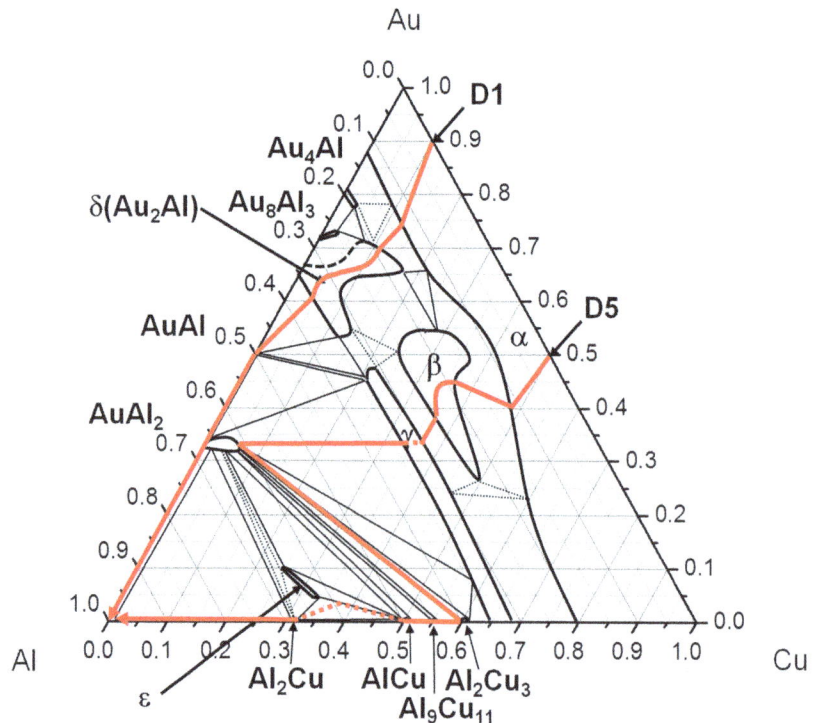

Fig. 7 The isotherm of the Au-Al-Cu at 500 °C determined by F.C. Levey et al. [9–11]

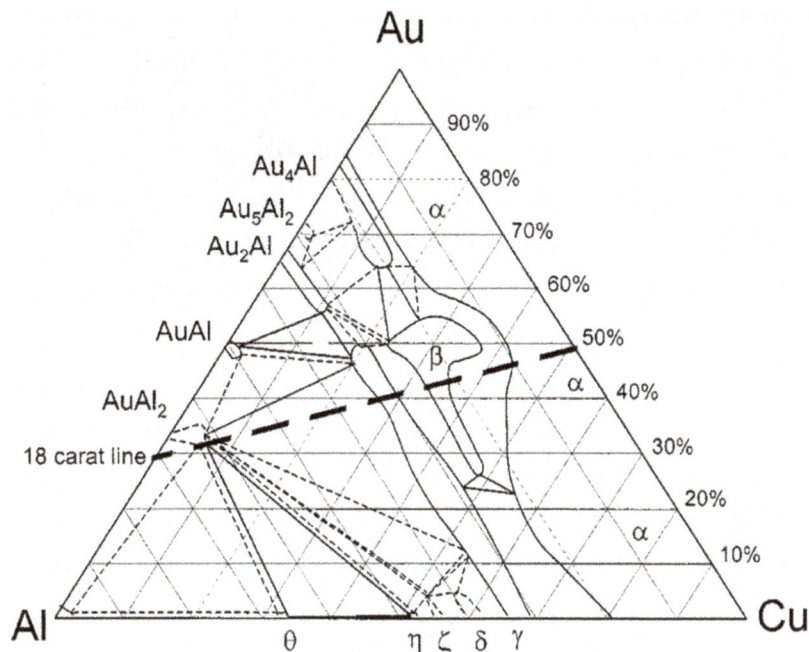

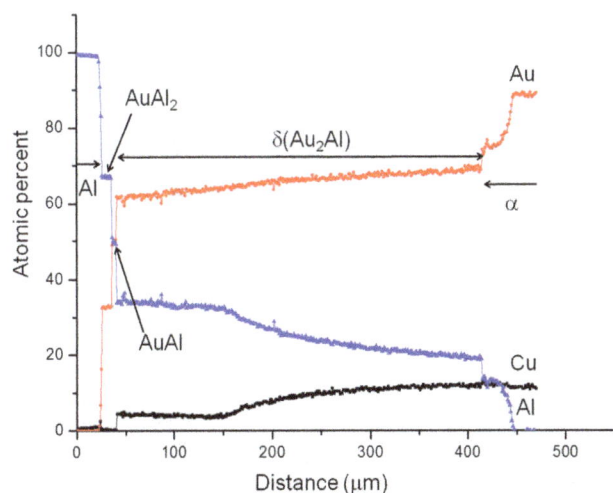

Fig. 8 A composition profile of the D1 sample measured by EPMA

original compositions after the diffusion couple annealing. If volume diffusion in a semi-infinite couple is a rate-limiting step, local equilibrium is assumed to exist, in which case the rules described previously can be used to relate the reaction zone morphology developed during isothermal diffusion to the phase diagram. The phase composition of the reaction zone is independent of time, whereas that of the diffusion path is fixed. A D8 diffusion couple was annealed for 5 days to compare it with the 1-day diffusion couple. No difference was observed among the phases formed in each diffusion layer,

and only the layer thickness differed. The versatility of this technique for constructing isothermal cross sections of ternary systems has been demonstrated repeatedly [14].

The D2 diffusion path was plotted on the Au-rich region (Fig. 12a), and its corresponding microstructure is shown in Figs. 12b, c. The solid a-b line crossed the α single-phase field, and the dashed b-c line crossed the δ (Au$_2$Al)-α-β three-phase region, indicating the presence of an interface in the diffusion structure, with equilibrium between the α single-phase and δ (Au$_2$Al)-β two-phase region. The microstructure of the corresponding three-phase region is shown in Fig. 12b and is marked as the interface b-c; c-d is the δ (Au$_2$Al)-β two-phase region. The solid d-e-f-g line in Fig. 12a shows that the diffusion path entered the δ (Au$_2$Al) single-phase region, returned to the δ (Au$_2$Al)-β two-phase region, and then entered the δ (Au$_2$Al) single-phase region again. The thickness of the δ (Au$_2$Al)-β two-phase region was approximately 150 μm and was contained within the δ (Au$_2$Al) single-phase region. Figure 12c shows the other side of this microstructure. This diffusion path could explain why the β phase formed within the δ (Au$_2$Al) phase layer.

The diffusion paths D8 and D9 crossed three three-phase regions (AlCu-AuAl$_2$-Al$_9$Cu$_{11}$, Al$_9$Cu$_{11}$-AuAl$_2$-Al$_2$Cu$_3$, and γ-AuAl$_2$-Al$_2$Cu$_3$) with AuAl$_2$ as the common matrix phase. Figure 13a shows the diffusion path near the Al-Cu binary region, and Fig. 13b is the plot displaying the equilibrium of these three phases along with their related two-phase regions. According to Fig. 13b, d-e is the γ-AuAl$_2$-Al$_2$Cu$_3$ three-phase

Fig. 9 The microstructure and tie lines of δ (Au$_2$Al) and β phases determined from A4, A8, and A9 alloy samples

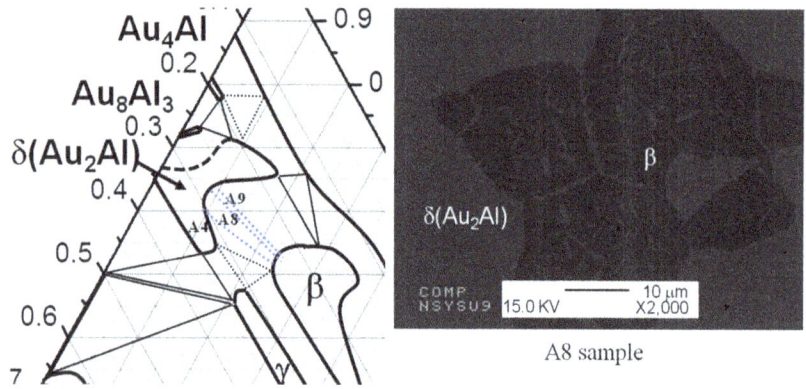

A8 sample

Sample	Phase	Composition (at.%)		
		Au	Cu	Al
A4	δ(Au$_2$Al)	60.22±0.25	9.21±0.28	30.57±0.19
	β	51.52±0.46	23.70±0.55	24.78±0.97
A8	δ(Au$_2$Al)	62.86±0.70	9.04±0.34	28.10±0.39
	β	51.46±1.15	24.69±0.79	23.86±0.89
A9	δ(Au$_2$Al)	62.71±0.89	10.29±0.38	27.00±0.69
	β	52.49±0.34	24.32±0.48	23.19±0.36

equilibrium, e-f is the AuAl$_2$-Al$_2$Cu$_3$ two-phase region, f-g is the Al$_9$Cu$_{11}$-AuAl$_2$-Al$_2$Cu$_3$ three-phase equilibrium, g-h is the Al$_9$Cu$_{11}$-AuAl$_2$ two-phase region, h-i is the AlCu-AuAl$_2$-Al$_9$Cu$_{11}$ three-phase equilibrium, i-j is AlCu-AuAl$_2$, j-k is the AlCu single-phase region, k-l is the ε-AlCu two-phase region, and l-m is the ε-AlCu-Al$_2$Cu three-phase equilibrium.

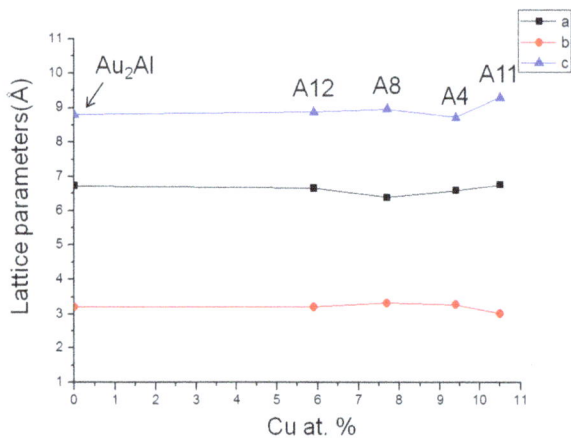

Sample	Phase	Lattice parameters(Å)		
		a	b	c
JC-PDF[(65-1504)]	Au$_2$Al	6.715	3.219	8.815
A12	δ(Au$_2$Al)	6.66±0.01	3.22±0.00	8.89±0.01
A8	δ(Au$_2$Al)	6.39±0.02	3.33±0.01	8.98±0.01
A4	δ(Au$_2$Al)	6.59±0.01	3.28±0.00	8.74±0.00
A11	δ(Au$_2$Al)	6.76±0.02	3.03±0.01	9.31±0.00

Fig. 10 The lattice parameters of δ (Au$_2$Al) phase

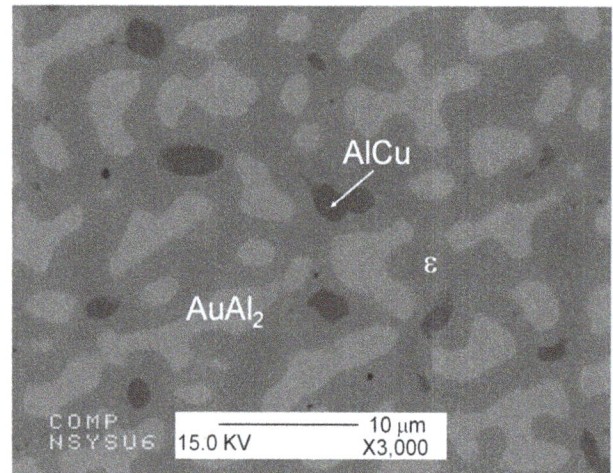

Color	Phase	Composition (at.%)		
		Au	Cu	Al
light gray	AuAl$_2$	31.30±0.82	4.27±0.37	64.42±0.64
dark gray	ε	15.24±1.03	20.46±1.22	64.30±0.42
black	AlCu	1.867±0.87	46.60±1.61	51.53±0.82

Fig. 11 The microstructure of A2 alloy sample including AuAl$_2$, ε and AlCu phases

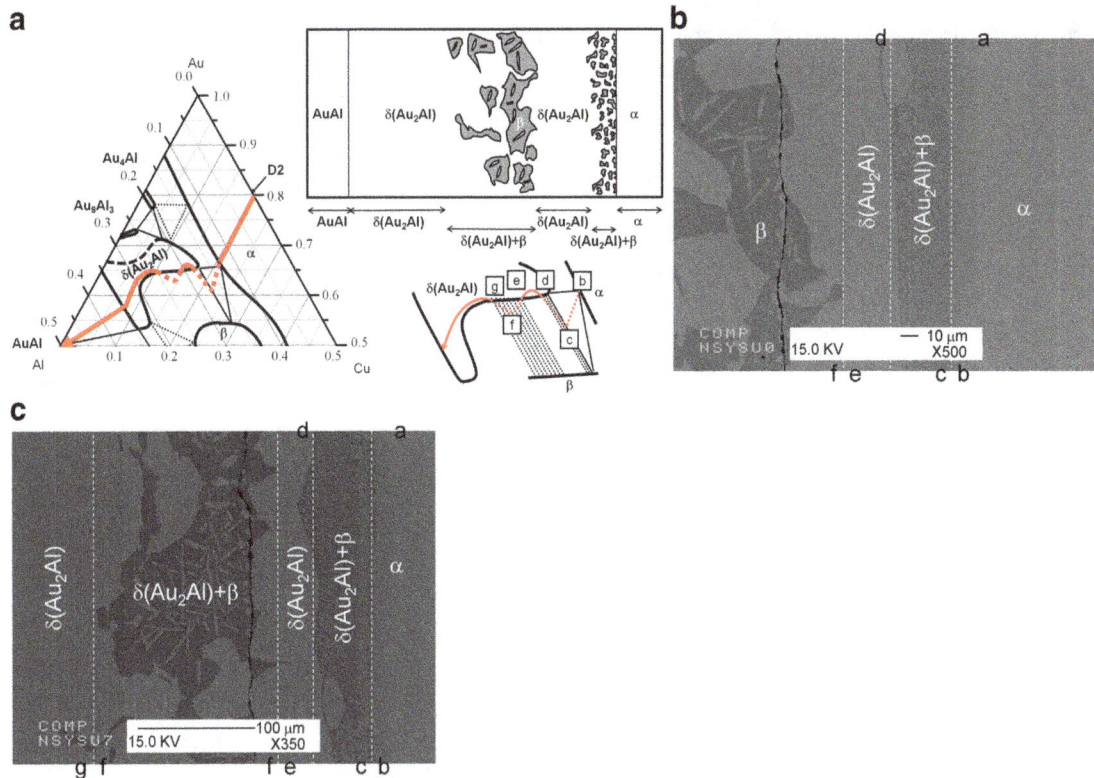

Fig. 12 **a** The diffusion path of D2 diffusion couple and sketch figure. **b** The microstructure of D2 diffusion couple with the label (*a* to *f*) **c** The microstructure of D2 diffusion couple shows β phase formed within δ (Au₂Al) phase with the label (*e* to *g*)

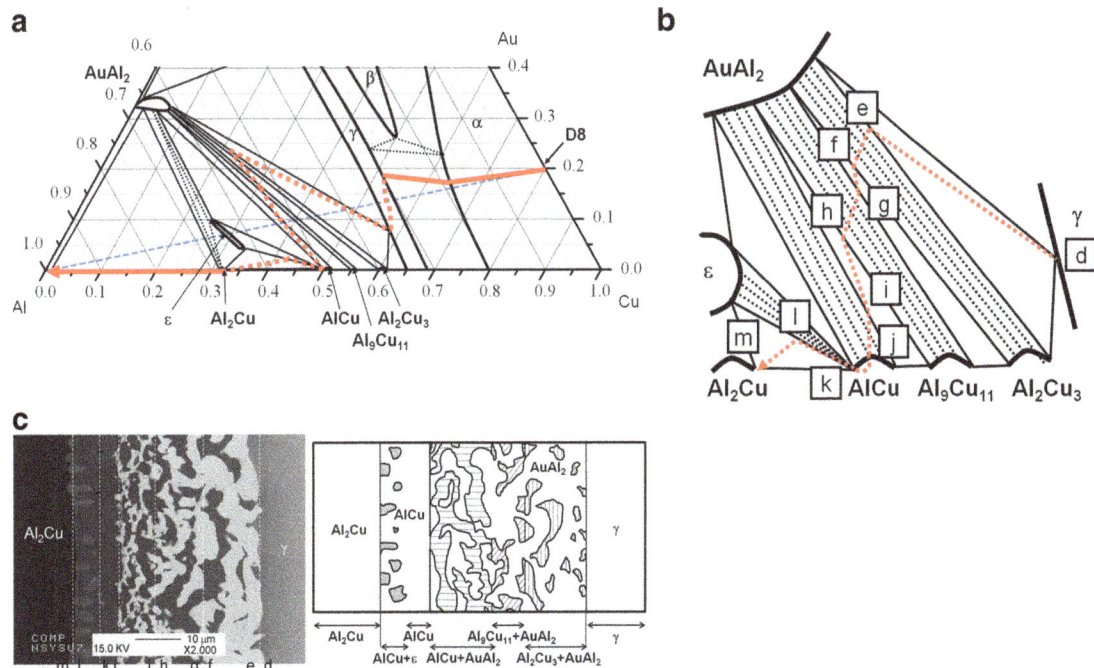

Fig. 13 **a** The diffusion path of D8 diffusion couple near the Al-Cu binary region. **b** The sketch figure of D8 diffusion path. **c** The microstructure of D8 diffusion couple with the label (*d* to *m*) and sketch figure

The microstructure and corresponding sketch are shown in Fig. 13c, and are labeled from d to m, as described previously.

In comparison with Levey [9–11], new ternary phase ε and their related three-phase equilibria were identified in this study. Ternary ε phase exhibited a composition similar to that of the binary Al_2Cu phase and contained 25 to 33.4 at.% of Cu and 62.4 to 65 at.% of Al. Three related three-phase equilibria were also observed, namely, $AuAl_2$-ε-AlCu, Al_2Cu-ε-AlCu, and $AuAl_2$-ε-Al_2Cu. The β phase reported by Levey was also identified. However, the phase relationships exhibited some differences, because the δ (Au_2Al) phase exhibited a wide range of solubility and a peculiarly shaped phase field in the Au-rich region. Therefore, three additional three-phase equilibria (β-δ (Au_2Al)-α, β-α-γ, and β-δ (Au_2Al)-γ) were identified in this study. Finally, the solubility of Au in the Al_9Cu_{11} and Al_2Cu_3 phases was limited.

Conclusion

The isothermal section of an Au-Al-Cu system annealed at 500 °C was presented in this study. The whole composition range included two ternary phases (ε and β) and 14 three-phase equilibria. The δ (Au_2Al) phase exhibited a wide range of solubility and a peculiar morphology. The ε phase contained 25 to 33.4 at.% of Cu and 62.4 to 65 at.% of Al. Ten three-phase equilibria were determined in this study, namely, Al-$AuAl_2$-Al_2Cu, Al_2Cu-ε-AlCu, $AuAl_2$-ε-AlCu, $AuAl_2$-AuAl-γ, AuAl-δ (Au_2Al)-γ, δ (Au_2Al)-α-β, Au_8Al_3-δ (Au_2Al)-Au_4Al, AlCu-$AuAl_2$-Al_9Cu_{11}, Al_9Cu_{11}-$AuAl_2$-Al_2Cu_3, and Al_2Cu_3-$AuAl_2$-γ. Four three-phase equilibria, namely, β-δ (Au_2Al)-γ, α-β-γ, α-δ (Au_2Al)-Au_4Al, and $AuAl_2$-ε-Al_2Cu were estimated to meet the criteria for phase relationships.

Acknowledgments We appreciate the financial support provided by the National Science Council, Grant No. NSC 98-2221-E-110-034-MY3.

References

1. Chang HS, Hsieh KC, Martens T, Yang A (2003) The effect of Pd and Cu in the intermetallic growth of alloy Au wire. J Electron Mater 32: 1182–1187
2. Gam SA, Kim HJ, Cho JS, Park YJ, Moon JT, Paik KW (2006) Effects of Cu and Pd addition on Au bonding wire/Al pad interfacial reactions and bond reliability. J Electron Mater 35:2048–2055
3. Cortie MB, Levey FC (2000) Structure and ordering of the 18-carat Al-Au-Cu β-phase. Intermetallics 8:793–804
4. Levey FC, Cortie MB (2001) Body-centred tetragonal martensite formed from $Au_7Cu_5Al_4$ β phase. Mater Sci Eng 303:1–10
5. Levey FC, Cortie MB, Cornish LA (2000) Displacive transformations in Au-18 Wt Pct Cu-6 Wt Pct Al. Metall Mater Trans A 31: 1917–1923
6. Okamoto H (1991) Al-Au (aluminum-gold). J Phase Equilib 12:114–115
7. Murray JL (1985) The aluminium-copper system. Int Mater Rev 30: 211–234
8. Okamoto H, Shakrabarti DJ, Laughlin DE, Massalski TB (1987) The Au-Cu (gold-copper) system. Bull Alloy Phase Diagr 8:454–473
9. Levey FC, Cortie MB, Cornish LA (2002) A 500 °C isothermal section for the Al-Au-Cu system. Metall Mater Trans A 33:987–993
10. Raghavan V (2008) Al-Au-Cu (aluminum-gold-copper). J Phase Equilib Diff 29:260–261
11. Bhatia VK, Levey FC, Kealley CS, Dowd A, Cortie MB (2009) The aluminium-copper-gold ternary system. Gold Bull 42:201–208
12. Bhatia VK, Kealley CS, Wuhrer R, Wallwork KS, Cortie MB (2009) Ternary β and γ phases in the Al-Au-Cu system at 750 °C. J Alloy Compd 488:100–107
13. Levey FC, Cortie MB, Cornish LA (2003) Determination of the 76 wt.% Au section of the Al-Au-Cu phase diagram. J Alloy Compd 354:171–180
14. Kodentsov AA, Bastin GF, Van Loo FJJ (2001) The diffusion couple technique in phase diagram determination. J Alloy Compd 320:207–217

22

Hardening effect of pre- and post-firing heat treatment for a firing-simulated Au-Pd-In metal-ceramic alloy

Byung-Wook Jeon · Sung-Min Kim · Hyung-Il Kim · Yong Hoon Kwon · Hyo-Joung Seol

Abstract The hardening effect of the pre- and post-firing heat treatments and their dual treatment was examined for a firing-simulated Au-Pd-In metal-ceramic alloy to determine if an additional post-firing heat treatment is effective in the hardening of an Au-Pd-In metal-ceramic alloy as well as to compare the hardening effects of pre- and post-firing heat treatments for a firing-simulated Au-Pd-In metal-ceramic alloy. The post-firing heat treatment was much more effective than the pre-firing heat treatment or dual treatments. The hardening effect of the pre- and post-firing heat treatments was caused by the induction of fine grain interior precipitation. In the pre-firing heat-treated specimen after casting, the apparent hardening was achieved during simulated porcelain firing, but an additional post-firing heat treatment did not introduce severe grain interior precipitation, resulting in very weak hardening. In the as-cast specimen without the pre-firing heat treatment, apparent hardening by grain interior precipitation was achieved only by post-firing heat treatment and not by simulated porcelain firing.

Keywords Pre-firing heat treatment · Post-firing heat treatment · Au-Pd-In metal-ceramic alloy · Porcelain firing simulation · Grain interior precipitation

Introduction

Metal-ceramic gold alloys are used to fabricate substructures of porcelain, and thus, the metal substructures are made thin enough to increase the aesthetic potential of semitranslucent

B.-W. Jeon · S.-M. Kim · H.-I. Kim · Y. H. Kwon · H.-J. Seol (✉)
Department of Dental Materials, Institute of Translational Dental Sciences, School of Dentistry, Pusan National University, Beomeo-Ri, Mulgeum-Eup, Yangsan-Si, Gyeongsangnam-Do 626-814, South Korea
e-mail: seolhyojoung@daum.net

porcelain superstructure [1, 2]. Metal-ceramic gold alloys have insufficient creep resistance and poor hardness after casting due to the high gold content. In addition, these alloys are heat treated several times at high temperatures for porcelain firing after casting. During this process, these alloys can be deformed by sag and thermal distortion [3, 4]. This causes serious problems, such as an inaccurate fit of the restoration [5, 6]. To resolve this problem, studies of heat treatments before porcelain firing have been conducted with a purpose to increase the hardness of metal-ceramic gold alloys before porcelain firing [7–9]. Koike reported that if a casting is pre-heated near the porcelain firing temperature before surface grinding, distortion from subsequent procedures can be controlled remarkably [7]. Liu and Wang reported that pre-firing heat treatment increased the hardness of Pd-free Au-Pt metal-ceramic gold alloy [8]. They attributed the strengthening to the homogenization of the microstructure and the precipitation of new fine particles. Fischer and Fleetwood reported that pre-firing heat treatment improved the processing of Au-Pt metal-ceramic frameworks by increasing the hardness, resistance to thermal distortion and proof stress [9].

The strength of metal ceramics depends largely on the strength of the metal substructure. Accordingly, porcelain fractures even by slight deformation of the metal substructure after the completion of fusion bonding [1]. Therefore, if the hardness of the metal-ceramic gold alloys is insufficient even after a pre-firing heat treatment, the fracture or separation of porcelain can occur during mastication. From such a viewpoint, it was considered that additional post-firing heat treatment at relatively lower temperatures might increase the hardness of the metal substructure without softening or damaging the porcelain. In the present study, the hardening effect of pre- and post-firing heat treatments and their dual treatment on a firing-simulated Au-Pd-In metal-ceramic alloy was examined. The aims of the present investigation were to determine if an additional post-firing heat treatment is effective in hardening an Au-Pd-In metal-ceramic alloy and to compare the hardening effects of pre- and post-firing heat treatments for a firing-simulated Au-Pd-In metal-ceramic alloy.

Materials and methods

Specimen alloy

The specimen alloy used in this study was a Cu-free Au-Pd-In metal-ceramic alloy (Aurolite 45, Aurium research, USA) with a white colour and fine grain, which belongs to type IV. The manufacturer reported the alloy to have a melting range of 1,190~1,290 °C and a casting temperature of 1,370 °C. Table 1 lists the chemical composition of the alloy. The specimens were cast using the lost wax casting technique in a standard broken arm centrifugal casting machine (Centrifugal casting machine, Osung, Korea). All castings were bench cooled to room temperature, followed by divesting and airborne-particle abrasion with 50 μm Al_2O_3. The specimens obtained were in the form of small square pieces, $10 \times 10 \times 0.8$ mm in size.

Heat treatment

The above-mentioned as-cast plate-like specimens were pre-firing heat treated at 980 °C for 10 min in a vertical electric furnace under an argon atmosphere to inhibit oxidation and then quenched rapidly by dropping it into ice brine to prevent hardening by atomic diffusion. The specimens were then heat treated in a ceramic furnace (Multimat 2 touch, Dentsply, Germany) to simulate the complete firing cycle according to Table 2. An additional post-firing heat treatment was performed in a ceramic furnace that was adjusted to 550 or 650 °C.

Hardness test

Hardness measurements were completed for the required heat-treated specimen using a Vickers microhardness tester (MVK-H1, Akashi Co., Japan) with 300 gf and dwell time of 10 s. All hardness results were recorded as the mean values of five measurements.

Table 1 Chemical composition of the specimen

Composition	Au	Pd	In	Ag	Ga	Ru
wt.%	45.00	40.00	8.50	4.90	1.50	0.10
at.%	30.61	50.37	9.92	6.09	2.88	0.13

Table 2 Simulated porcelain firing cycle

Firing stage	Start temp. (°C)	Pre-drying (min)	Heat rate (°C/min)	Final temp. (°C)	Time at final temp. (min)	Vacuum time (min)
Degassing	650	0	70	1,010	0	0
Wash	550	2	70	980	1	7:09
Opaque	550	3	70	970	1	7:00
Main bake	550	5	70	960	1	6:51
Correction	550	4	70	950	1	6:43
Glaze	550	1	70	910	1	0

Field emission scanning electron microscopy (FE-SEM)

For FE-SEM (JSM-6700 F, Jeol, Japan), the as-cast plate-like specimens were subjected to the required heat treatment and prepared using a standard metallographic technique. Subsequently, a freshly prepared aqueous solution of 10 % potassium cyanide (KCN) and 10 % ammonium persulfate $((NH_4)_2S_2O_8)$ was used for final etching of the specimens. The specimens were examined by FE-SEM at 15 kV.

X-ray diffraction (XRD) study

For XRD (XPERT-PRO, Philips, Netherlands), the as-cast plate-like specimens were subjected to the required heat treatment and prepared using a standard metallographic technique. The XRD profile was recorded at 40 kV and 30 mA using nickel-filtered Cu Kα radiation as the incident beam. The scanning rate of a goniometer was 1° (2θ/min).

Energy dispersive spectrometer (EDS) analysis

The changes in the element distribution of the heat-treated plate-like specimens with heat treatment were examined using an energy dispersive X-ray spectrometer (INCA x-sight, Oxford Instruments Ltd., UK) attached to the FE-SEM at 15 kV.

Results and discussion

Hardness changes at various cooling rates during porcelain firing simulation

Metal-ceramic gold alloys can be hardened during cooling after porcelain firing [9, 10], and the hardness of a metal-ceramic gold alloy can be increased significantly by controlling the cooling rate. To identify the most effective cooling rate for hardening of the alloy, the specimens underwent cooling at various cooling rates during the degassing treatment, which is the first stage of the porcelain firing cycle. Tables 2 and 3 list

Table 3 Cooling rate during simulated firing

Cooling rate	Ice quenching	Quick cooling	Stage 0	Stage 1	Stage 2	Stage 3
Condition	Quenching into ice brine	Firing chamber moves immediately to upper end position and air cooled	Firing chamber moves immediately to upper end position	Firing chamber opens about 70 mm	Firing chamber opens about 50 mm	Firing chamber remains closed

the simulated porcelain firing cycle and each cooling rate, respectively. The degassing schedule was directed by the alloy manufacturer. To obtain a single-phase specimen before the firing simulation, as-cast specimens were pre-firing heat treated at 980 °C for 10 min and ice quenched. The hardness was 207.0 ± 1.18 HV in this state. The specimens were then degassing treated and cooled to 600 °C at six cooling rates in a porcelain furnace and then bench cooled to room temperature. Table 4 lists the hardness at each cooling rate during the simulated degassing treatment. The specimen cooled by ice quenching showed the lowest hardness, 202.6 HV, which is similar to that of the pre-firing heat-treated specimen. Therefore, the ice-quenched specimen after degassing was believed to be single phase. The specimen cooled by quick cooling showed the highest hardness, 255.1 HV. As the cooling rate becomes slower than quick cooling, the increasing rate in hardness decreased gradually. Therefore, the cooling rate during simulated porcelain firing was fixed to quick cooling in the present study.

Hardening effect of the pre- and post-firing heat treatments

To observe the effects of a pre-firing heat treatment on the hardness of a porcelain firing-simulated specimen, the changes in hardness by simulated porcelain firing of the as-cast and pre-firing heat-treated specimens were examined. Figure 1 shows the changes in hardness during simulated porcelain firing at a fixed cooling rate (quick cooling) in the as-cast and pre-firing heat-treated specimens. In the as-cast state, the hardness was approximately 220 HV, which was not changed apparently by the degassing and wash treatments. Subsequently, the hardness increased to the maximum value, 252.0 HV, by an opaque treatment. By the subsequent simulated firing schedule of the main bake, correction and glaze, the hardness decreased gradually and finally became similar to that of the as-cast state (226.9 HV). From the above,

the porcelain firing full cycle did not result in hardening of the as-cast specimen [11], unlike the case of the Au-Pt and Au-Pd-Ag metal-ceramic alloys [9, 10, 12].

After the pre-firing heat treatment after casting, the hardness decreased to 207 HV [10, 13]. On the other hand, the hardness increased to 255 HV after the degassing treatment, the first firing stage, and the high hardness was maintained during the subsequent firing stages. Therefore, the pre-firing heat treatment itself softens the as-cast specimen but results in apparent hardening from the first stage of firing with softening restrained until the final firing stage. From a clinical point of view, the hardening of a metal-ceramic alloy during porcelain firing is advantageous because it improves the fit of the restoration by reducing the level of sag [9], which can be obtained by a pre-firing heat treatment.

To determine the hardening effect of the post-firing heat treatment on firing the simulated specimen, the as-cast specimen and pre-firing heat-treated specimen were post-firing heat treated in a porcelain furnace after firing. Figure 2 shows the changes in hardness of the as-cast and pre-firing heat-treated specimen during post-firing heat treatment at 550 and 650 °C. Here, 550 °C is the hardening temperature recommended by the alloy manufacturer. In the as-cast specimen marked by black, the hardness increased by 54 HV within 20 min at both temperatures. After that, however, the hardness decreased with increasing heating time. The decreasing rate was faster at 650 °C than at 550 °C due to the faster atomic diffusion at a higher temperature [14].

In the pre-firing heat-treated specimen after casting, which is marked as white, the hardening effect of the post-firing heat treatment was relatively weak. In Fig. 2, the hardness increased slightly by 13 HV within 15 min and the hardness then decreased gradually. The decreasing rate was faster at 650 °C than at 550 °C, as it was in the as-cast specimen without the pre-firing heat treatment. These results show that the post-firing heat treatment was much more effective than

Table 4 Hardness at each cooling rate during simulated degassing treatment

Cooling rate	Ice quenching	Quick cooling	Stage 0	Stage 1	Stage 2	Stage 3
Hardness (HV)	202.6 (±2.67)	255.1 (±4.36)	243.6 (±2.97)	237.9 (±2.35)	224.1 (±4.77)	210.4 (±4.67)

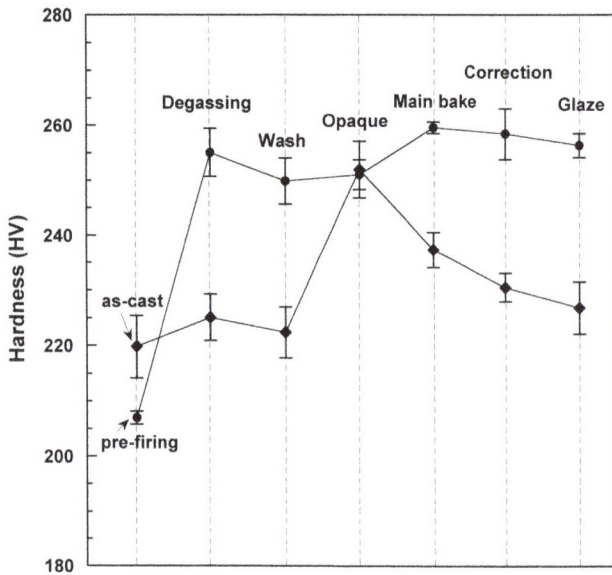

Fig. 1 Hardness changes in the as-cast and pre-firing heat-treated specimen during simulated complete firing

the pre-firing heat treatment (Fig. 1) or pre- and post-firing dual heat treatment (Fig. 2, white mark) for hardening of the Au-Pd-In metal-ceramic alloy.

Microstructural changes by simulated porcelain firing and post-firing heat treatment

The microstructural changes of the as-cast and pre-firing heat-treated specimens by simulated porcelain firing were observed by FE-SEM. Figure 3 shows FE-SEM images of the as-cast specimen (a), as-cast and then porcelain firing-simulated

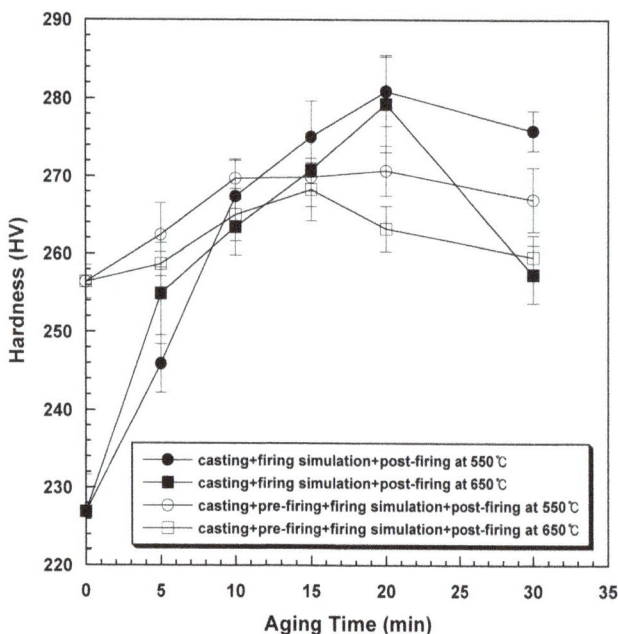

Fig. 2 Hardness changes in the as-cast and pre-firing heat-treated specimen during post-firing heat treatment at 550 and 650 °C

specimen (b), pre-firing heat-treated specimen after casting (c) and the as-cast and then pre-firing and porcelain firing-simulated specimen (d) at magnifications of ×2,500 (1), ×8,000 (2) and ×35,000 (3). In the as-cast specimen (a), in addition to the equiaxed structure and grain boundary precipitates, lamellar and cubic- or particle-like structures were observed adjacent to the grain boundaries. By subsequent porcelain firing simulation (b), the grain boundary precipitates became partly narrow, and the lamellar structures observed adjacent to the grain boundaries disappeared. The fine cubic- or particle-like structures became smaller and spread toward the grain interior. Grain interior precipitates were not observed in the matrix marked by the circle in (b-3).

In the pre-firing heat-treated specimen after casting (c), there was no precipitate in the grain interior and grain boundaries, and only a single-phased equiaxed structure was observed, resulting in a 15HV decrease in hardness from the as-cast state. By subsequent porcelain firing simulation (d), precipitates formed along the grain boundaries and cubic- or particle-like structures were observed in the whole grain interior, as it was in the porcelain firing-simulated specimen after casting (b). On the other hand, unlike the microstructures of (b), fine grain interior precipitates were observed in the matrix marked by the circle at (d-3). Therefore, the hardening effect of the pre-firing heat treatment on simulated porcelain firing was caused by the induction of the grain interior precipitation of fine scale, as can be seen by a comparison of (d-3) with (b-3).

Microstructural changes by post-firing heat treatment of the as-cast and pre-firing heat-treated specimens were observed. Figure 4 shows FE-SEM images of the porcelain firing simulated and then post-firing heat-treated specimens at 550 °C for 30 min after casting (a) and after casting and pre-firing heat treatment (b) at magnifications of ×2,500 (1), ×8,000 (2) and ×35,000 (3). In specimen (a), the grain boundary precipitates and cubic- or particle-like structures in the grain interior became coarse compared to those before the post-firing heat treatment in Fig. 3(b), and the lamellar structures appeared. Therefore, the microstructure became partly similar to that of the as-cast specimen in Fig. 3(a) except for the spread of cubic or particle-like structures to the grain interior. The decisive difference in microstructure by post-firing was that the fine grain interior precipitates covered the entire matrix region, as shown in a circle mark of Fig. 4(a-3). This must be the main reason for the apparent increase in hardness by post-firing heat treatment in the firing-simulated specimen after casting. This hardening mechanism by grain interior precipitation was similar to that by the pre-firing heat treatment, as shown in the circle mark of Fig. 3(d-3).

In the pre-firing heat-treated specimen after casting (b), no obvious microstructural differences, particularly in the matrix, were observed after the post-firing heat treatment, as can be seen by a comparison of the circle-marked region in

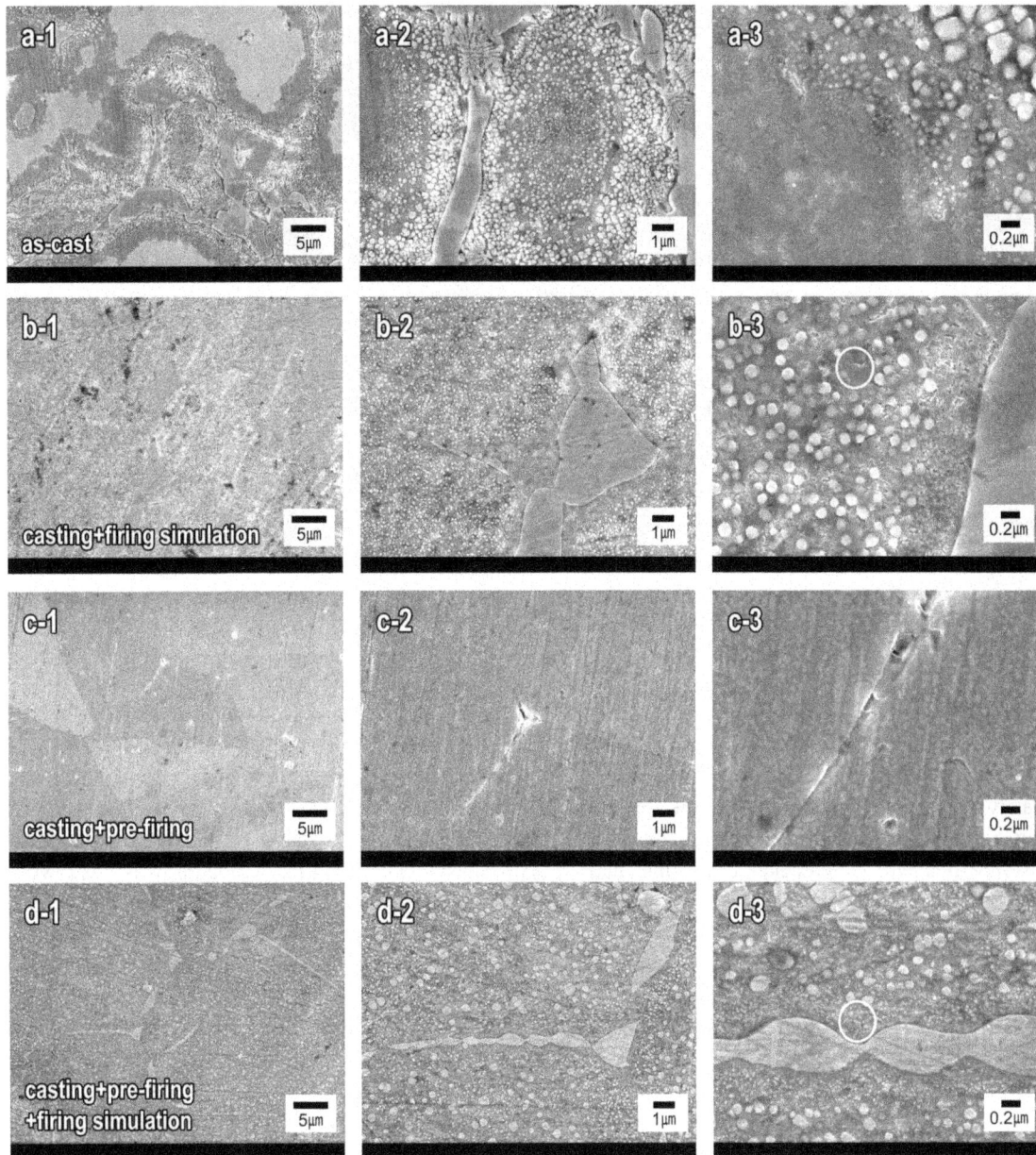

Fig. 3 Microstructural changes in the as-cast and pre-firing heat-treated specimen by simulated complete firing: as-cast (*a*), firing simulated after casting (*b*), pre-firing heat-treated after casting (*c*) and firing simulated after casting and pre-firing heat treatment (*d*) at magnifications of ×2,500 (*1*), ×8,000 (*2*) and ×35,000 (*3*)

Fig. 4(b-3) with that in Fig. 3(d-3). In the pre-firing heat-treated specimen after casting, apparent hardening by grain interior precipitation was achieved during simulated porcelain firing, but an additional post-firing heat treatment did not introduce severe grain interior precipitation, resulting in very weak hardening.

Phase transformation and elemental distribution

Figure 5 shows the changes in the XRD patterns of the as-cast and pre-firing heat-treated specimen after simulated complete firing and post-firing heat treatment at 550 °C for 30 min (a, as

cast; b, pre-firing heat treated after casting; c, simulated complete firing; d, post-firing heat treated). In the as-cast specimen (a), a face-centred cubic (f.c.c.) Pd-Au-rich (α) phase with a lattice parameter of a_{200}=3.9958 Å was the main phase and another phase marked by arrows was a minor phase due to the precipitates. The XRD pattern of the as-cast specimen did not change by subsequent simulated complete firing, (a + c) and post-firing heat treatment, (a + c + d). In the pre-firing heat-treated specimen after casting (b), a single f.c.c. Pd-Au-rich (α_0) phase with a lattice parameter of a_{200}=3.9931 Å was obtained. By subsequent simulated complete firing, (b + c), and post-firing heat treatment, (b + c + d), a minor phase

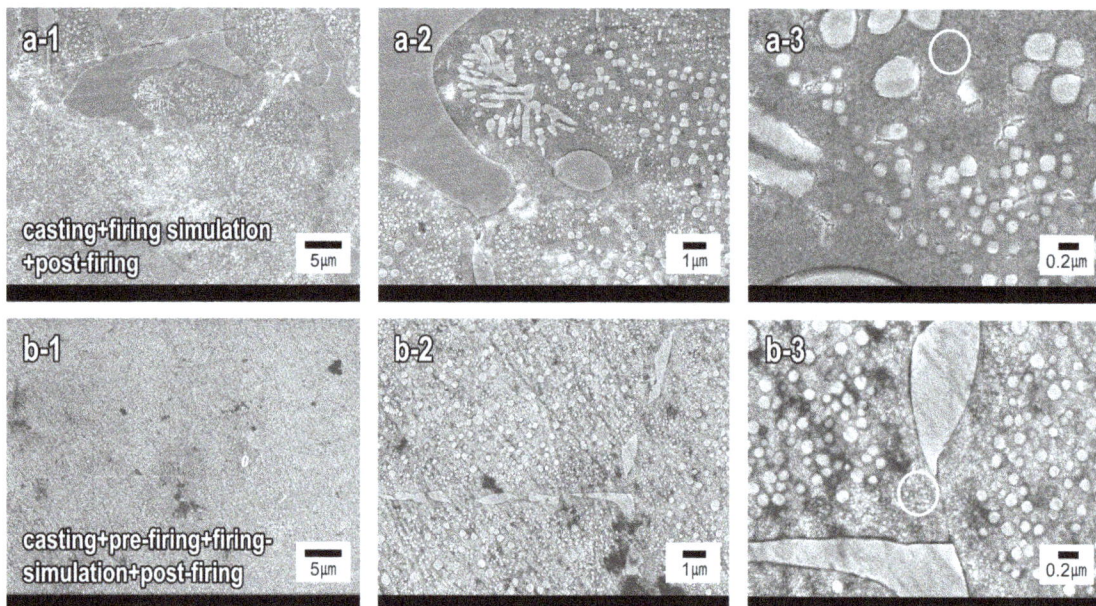

Fig. 4 Microstructural changes in the as-cast and pre-firing heat-treated specimen by post-firing heat treatment at 550 °C for 30 min: post-firing heat treated after casting and firing simulation (*a*), post-firing heat treated after casting, pre-firing and firing simulation (*b*) at magnifications of ×2,500 (*1*), ×8,000 (*2*) and ×35,000 (*3*)

marked by the arrows was newly formed by precipitation. Its peak position was same as that of the as-cast specimen (a), indicating it to be the same phase.

EDS analysis was performed to identify the precipitated elements during the above-mentioned heat treatment. Figure 6 presents an FE-SEM image of the porcelain firing simulated and the post-firing heat-treated specimen at 650 °C for 30 min after casting and pre-firing heat treatment (M, matrix; P, grain interior precipitate; Gb, grain boundary precipitate at a magnification of ×7,000). Table 5 lists the results of EDS point analysis for each region marked with the arrow in Fig. 6.

In the solute-depleted matrix region, which is covered by grain interior precipitates, (M + P), the composition was similar to the alloy composition in Table 1. In the grain boundary precipitate (Gb), Au and Ag decreased and Pd, Ga and In increased compared to the alloy composition. In particular, the Ga content increased to more than twice that of the alloy composition. By considering the elemental distribution in the grain boundary precipitate, the precipitated phase was possibly the ordered Ga_2Pd_5 phase containing In [15]. From the above, it was believed that precipitation and then ordering of Pd, Ga and In occurred by simulated porcelain firing and post-firing heat treatment, resulting in apparent hardening.

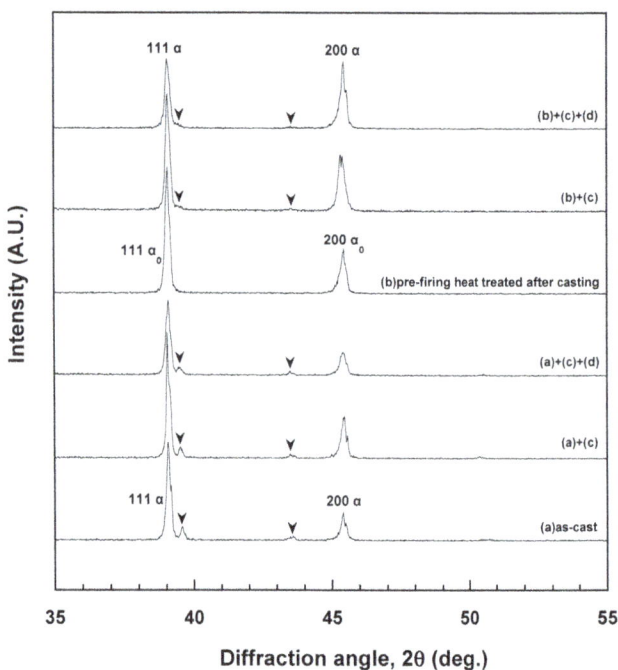

Fig. 5 Changes in the XRD patterns of the as-cast and pre-firing heat-treated specimen after simulated complete firing and post-firing heat treatment at 550 °C: as-cast (**a**), pre-firing heat treated after casting (**b**), simulated complete firing (**c**), post-firing heat treated (**d**)

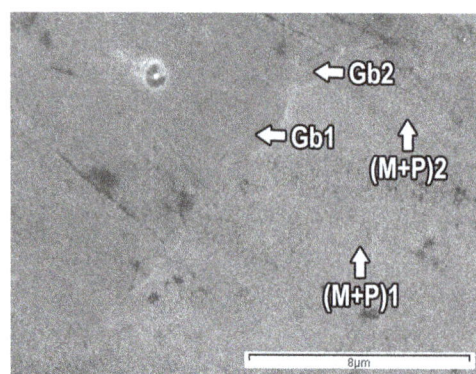

Fig. 6 FE-SEM images of the post-firing heat-treated specimens at 650 °C for 30 min for EDS analysis

Table 5 EDS analysis at the regions marked in Fig. 6

at.%	Au	Pd	In	Ag	Ga	Ru
(M + P)1	32.54	51.22	9.08	4.79	2.37	0
(M + P)2	30.76	51.54	10.58	3.73	3.39	0
Gb1	22.38	58.47	11.54	0	7.61	0
Gb2	23.03	58.10	12.01	0	6.86	0

Conclusions

The hardening effect of the pre- and post-firing heat treatments and their dual treatment was examined for a firing-simulated Au-Pd-In metal-ceramic alloy to determine if an additional post-firing heat treatment is effective in the hardening of an Au-Pd-In metal-ceramic alloy as well as to compare the hardening effects of pre- and post-firing heat treatments for a firing-simulated Au-Pd-In metal-ceramic alloy.

(1) The post-firing heat treatment was much more effective than the pre-firing heat treatment or pre- and post-firing dual heat treatment for the hardening of an Au-Pd-In metal-ceramic alloy.

(2) The hardening effect of pre- and post-firing heat treatments was caused by the induction of the grain interior precipitation of fine scale.

(3) In the pre-firing heat-treated specimen after casting, the apparent hardening was achieved during simulated porcelain firing, but an additional post-firing heat treatment did not introduce severe grain interior precipitation, resulting in very weak hardening.

(4) In the as-cast specimen without the pre-firing heat treatment, apparent hardening by grain interior precipitation was achieved only by post-firing heat treatment and not by simulated porcelain firing.

Acknowledgments This research was supported by Basic Science Research Program through the National Research Foundation of Korea (NRF) funded by the Ministry of Education, Science and Technology (grant number: 2011-0010995)

References

1. Yamamoto M (1985) Metal-ceramics: principle and methods of Makoto Yamamoto. Quintessence Publishing Co Inc, Chicago, pp 23–25, Chap 1
2. Straussberg G, Katz G, Kuwata M (1966) Design of gold supporting structures for fused porcelain restorations. J Prosthet Dent 16:928–936
3. Chew CL, Norman RD, Stewart GP (1990) Mechanical properties of metal-ceramic alloys at high temperature. Dent Mater 6:223–227
4. O'Brien WJ (2002) Dental materials and their selection, 3rd edn. Quintessence Publishing Co Inc, Chicago, p 204, Chap 14
5. Tuccillo JJ, Nielsen JP (1967) Creep and sag properties of a porcelain-gold alloy. J Dent Res 46:579–583
6. Iwashita H, Kuriki H, Hasuo T, Ishikawa K, Hashimoko K, Harada H, Uochi T, Hata Y (1977) Studies on dimensional accuracy of porcelain fused to precious metal crown: the influence of the porcelain to the metal coping on the porcelain fusing procedure. Shigaku 65:110–125
7. Koike K (1997) Fabrication of ceramo-metal crowns with accurate fitness: deformation of casting and its remedies. Shika Giko 5:31–41
8. Wang JN, Liu WB (2006) A Pd-free high gold dental alloy for porcelain bonding. Gold Bull 39:114–120
9. Fischer J, Fleetwood PW (2000) Improving the processing of high-gold metal-ceramic frameworks by a pre-firing heat treatment. Dent Mater 16:109–113
10. German RM (1980) Hardening reactions in a high-gold content ceramo-metal alloy. J Dent Res 59:1960–1965
11. Li D, Baba N, Brantley WA, Alapati SB, Heshmati RH, Daehn GS (2010) Study of Pd-Ag dental alloys: examination of effect of casting porosity on fatigue behavior and microstructural analysis. J Mater Sci Mater Med 21:2723–2731
12. Vermilyea SG, Huget EF, Vilca JM (1980) Observations on gold-palladium-silver and gold-palladium alloys. J Prosthet Dent 44:294–299
13. Watanabe I, Watanabe E, Cai Z, Okabe T, Atsuta M (2001) Effect of heat treatment on mechanical properties of age-hardenable gold alloy at intraoral temperature. Dent Mater 17:388–393
14. Jeon GH, Kwon YH, Seol HJ, Kim HI (2008) Hardening and overaging mechanisms in an Au-Ag-Cu-Pd alloy with In additions. Gold Bull 41:257–263
15. Khalaff K, Schubert K (1974) Kristallstruktur von Pd₅Ga₂. J Less-Common Metals 37:129–140

23

Effect of gold addition on the microstructure, mechanical properties and corrosion behavior of Ti alloys

Yong-Ryeol Lee · Mi-Kyung Han · Min-Kang Kim · Won-Jin Moon · Ho-Jun Song · Yeong-Joon Park

Abstract We performed a systematic investigation of the Ti–xAu (x=5, 10, 15, 20, and 40 wt%) alloys to assess the effect of addition of element Au on the microstructure, mechanical properties, and corrosion behavior of commercially pure titanium (cp-Ti). The phase and microstructure were characterized using X-ray diffraction (XRD), optical microscopy, scanning electron microscopy (SEM), and transmission electron microscopy (TEM). The results indicated that the Ti–xAu alloys containing up to 15 wt% Au showed a hexagonal close-packed α-Ti structure, whereas the Ti–xAu alloys containing more than 20 wt% Au were mainly composed of the α-Ti phase and Ti$_3$Au intermetallic phase. We also investigated the effect of alloying element Au on the mechanical properties (including Vickers hardness and modulus) and corrosion behavior of Ti–xAu binary alloys. The addition of gold to Ti improved its hardness. Electrochemical results showed that the Ti–xAu alloys exhibited improved corrosion resistance than cp-Ti.

Y.-R. Lee · M.-K. Han · M.-K. Kim · H.-J. Song · Y.-J. Park (✉)
Department of Dental Materials, MRC for Biomineralization Disorders, School of Dentistry, Chonnam National University, Gwangju 500-757, Korea
e-mail: yjpark@jnu.ac.kr

Y.-R. Lee
e-mail: lyr1366@empal.com

M.-K. Han
e-mail: mikihan@jnu.ac.kr

M.-K. Kim
e-mail: sniky785@jnu.ac.kr

H.-J. Song
e-mail: songhj@jnu.ac.kr

W.-J. Moon
Korea Basic Science Institute, Gwang-Ju Center, 300 Yongbong-dong, Buk-gu, Gwangju 500-757, Korea
e-mail: wjmoon@kbsi.re.kr

Keywords Ti–Au alloys · Dental material · Mechanical properties · Microstructure · Corrosion behavior

Introduction

Favorable mechanical properties including high-specific strength, excellent corrosion resistance both in air and in biological fluids, and high biocompatibility have made titanium (Ti) and titanium alloys suitable for application in dental and medical fields [1, 2]. However, cp-Ti has several disadvantages, such as low deformability, low wear resistance, and difficulty in manufacturing, welding, and machining [3]. Other drawbacks of Ti are a high melting point (±1,700 °C) and a high reactivity with surrounding impurities such as oxygen and nitrogen at elevated temperatures [4]. Therefore, titanium is still unsatisfactory for practical use. Attempts have been made to develop new titanium alloys with improved mechanical properties and castability by alloying Ti with a variety of elements. The TiNi (Nitinol) and Ti–6Al–4 V alloy are the most widely used materials [5, 6]. However, the metal ions, such as Ni, Al, and V, released from the alloy cause allergic and cytotoxic effects and neurological disorders [7]. Therefore, the search continues for new Ti-based alloys with the desired properties, yet lacking any toxic elements, for dental applications.

Gold is classified as a precious metal and has a much higher standard reduction potential than titanium (+1.498 V vs. normal hydrogen electrode at 25 °C) [8]. It has been recognized to be a nontoxic and nonallergic element. Historically, gold has been used in dentistry for more than 4,000 years due to its good biocompatibility and versatility [9–12]. The use of Au as an alloying element may help to improve the corrosion resistance and biocompatibility of Ti. Recently, Rosalbino et al. reported the positive influence of Au addition on the corrosion behavior of Ti [13]. They showed that the corrosion resistance of Ti alloy containing 3.6 wt% Au was similar to or better than that of the Ti–6Al–7Nb alloy currently used as a biomaterial. Takahashi et al. examined the mechanical properties of Ti–Au alloys and found that the Ti–Au alloys had higher yield strength, tensile strength, and hardness [14]. Oh et al. reported that the cytotoxicities of the Ti–Au alloys were similar to that of pure Ti [15]. They also reported that the fusion temperature was lowered by alloying Ti with Au, thus facilitating the casting process. Therefore, it was reasonable to employ gold as a strengthening alloying element to improve the clinical performance of cp-Ti. Not only for the mechanical properties, but also the galvanic corrosion is a concern for dental implant systems which are consisted of cp-Ti fixture and noble gold alloy crowns. Therefore, it is important to develop gold-containing Ti alloys which better harmonize with the gold crowns and have the favorable bone-bonding ability. Despite the fact that there exists an extensive literature on the fundamental physical metallurgical aspects of Ti–Au alloys, studies on the relationship between phase/microstructure and various properties of Ti–xAu alloys are still relatively rare, and hence further investigations are needed to provide an in-depth understanding of their mechanical properties.

In the present study, for developing a dental titanium alloy with better mechanical properties and corrosion resistance than cp-Ti, the effect of addition of 5, 10, 15, 20, and 40 wt% alloying element Au on the microstructure, mechanical properties, and corrosion behavior of Ti–Au binary alloys was investigated. In this work, "Ti–xAu" will henceforth stand for "Ti–wt% Au".

Experimental

Material preparation

A commercially available cp-Ti (ASTM grade 2, Daito Steel Co. Ltd., Japan) was used as the control titanium material. Experimental Ti–Au alloys (5, 10, 15, 20, and 40 wt% Au) were prepared by arc melting of the stoichiometric quantities of elements placed on a water-cooled copper hearth using a tungsten electrode under high-purity argon atmosphere. The starting materials (Ti sponge, Alfa Aesar, USA, 99.95 %; Au ingot, LS-Nikko, Korea, 99.99 %) were used without purification. During the arc-melting procedure, a titanium getter was heated prior to melting the reactant mixture to further purify the argon atmosphere. The samples were remelted several times to promote sample homogeneity. Subsequently, the samples were heat treated using tube furnace under argon atmosphere (99.9999 %) for 4 h at 150 °C below the respective solidus temperatures followed by cooling in furnace at rate of 10 °C/min down to 600 °C and air cooling to room

temperature. These heat-treatment conditions were chosen in accordance with the binary Ti–Au phase diagrams [16].

Material characterization

Phase analysis and structural characterization were performed by X-ray diffraction. The X-ray diffraction (XRD) patterns were collected for bulk samples using a X'Pert PRO Multi Purpose X-ray Diffractometer (40 kV and 40 mA) with Cu K_α ($\lambda = 1.54056$ Å). The scanning speed was 2°/min, and the scanning 2θ angle ranges from 20° to 80°. The cp-Ti was used as a control. The lattice parameters were obtained by least squares refinement of data in the 2θ range of 20–80° with the aid of a Rietveld refinement program [17]. The phase transformation of Ti–xAu alloys was investigated by heating approximately 200 mg of the sample to 1,000 °C at a rate of 20 °C/min using differential scanning calorimetry (DSC; DSC 404 C, Netzsch, Germany). The microstructures of samples were examined using metallurgical microscope (Epiphot FX-35WA, Nikon, Japan), scanning electron microscope (SEM; Hitachi, S-3000 N, Japan), high-resolution transmission electron microscopy (HRTEM; Philips, TECHNAI-F20, Netherlands), selected area energy diffraction (SAED), and energy dispersive X-ray analysis (EDX; EMAX, Horiba, Japan). The elemental distribution of the Au and Ti in the alloys was mapped using the electron probe microanalysis (EPMA; Shimadzu, EPMA-1600, Japan). The microhardness of polished alloys was measured using a Vickers microhardness tester (Zwick, Postfach4350, Germany) with a 500-g load for 30 s. Elastic modulus was measured using a Nanoindenter XP (MTS Co, USA) with a maximum indentation depth of 2 μm.

To observe the corrosion behavior of Ti–xAu alloys, potentiodynamic anodic polarization tests were conducted at a scan rate of 5 mV/s from −1.5 V saturated calomel electrode (SCE) to +1.5 V SCE using a potentiostat (WAT100, WonA Tech Co., Ltd, Korea) in a 0.9 % NaCl solution at 37±1 °C. A three-electrode cell was used. The counter electrode was a high-density graphite electrode, and the reference electrode was a SCE. Before immersing the test sample, the electrolyte was bubbled with Ar gas at 150 mL min^{-1} for more than 20 min to eliminate the residual oxygen in the electrolyte. The used electrolyte was replaced with the fresh electrolyte before each measurement. Both corrosion density and potential were estimated by Tafel plots using both the anodic and cathodic branches. At least three samples were tested to confirm the repetition of the experimental results.

The galvanic current density of various Ti–xAu/cp-Ti galvanic pairs were measured over a 20-min period using a potentiostat/galvanostat under deaerated conditions (ZIVE SP2, WonA Tech). The experimental setup for the electrochemical measurements consisted of a three-electrode cell with the sample as a working electrode with exposed area of 0.79 cm^2, a SCE as a reference electrode, and a cp-Ti as a counter electrode. The distance between the cp-Ti and the samples was maintained at 1 cm.

Results and discussion

Phase and microstructure

X-ray diffraction (XRD) analysis was conducted for characterizing the phase of the cast alloy by matching each characteristic peak with the JCPDS files. The X-ray diffraction patterns as a function of x for the Ti–xAu (x=5, 10, 15, 20, and 40 wt%) samples are shown in Fig. 1, and are compared with those of cp-Ti. The cp-Ti showed a hexagonal close-packed (hcp) crystal α phase structure whose unit cell constants were a=2.959(1)Å and c=4.703(1) Å (c/a ratio= 1.589), and they corresponded well with those in the literature (JCPDS card no. 44-1294). It could be clearly seen that the phases/crystal structure of the binary Ti–xAu alloy was sensitive to the Au contents in the cast alloy. The Ti–xAu (x=5, 10, and 15 wt%) alloys were comprised mainly of α-Ti phase. When the Au content was increased to 20 wt% or greater, the phase constitution was changed from a single α-Ti phase to a mixture of α-Ti phase and the secondary phase of Ti$_3$Au. As indicated by the intensity of the Ti$_3$Au peaks in the XRD

patterns, the secondary phase of Ti$_3$Au increased with the increase of Au contents. According to the equilibrium-phase diagram of the Ti–Au system, the Ti–Au alloy has a eutectoid point of α-Ti and Ti$_3$Au at 15.3 % Au [4, 16]. Therefore, the precipitation of the intermetallic compound Ti$_3$Au was expected to occur at an Au concentration of approximately 15 wt% or more. Even though the Ti$_3$Au phase was not detected in the Ti–15Au sample using XRD due to the detection limit of this instrument, the presence of Ti$_3$Au phase was confirmed by SEM and transmission electron microscopy (TEM) at this Au concentration.

It is widely known that the mechanical properties of Ti alloys depend essentially on the microstructure. A good understanding of the microstructure of Ti alloys is therefore a prerequisite for controlling their properties. To investigate the influence of the addition of Au on the microstructures of Ti, samples were investigated using metallurgical microscope and scanning electron microscope (SEM). Figure 2 shows the microstructures of Ti–xAu alloys with different Au contents (5, 10, 15, 20, and 40 wt%). The Ti–5Au and Ti–10Au alloys exhibited typical lath-type morphologies with a hexagonal crystal structure of the α phase. When the Au content was 15 wt%, the alloy had an equiaxed structure and the secondary phase of Ti$_3$Au was observed in alloys. From Fig. 2d, e, it was observed that Ti$_3$Au was homogeneously distributed in the α-Ti matrix, and the volume fraction of Ti$_3$Au phase increased with the increase in Au content. The compositional analyses were performed using EDX at various areas marked in Fig. 2, and the results are summarized in Table 1. With the aid of EDX and TEM analysis, the bright region in the SEM micrographs of Fig. 2 was associated with the Au-rich phase (Ti$_3$Au phase), and the dark region was the Ti matrix (α-Ti phase).

At a lower Au concentration, only the α-Ti phase was detected using the XRD. However, when the minimal amount of intermetallic compounds is existed in the alloy, it is difficult to detect the phases by using XRD due to instrumental detection limit. Therefore, in order to study the homogeneity of the microstructure of the Ti–5Au and Ti–10Au alloys, the elemental distribution of Au and Ti in the alloys was mapped by using the EPMA. Figure 3 shows the SEM backscattered electron image of Ti–xAu (x=5, 10, and 15) alloys and the EPMA chemical analysis of Au elements. Imaging the polished Ti–xAu alloys using backscattered electrons revealed the presence of two different phases. As darker spots represented the lighter elements in the backscattered image, Ti-rich phases were observed as dark spots, whereas the bright area was the Au-rich phase. EPMA mappings of the polished sample surfaces were also carried out to show the distribution of alloying elements in the phases. As the brighter area indicated a high concentration region of Au element, elemental

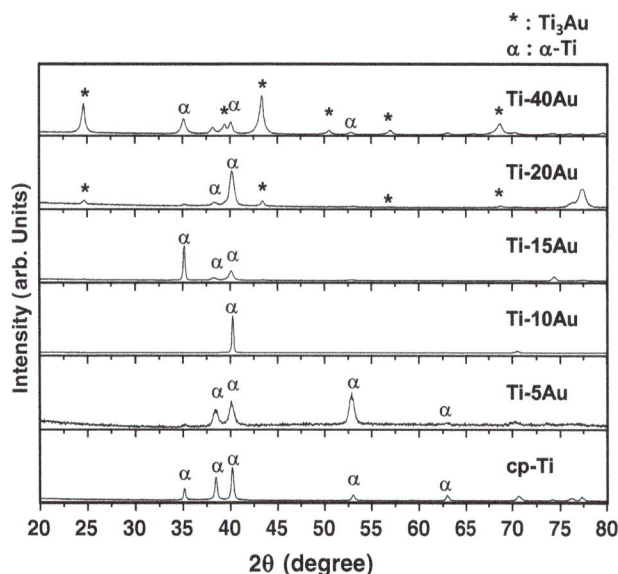

Fig. 1 XRD patterns of cp-Ti and the series of binary Ti–xAu alloys (α and * represent diffraction peaks of the α-Ti and Ti$_3$Au, respectively)

Fig. 2 Micrographs of Ti–xAu alloys **a** Ti-5Au, **b** Ti-10Au, **c** Ti-15Au, **d** Ti-20Au, and **e** Ti-40Au. In each micrograph, the *left images* were metallurgical micrographs (×400; *scale bar* 10 μm) and the *right images* were SEM micrographs (*scale bar* 50 μm). The *boxed letters* denote the areas for compositional analyses using EDX.

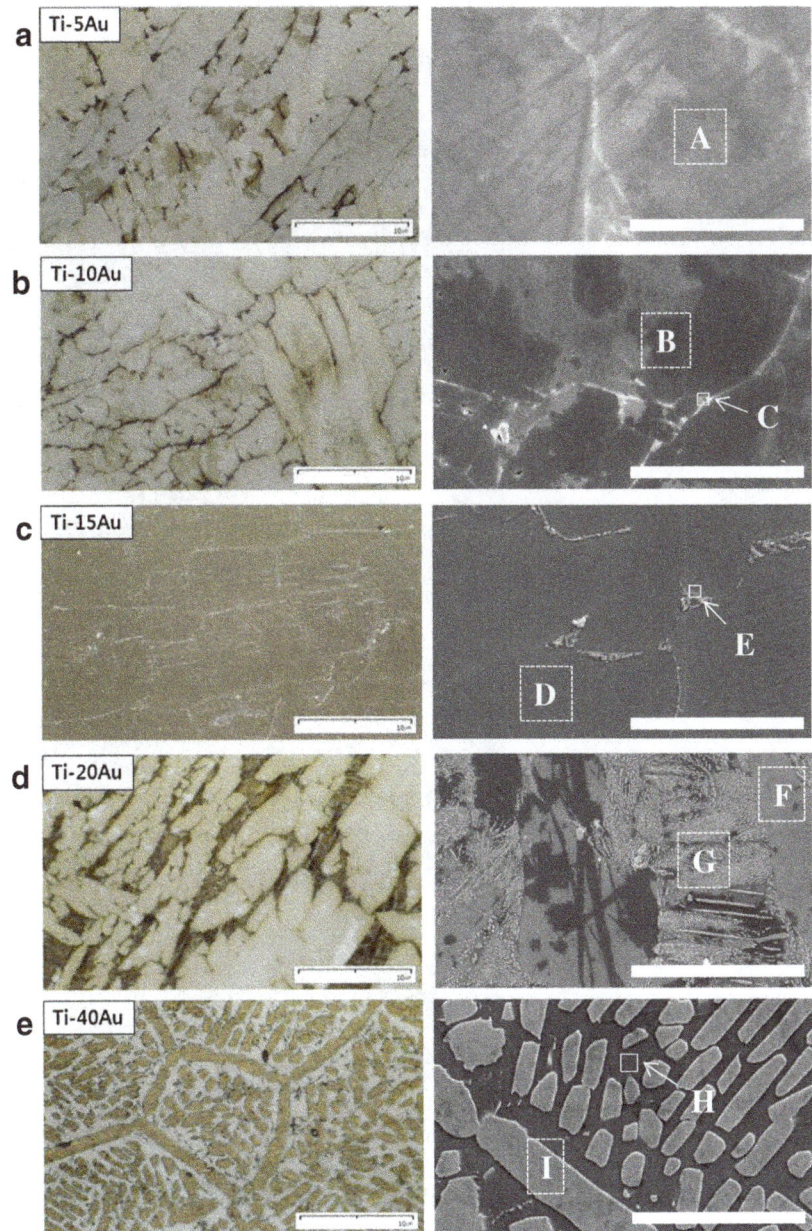

Table 1 Compositional analysis at areas shown in Fig. 2 for Ti–xAu alloys

Alloy code	Area	Au (wt%)
Ti-5Au	A	5.7
Ti–10Au	B	10.8
	C	16.9
Ti–15Au	D	12.8
	E	72.9
Ti–20Au	F	22.1
	G	49.5
Ti–40Au	H	16.5
	I	65.9

mapping images showed that the Au-rich phase was concentrated at the interface.

Detailed analysis of phases was conducted using TEM. Typical representative TEM images and SAED patterns of the Ti–10Au and Ti–20Au alloys are shown in Fig. 4. The SAED patterns acquired from the regions marked as "A" and "B" in the TEM image of Ti–10Au alloy were obtained from $[413]_\alpha$ zone axis of a hexagonal α-Ti structure, indicating that the Ti–10Au alloy was comprised mainly of α-Ti phase. The SAED patterns acquired from the regions marked as "C" and "D" in the TEM image of Ti–20Au alloy were obtained from $[213]_\alpha$ and $[100]_\alpha$ zone axis of a hexagonal α-Ti structure, respectively. Whereas, the SAED pattern acquired from the

Fig. 3 SEM backscattered electron images (BEI) and EPMA elemental maps for Au of **a** Ti-5Au, **b** Ti-10Au, and **c** Ti-15Au alloys

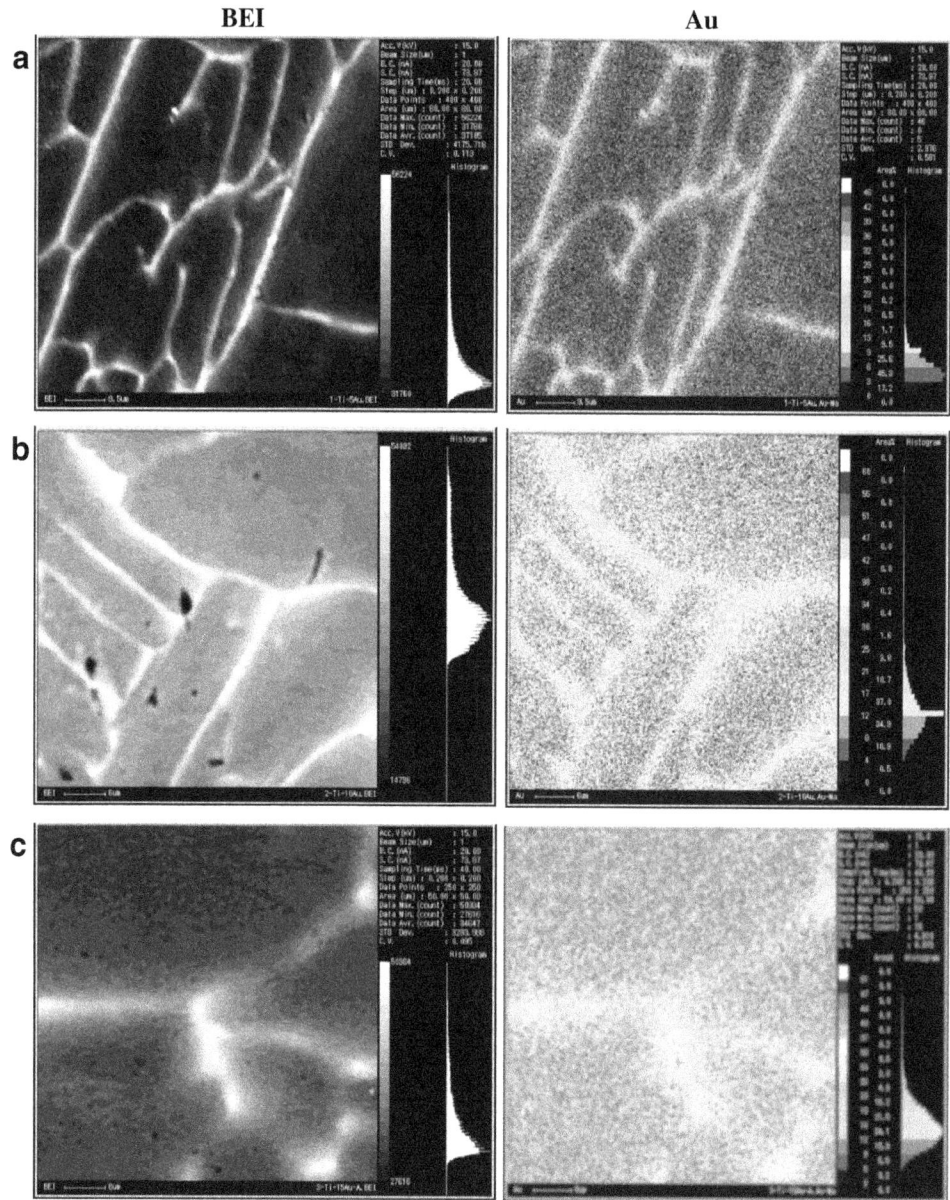

"E" region consisted of characteristic single-crystalline bcc Ti$_3$Au phase, and this finding was in good agreement with the XRD and EDX analysis results.

DSC analysis was performed to detect any abnormal thermal effects from room temperature to 1,000 °C which could be representative of the presence of free segregation of Au or intermetallic precipitates. As shown in Fig. 5, very smooth temperature-dependent traces were found at about 150 °C for all Ti–xAu alloys with the content up to 10 wt%. This peak resulted from stabilization of a thermally unstable structure such as grain boundary relaxation or grain boundary reordering. The second exothermic peak at about 900 °C occurred due to the martensitic transformation of α-Ti to β-Ti. Other than those two peaks, no extraneous exothermic or endothermic peaks were detected. The martensitic

transformation temperature decreased with increasing Au content as shown in Fig. 5. This implies that Au enhanced the stability of the β-phase, thus lowering the transformation temperature. When the alloy contained 20 wt% or more Au, the alloy showed an extraneous exothermic peak which resulted from the formation of the Ti$_3$Au phase. There was a gradual decrease in enthalpy of the exothermic peaks in DSC with the addition of Au in Ti alloys. The decrease in enthalpy may be understood as a result of short-range chemical ordering in the Au containing alloys.

Mechanical properties

The effect of alloying element Au on the mechanical properties and corrosion behavior of Ti was investigated. As shown

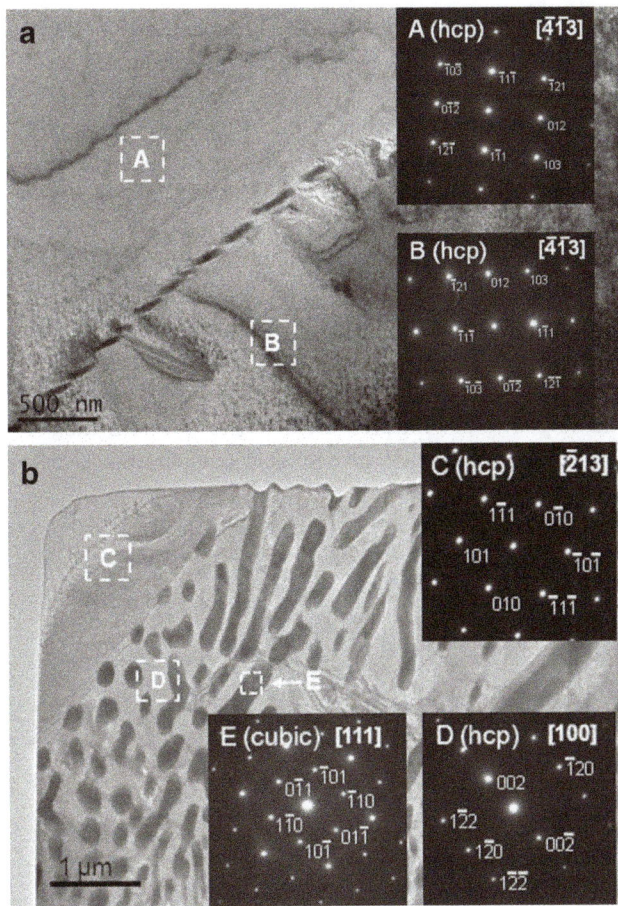

Fig. 4 TEM images and SAED patterns of **a** Ti-10Au and **b** Ti-20Au alloys

Table 2 Hardness and elastic modulus values of Ti–xAu alloys compared with cp-Ti ($n=5$)

Alloy code	Vickers hardness (VHN)	Elastic modulus (GPa)
cp-Ti	164.54 (3.54)[a] a	132.35 (12.22) a
Ti–5Au	375.40 (8.71) d	130.51 (2.57) a
Ti–10Au	343.40 (15.34) c	127.40 (7.61) a
Ti–15Au	318.60 (10.01) b	134.17 (2.61) a
Ti–20Au	366.40 (30.05) d	128.55 (3.87) a
Ti–40Au	433.20 (15.77) e	162.40 (4.66) b

Different letters within same column indicate statistically significant difference at 5 % ($p<0.05$)

[a] Mean (S.D.)

Au, the microhardness values increased again. This increase in hardness was probably caused by the formation of a eutectoid structure consisting of α-Ti and the intermetallic Ti₃Au compound. A similar precipitation hardening was observed previously in the Ti–7Cr system [18] and Ti–30Nb system [19].

The results of elastic modulus measured by nanoindentation are also shown in Table 2. The elastic modulus for cp-Ti was ~130 GPa. In the present study, the elastic moduli values for Ti–xAu ($x=5$, 10, 15, and 20) alloys were in the range of 124–132 GPa. For these compositions, there was no statistically significant difference between Ti–xAu alloys ($p<0.05$) and cp-Ti. However, the Ti-40Au alloy exhibited significantly high values of the elastic modulus (164 GPa) due to the presence of the intermetallic Ti₃Au alloy.

Corrosion behavior

The purpose of this study was to evaluate the corrosion resistance of experimental Ti–Au alloys. Potentiodynamic

in Table 2, all of the Ti–Au alloys had significantly higher ($p<0.05$) hardness than the cp-Ti. The hardness value tended to rise with increasing Au content up to 10 wt%, which was mainly due to the solid-solution strengthening of the α-phase. The VHN values for the alloys containing 5 and 10 wt% Au were 375 and 343, respectively. As for the Ti–15Au alloy, its microhardness was slightly decreased due to the formation of equiaxed structure. When the alloy contained 20 wt% or more

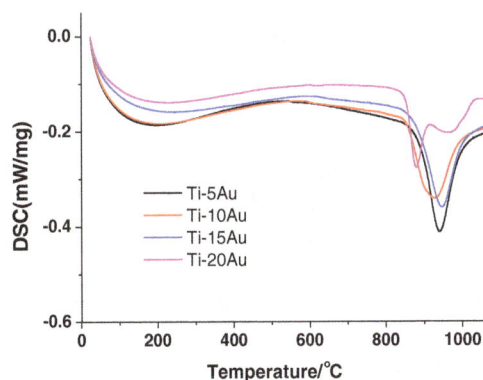

Fig. 5 DSC curves of the series of binary Ti–xAu alloys

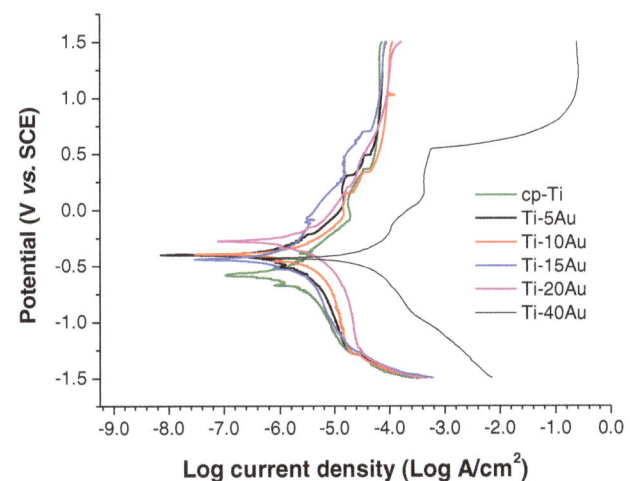

Fig. 6 Representative potentiodynamic polarization curves for cp-Ti and Ti–xAu alloys

Table 3 Corrosion potential (E_{corr}) and corrosion current density (I_{corr}) of cp-Ti and Ti–xAu alloys ($n=3$)

Alloy code	E_{corr} (mV)	I_{corr} (μA/cm^2)
cp-Ti	−575.00 (40.95)[a] a	0.61 (0.17) a
Ti–5Au	−462.67 (59.76) b	0.29 (0.08) a
Ti–10Au	−395.00 (23.64) b	0.43 (0.11) a
Ti–15Au	−472.67 (90.01) b	0.36 (0.14) a
Ti–20Au	−278.33 (20.43) c	0.94 (0.70) a
Ti–40Au	−425.73 (63.40) b	7.78 (6.51) b

Within the same column, mean values with the same letters were not statistically different at 5 % ($p<0.05$)

[a] Mean (S.D.)

polarization and galvanic couple technique were used in order to investigate the effect of Au content on corrosion resistance. The potentiodynamic polarization curves of cp-Ti and Ti–xAu alloy were recorded at a sweep rate of 0.005 V/s, between −1.5 and 1.5 V potential ranges in 0.9 % NaCl solutions, and the results are shown in Fig. 6. All curves exhibited active–passive behavior with apparently small oscillation of current density. These regions might be associated with the formation of one or more protective films. Except for the case of 40 wt% Au addition, all of the Ti–xAu alloys revealed that the initial passive current densities occurred in the range of 1.5 to 2.5× 10^{-5} A/cm^2. These current density values were similar to that of cp-Ti (1.8×10^{-5} A/cm^2). At the potential of 0.55 V, the current densities were in the following order: Ti–40Au (5.7×10^{-4} A/cm^2)>Ti–10Au (6.8×10^{-5} A/cm^2)>cp-Ti≈ Ti–5Au (5.0×10^{-5} A/cm^2)>Ti–20Au (4.0×10^{-5} A/cm^2)> Ti–15Au (2.0×10^{-5} A/cm^2). This trend of abrupt increase of current density in Ti–40Au at approximately 0.55 V (vs. SCE) agreed with the result of previous report [14].

Using the Tafel extrapolation method, we calculated corrosion potential (E_{corr}) and corrosion current density (I_{corr}) of cp-Ti and Ti–xAu alloys from both the anodic and the cathodic branches of the potentiodynamic polarization curves, as listed

in Table 3. Positive shifts in E_{corr} were observed in all of the Ti–xAu alloys, indicating a better corrosion resistance of the coated samples. The results showed that the Ti–20Au alloy had the most positive E_{corr} value (−0.278 V). In comparison with cp-Ti, Ti–xAu alloys (x=5, 10, and 15 wt%) exhibited slightly reduced current density values. When Au content was increased to more than 20 wt%, the I_{corr} increased from 0.94 μA/cm^2 (Ti–20Au) to 7.78 μA/cm^2 (Ti–40Au) due to the preferential attack of the Ti$_3$Au phase. This finding was consistent with the previous report showing that the corrosion resistance of the Ti–Au alloys was deteriorated due to the preferential dissolution of Ti$_3$Au [20]. From this experiment, it could be known that the presence of Ti$_3$Au in Ti–Au alloy not only cause an increase of corrosion potential but also weaken the passive film.

The passivation phenomenon might be studied better by the galvanic couple technique. Mean values of galvanic current densities versus the time of coupling in cp-Ti/Ti–xAu alloys are shown in Fig. 7. Ti–xAu alloys apparently exhibited negative current values, indicating that the electron flows from cp-Ti to Ti–xAu alloy, thus causing corrosion of the cp-Ti, because it lost electrons. The galvanic current density increased with time and reached a stable value. This increase in the galvanic current density could be attributed to the formation of a TiO$_2$ product film on the cp–Ti surface which grew with time and partially protected this material. The Ti–10Au alloy showed the fastest cathodic behavior among the Ti–xAu alloys, since the time taken by the Ti–10Au alloys to attain the stable resting potential was less as compared to that taken by other Ti–xAu alloys.

Conclusions

The effect of Au alloying element on the microstructure, mechanical properties, and corrosion behavior of commercially pure titanium (cp-Ti) was investigated. Based on the results of XRD, EPMA, and optical microscopy, the Ti–xAu alloys with up to 10 wt% Au content showed hcp α structures, and the Ti–xAu alloys (x=15, 20, and 40 wt%) were composed of α-Ti and Ti$_3$Au intermetallic phases. As results of solid-solution strengthening of α-Ti and precipitation hardening with Ti$_3$Au intermetallic compound, the Ti–xAu alloys showed better hardness than the cp-Ti. The Ti–xAu alloys showed a higher corrosion resistance. After considering all the mechanical properties and corrosion behavior of the Ti–xAu alloys, the Ti alloy with 10 wt% Au is a good candidate for dental casting materials. Due to the fact that Ti–Au alloys have better hardness and corrosion resistance compared to conventional titanium, further investigation on the effects of compositional variations with less Au content in Ti–xAu alloy systems for the possible application as implant materials might be needed.

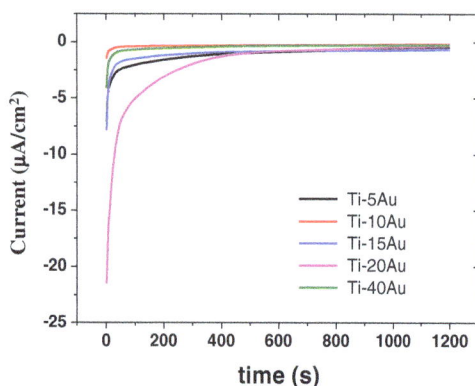

Fig. 7 Mean values of galvanic current density versus time of the couplings cp-Ti/Ti–xAu alloys

Acknowledgments This research was supported by the National Research Foundation of Korea (NRF) grant funded by the Korea government (MSIP) (No. 2011-0030762) and by the Basic Science Research Program through the NRF funded by the Ministry of Education, Science and Technology (2011-0002706).

References

1. Leyens C, Peters M (2003) Titanium and titanium alloys. Wiley, New York
2. Lloyd C, Scrimgeour S, Brown D, Clarke R, Curtis R, Hatton P, Ireland A, McCabe J, Nicholson J, Setcos J (1997) Dental materials: 1995 literature review. J Dent 25:173–208
3. Ezugwu E, Wang Z (1997) Titanium alloys and their machinability—a review. J Mater Process Technol 68:262–274
4. Niinomi M (1998) Mechanical properties of biomedical titanium alloys. Mater Sci Eng A 243:231–236
5. Rocher P, El Medawar L, Hornez J, Traisnel M, Breme J, Hildebrand H (2005) Biocorrosion and biocompatibility of NiTi alloys. Eur Cell Mater 9:23–24
6. Eisenbarth E, Velten D, Müller M, Thull R, Breme J (2004) Biocompatibility of β-stabilizing elements of titanium alloys. Biomaterials 25:5705–5713
7. Cremasco A, Messias AD, Esposito AR, Duek EAR, Caram R (2011) Effects of alloying elements on the cytotoxic response of titanium alloys. Mater Sci Eng C 31:833–839
8. Lide DR (1995) CRC handbook of chemistry and physics, 76th edn. CRC, Boca Raton
9. Knosp H, Holliday RJ, Corti CW (2003) Gold in dentistry: alloys, uses and performance. Gold Bull 36:93–102
10. Corti CW, Holliday RJ (2004) Commercial aspects of gold applications: from materials science to chemical science. Gold Bull 37:20–26
11. Takada Y, Ito M, Kimura K, Okuno O (2005) Electrochemical properties and released ions of Au-1.6 mass% Ti alloy. Dent Mater J 24:153–162
12. Fischer J (2002) Effect of small additions of Ir on properties of a binary Au–Ti alloy. Dent Mater 18:331–335
13. Rosalbino F, Delsante S, Borzone G, Scavino G (2012) Influence of noble metals alloying additions on the corrosion behaviour of titanium in a fluoride-containing environment. J Mater Sc-Mater Med 23:1129–1137
14. Takahashi M, Kikuchi M, Okuno O (2004) Mechanical properties and grindability of experimental Ti-Au alloys. Dent Mater J 23:203–210
15. Oh KT, Kang DK, Choi GS, Kim KN (2007) Cytocompatibility and electrochemical properties of Ti–Au alloys for biomedical applications. J Biomed Mater Res Part B 83:320–326
16. Murray JL (1987) Phase diagrams of binary titanium alloys. ASM International, Geauga County
17. Hunter B, Howard C (2000) Rietica; Australian Nuclear Science and Technology Organization: Menai. Australia
18. Koike M, Okabe T, Itoh M, Okuno O, Kimura K, Takeda O, Okabe TH (2005) Evaluation of Ti-Cr-Cu alloys for dental applications. J Mater Eng Perform 14:778–783
19. Kikuchi M, Takahashi M, Okuno O (2003) Mechanical properties and grindability of dental cast Ti-Nb alloys. Dent Mater J 22:328–342
20. Takahashi M, Kikuchi M, Takada Y, Okuno O, Okabe T (2004) Corrosion behavior and microstructures of experimental Ti-Au alloys. Dent Mater J 23:109–116

24

Extended reliability of gold and copper ball bonds in microelectronic packaging

Chong Leong Gan · Classe Francis ·
Bak Lee Chan · Uda Hashim

Abstract Wire bonding is the predominant mode of interconnection in microelectronic packaging. Gold wire bonding has been refined again and again to retain control of interconnect technology due to its ease of workability and years of reliability data. Copper (Cu) wire bonding is well known for its advantages such as cost-effectiveness and better electrical conductivity in microelectronic packaging. However, extended reliabilities of Cu wire bonding are still unknown as of now. Extended reliabilities of Au and Pd-coated Cu (Cu) ball bonds are useful technical information for Au and Cu wire deployment in microelectronic packaging. This paper discusses the influence of wire type and mold compound effect on the package reliability and after several component reliability stress tests. Failure analysis has been conducted to identify its associated failure mechanisms after the package conditions for Au and Cu ball bonds. Extended reliabilities of both wire types are investigated after unbiased HAST (UHAST), temperature cycling (TC), and high-temperature storage life test (HTSL) at 150, 175, and 200 °C aging temperatures. Weibull plots have been plotted for each reliability stress. Obviously, Au ball bond is found with longer time to failure in unbiased HAST stress compared to Cu ball bonds for both mold compounds. Cu wire exhibits equivalent package and or better reliability margin compared to Au ball bonds in TC and HTSL tests. Failure mechanisms of UHAST and TC have been proposed, and its mean time to failure (t_{50}), characteristic life ($t_{63.2}$, η), and shape parameter (β) have been discussed in this paper. Feasibility of silver (Ag) wire bonding deployment in microelectronic packaging is discussed at the last section in this paper.

Keywords Gold bonding wire · Copper bonding wire · Silver wire · Extended reliability · Microelectronic packaging

Introduction

Gold bonding wire has been extensively used for the fabrication of integrated circuits because of its good electrical conductivity and mechanical stability with a diameter of 20 µm or less. With significant increases in gold price, gold ball bonding has become a more costly process that has a considerable economic effect on the assembly of packages used in consumer electronics. An alternative wire material to gold is copper, which is much cheaper, has several technical benefits including better electrical conductivity, and has been widely used in discrete and power devices with wire diameters typically larger than 30 µm in diameter for many years. The potentials and cost considerations of finding an alternative to replace gold wire bonding in microelectronic packaging are driven by new technologies coming to the market [1]. Copper wire bonding appears to be the alternate materials, and various engineering studies on copper wire deployment have been reported [2, 3]. The Au–Al intermetallic compound (IMC) growth is widely characterized and analyzed [4–6]. Zulkifli MN et al. [7] suggested new approaches: examining the effect of individual phase and surroundings on the strengthening produced by the Au–Al intermetallic compound, combining FEA based on friction and wire-bonding parameters, and correlating TEM results with results obtained from other techniques should enable a more detailed understanding of the bondability and strength of thermosonic gold wire bonds. Key technical barriers such as intermetal dielectric cracking due to excessive bonding, copper ball bond corrosion under moist environments, and extended reliability of copper ball bonds are identified accordingly [8–10]. Copper ball bond is

C. L. Gan (✉) · C. Francis · B. L. Chan
Spansion (Penang) Sdn. Bhd., Bayan Lepas,
11900 Penang, Malaysia
e-mail: chong-leong.gan@spansion.com

C. L. Gan · U. Hashim
Institute of Nano Electronic Engineering (INEE),
Universiti Malaysia Perlis, 01000 Kangar, Perlis, Malaysia

more susceptible to moisture corrosion compared to gold ball bonds and undergoes different corrosion mechanisms in microelectronic packaging [11, 12]. Our previous studies indicate that Pd-coated copper ball bond outperforms gold ball bonds in biased HAST wearout reliability [13]. Extended reliability of high-temperature storage life (HTSL) of copper ball bonds in TSOP package is found with apparent activation energy (E_{aa}) of ~0.70 eV compared to gold ball bonds [14]. Blish et al. [15] investigated E_{aa} of typical Au–Al IMC of 1.0~1.5 eV, which is Al thickness dependent. Hence, extended reliability is crucial to determine the lifetime of gold and copper ball bonds in microelectronic packaging. The Cu–Al and Au–Al IMC growth kinetics were studied, and it was found that Cu–Al growth is at least 5× slower than Au–Al IMC [16]. However, copper wire bonding still pose reliability challenges and complex failure mechanisms which could be the main barriers to entirely replace gold wire bonding [17]. In this study, we have prepared FBGA 64 package assembled with gold and Pd-coated copper wire and load for unbiased HAST (UHAST), temperature cycling (TC), and HTSL tests.

Will Cu or Ag wires entirely replace Au wire bonding?

In general, Cu wire is not an ultimate bonding wire solution in semiconductor packaging. Cu wire bonding is more suitable to be deployed in low-pin-count semiconductor packaging, flash memory packaging, or high-power devices which utilize a larger diameter for bonding wire. The various considerations such as its long-term extended reliability performance and bondpad cratering challenges still pose a showstopper for full sweep of copper wire bonding in semiconductor packaging. Undeniably, the improved N_2 kit (which is installed on wire bonder) will improve the wire-bonding process with an inert environment since Cu wire is vulnerable to corrosion and oxidation in production floor. Cu wire will not entirely replace conventional Au wire bonding in semiconductor packaging but rather another option of packaging methods other than Au, Al, and possibly Ag wire bonding.

Au, Cu, or Ag wire alloys for semiconductor packaging?

Cu as new interconnect material has increasingly gained popularity due its lower cost and good thermal and mechanical properties compared to Au wire. Table 1 shows some general material properties, where the similar mechanical properties of Ag and Au can be seen and Ag is more superior in terms of electrical and thermal conductivity. When compared to Cu, Ag is similar in conductivity, but softer in terms of mechanical properties. These properties will change as it is dependent on the purity of the metals.

Au wire exhibits excellent UHAST extended reliability and more stable assembly processes (in terms of shear strength and wire pull strength in the as-bonded stage). This is the most

pivotal deciding factor for keeping Au wire bonding in some of the customer-end field applications such as medical, automotive, and military market segments. Silver (Ag) wire bonding is still a new interconnect method in semiconductor packaging and yet to be widely adopted by major semiconductor companies due to lack of reliability data and further engineering evaluations. Many claims on its advantages of moderate AgAl IMC formation and growth rate, and easier pluck-and-play for mass production are key success factors for Ag wire to replace Au or Cu wire bonding. Cho et al. [18] reported that Pd alloying of the Ag wire was effective in improving the reliability of Ag ball bond. The lifetime in PCT increased with increasing Pd concentration in the Ag wire. Free air ball formation is found better in Ag–Au–Pd compared to 2 N Ag wire alloy [19]. The bonding process of Ag wire bonding is pretty similar to Au wire bonding [20]. Another bondability study is conducted on Au–Ag wire alloy, and caution should be given to bonding temperature and first ball bond parameter setting [22]. This observation convinces the great opportunity of using Ag–Au–Pd alloy instead of bare 2 N Ag wire in microelectronic packaging. Another reliable Ag–8Au–3Pd wire alloy is found with high reliability and low electrical resistivity, which is processed with annealing twins [21, 23, 24]. Bare Ag wire or tertiary Ag alloy (such as Ag–8Au–3Pd or Ag–Au–Pd) are identified as next potential candidates of microelectronic packaging. In a nutshell, Au, Ag, and Cu (bare or Pd-coated Cu) wire alloys will exist as three alternatives of wire-bonding techniques in semiconductor

Table 1 Material properties of bare Ag, Au, and Cu

Material properties	Units	Ag	Au	Cu
Thermal conductivity	W/mK	430	320	400
Electrical resistivity	10^{-8} Ωm	1.63	2.2	1.72
Young's modulus	GPa	83	78	130
Vickers hardness	MPa	251	216	369

Table 2 Summary of experimental matrix (for Au and Cu wires)

Mold compound type	Extended reliability test	Test conditions	Sample size
A	UHAST	85%RH, 130 °C	77
A	TC	-40 to 150 °C	77
A	HTSL	150 °C	45
A	HTSL	175 °C	45
A	HTSL	200 °C	45
B	UHAST	85%RH, 130 °C	77
B	TC	-40 to 150 °C	77
B	HTSL	150 °C	45
B	HTSL	175 °C	45
B	HTSL	200 °C	45

Table 3 Summary of extended reliability results (mold compound types A and B of Au and Cu wires)

Mold compound	Wire type	Test type	Test conditions	t_{first} (h/cyc)	t_{50} (h/cyc)	$t_{63.2}$ (η)	β
A	Cu	UHAST	85%RH, 130 °C	3,000	6,610	7,061	5.53
A	Cu	TC	-40 to 150 °C	9,500	16,756	17,504	7.23
A	Au	UHAST	85%RH, 130 °C	4,000	8,681	9,541	3.67
A	Au	TC	-40 to 150 °C	9,000	15,060	15,691	8.32
B	Cu	UHAST	85%RH, 130 °C	1,248	9,503	11,331	1.92
B	Cu	TC	-40 to 150 °C	11,000	18,175	18,919	8.35
B	Au	UHAST	85%RH, 130 °C	2,000	7,086	7,911	3.22
B	Au	TC	-40 to 150 °C	10,500	15,251	15,251	11.43

packaging based on its customer-end field applications and packaging cost considerations. The packaging cost of Ag wire bonding is moderate and in between of Au and Cu wire bonding.

Future of Au wire bonding in semiconductor packaging

Au wire bonding will still exist in microelectronic packaging in view of its process stability and higher moisture reliability margin compared to recent penetration of Cu wire bonding. The primary motivation for uptake of copper wire bonding is, however, more strongly cost driven rather than motivated by very clear and distinct process, performance, or reliability advantages [3]. In our extended reliability study, Au ball bond exhibits higher UHAST wearout reliability margin compared to Cu ball bonds for epoxy mold compound (EMC) A. This is one of the most important factors to keep Au wire bonding in microelectronic packaging, especially for high-pin-count semiconductor packaging. Cu is known for more susceptible corrosion activity under moist environments. End customers such as automotive, military, and medical industries had expressed concerns over massive transition from Au to Cu wire bonding. Ag wire bonding, however, is not a mature

assembly process, and it requires further engineering studies on IMC formation, extended reliability, and test yield analysis on flawless high-volume manufacturing process to validate the feasibility of Ag wire bonding in microelectronic packaging. There is clearly a place for copper (bare copper or Pd-coated copper wire) or silver (bare silver or Ag–Au–Pd alloy) wire bonding in microelectronic packaging, but it is likely that rather than replacing gold wire entirely, copper or silver wire bonding will become another option alongside gold wire bonding which microelectronic package designers can consider for package assembly.

Experimental Procedure

Materials used include 0.8 mil Pd-coated Cu wire (Cu), 4 N (99.99 % purity) gold (Au) wire, and 110-nm flash devices packaged into fortified fine pitch BGA packages, with green (<20 ppm chloride in content) molding compound and substrate. In this Cu wire wearout reliability study, there are a total of four legs comprised of Pd-coated Cu wire and 4 N Au wire bonded on fine pitch 64-ball BGA packages on a 2 L substrate. Sample size used is 77 units for UHAST and TC

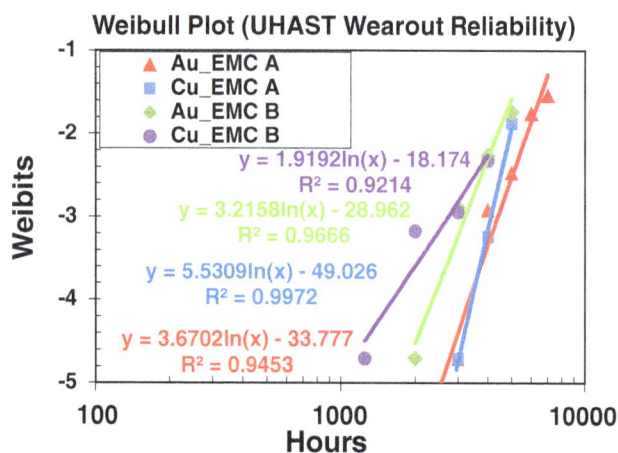

Fig. 1 Extended UHAST reliability plots of Au and Cu ball bonds with different EMCs

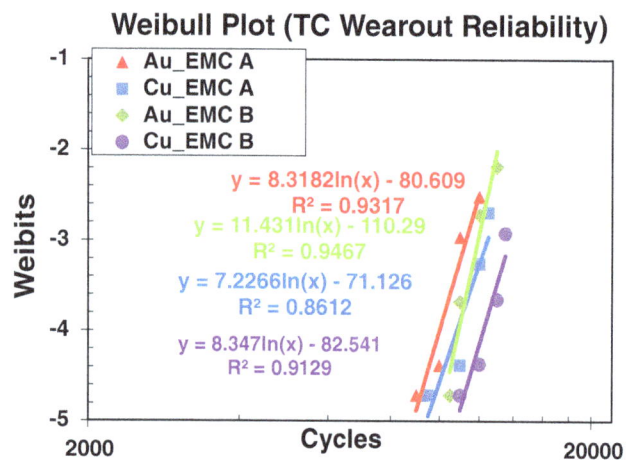

Fig. 2 Extended TC reliability plots of Au and Cu ball bonds with different mold compounds A and B

Table 4 Key material characteristics of EMCs A and B

Material characteristics	Units	EMC A	EMC B
Linear coefficient of thermal expansion 1 (CTE, α_1)	$10^{-5}/°C$	1.1 ± 0.3	0.7 ± 0.3
Linear coefficient of thermal expansion 2 (CTE, α_2)	$10^{-5}/°C$	4.5 ± 1.0	3.0 ± 1.0
Glass transition temperature (T_g)	°C	125 ± 15	125 ± 15
Cl$^-$ content	ppm	<10	<10
Na content	ppm	<10	< 10

stresses. The corresponding stress test conditions are tabulated in Table 2. After electrical test, good samples were then subjected for preconditioning and 3 times reflow at 260 °C as described in JEDEC IPC-STD 020 standard, followed by UHAST stress testing per JESD22-A118 at 130 °C/85%RH) [5] and TC per JESD22-A104 at -40 to 150 °C. Electrical testing was conducted after several readpoints of stress as well as to check Au and Cu ball bond integrity in terms of its moisture and thermomechanical reliability.

Another set of materials was used to estimate the apparent activation energies (E_{aa}) of Au and Cu ball bonds assembled with EMC A and EMC B. The key materials used include 0.8 mil Pd-coated Cu wire and 4 N (99.99 % purity) Au wire, fine pitch BGA package, and 110-nm device which is packed in fortified fineline BGA package, green (<20 ppm chloride in content) in molding compound and substrate. All direct material used in this evaluation study for the 110 nm, flash device (with top Al metallization bondpad) for packaging purpose. A total of six legs of 45 units of Au and Pd-coated Cu wire bonded on fine pitch 64-ball BGA packages are subjected to 150, 175, and 200 °C

aging temperatures. Electrical testing was conducted after each hour and cycle of stress to check Au and Cu ball bond integrity in terms of its ball bond HTSL reliability with various aging conditions.

Result and Discussion

The extended reliability results of UHAST and TC tests of Cu and Au ball bonds used in encapsulated mold compound types A and B are given in Table 3. All extended reliability curves belong to wearout reliability since its β (shape parameter) is more than 1.0 in all UHAST and TC reliability curves. In reliability testing, t_{first} denotes time to first occurrence of failure in reliability stresses, t_{50} (also known as MTTF) refers to time to failure of 50 % of the tested sample, and $t_{63.2}$ (also known as characteristic life) refers to time to failure of 63.2 % of the tested sample in reliability testing. Mold compound B reveals higher hours and cycles to failure (t_{first}, t_{50}, and $t_{63.2}$) in TC test compared to mold compound A, although they are with pretty similar technical specifications. Au ball bonds

Fig. 3 The representative SEM image shows Cu ball bond microcracking along the Au–Al interface after UHAST 3,000 h

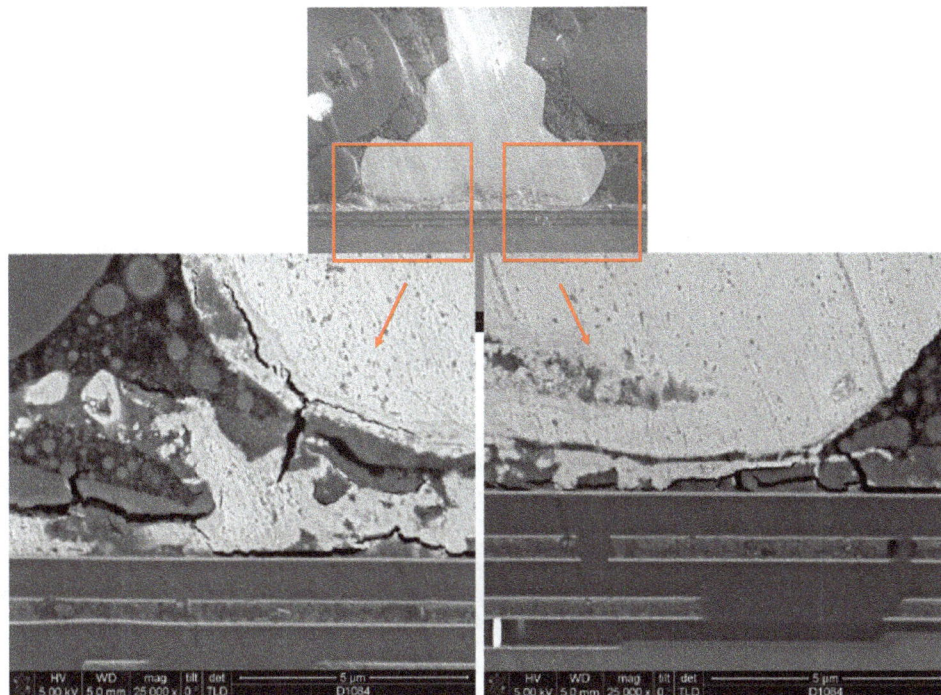

Fig. 4 Proposed UHAST failure mechanism of Cu ball bond

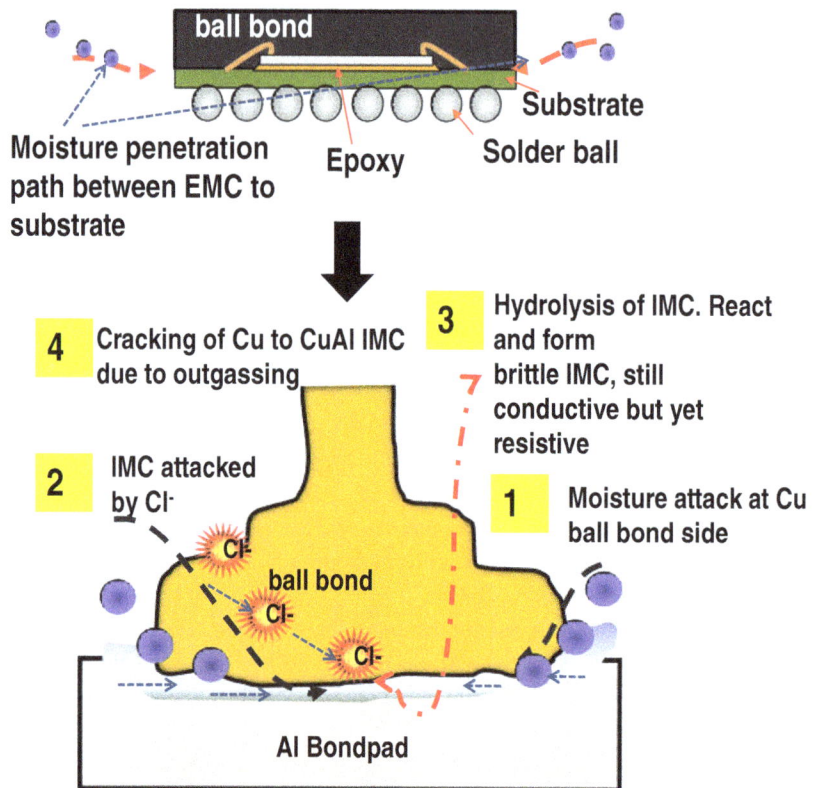

show superior extended UHAST reliability than Cu ball bonds for mold compound A but smaller t_{50} and $t_{63.2}$ for mold compound B (see Fig. 1 and Table 3). This could be due to a more stable Au and the higher corrosion resistance under moisture UHAST conditions compared to Cu ball bonds in mold compound A. The microstructure analysis of Au ball bond after extended UHAST test is shown in Fig. 8.

However, we observed Cu ball bonds with higher TC extended reliability performance (higher t_{first}, t_{50}, and $t_{63.2}$)

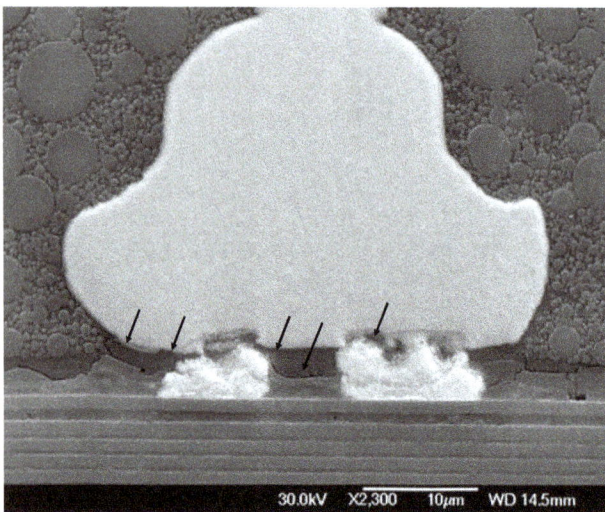

Fig. 5 The representative SEM image shows Au ball bond microcracking along the Au–Al interface after UHAST 2,000 h)

compared to Au ball bonds in FBGA 64 package of both mold compounds A and B (Fig. 2). The key material characteristics of epoxy mold compounds A and B are given in Table 4. The effect of wire type is not the key factor affecting the TC reliability performance, but we observed a higher extended reliability performance in Cu ball bonds compared to Au ball bonds. Again, the coefficients of thermal expansion (CTE) of mold compounds A and B are pretty similar, except that the CTE is different between Au and Cu to the silicon die. Overall, EMC B (with α_2 of 3.0×10^{-5}/°C which is closer to silicon die CTE of 3.0 ppm/°C compared to EMC A) shows higher TC reliability margin and longer cycles to failure for Au and Cu wires.

Failure analysis and mechanisms of Au and Cu ball bonds

Unbiased HAST

Typical failure mechanisms of both unbiased HAST (UHAST) of gold (Au) and copper (Cu) ball bonds are IMC microcracking along the Au–Al and Cu–Al interfaces. The stress-induced microcracking occurs in the event of Au and Cu ball bond corrosion in high temperature and moisture environment (130 °C, 85%RH). Insulative corrosion products will cause high resistance or open failure after long durations of UHAST stressing. Figure 3 shows representative Cu ball bond microcracking and opening after 3,000 h in UHAST.

Fig. 6 The representative SEM
image shows that Cu ball bond
microcracking starts from the
edge of the Cu ball bond
(larger gapping) toward the
center of the ball bond
(narrower gapping)

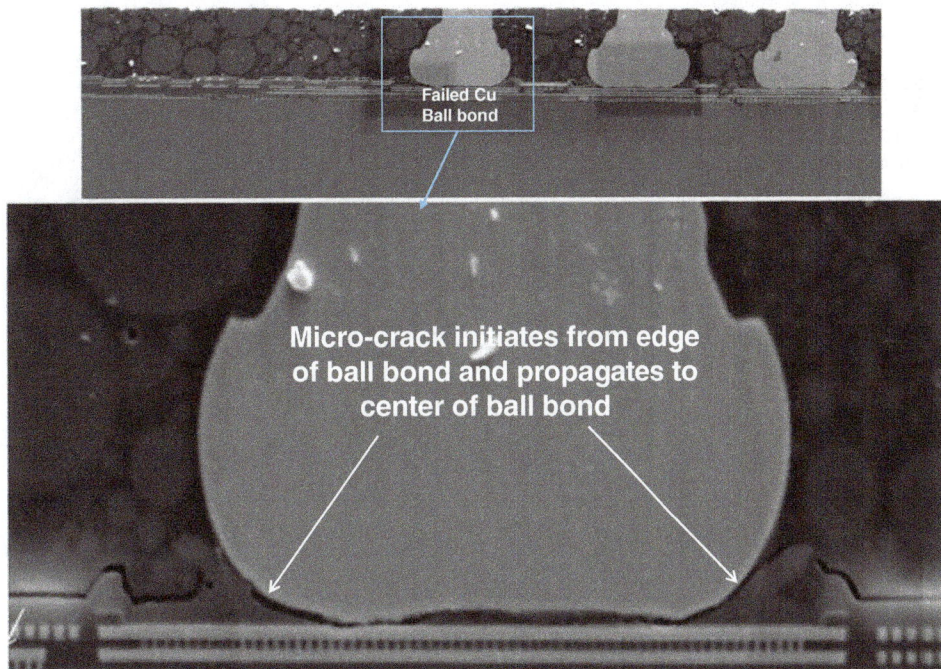

We observed that the crack starts from the edge of Cu ball bonds, and it might propagate toward the center of the ball bond (see Fig. 6).

Figure 4 shows the proposed Cu ball bond corrosion mechanism in the UHAST test. The proposed corrosion mechanism is described in Eqs. 1 to 3 below.

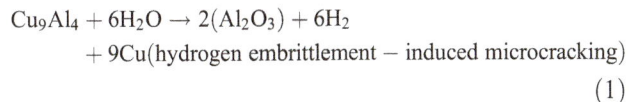

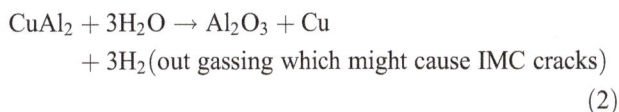

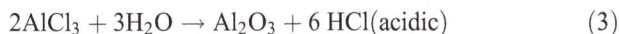

$$Cu_9Al_4 + 6H_2O \rightarrow 2(Al_2O_3) + 6H_2$$
$$+ 9Cu(\text{hydrogen embrittlement} - \text{induced microcracking}) \quad (1)$$

$$CuAl_2 + 3H_2O \rightarrow Al_2O_3 + Cu$$
$$+ 3H_2(\text{out gassing which might cause IMC cracks}) \quad (2)$$

$$2AlCl_3 + 3H_2O \rightarrow Al_2O_3 + 6\,HCl(\text{acidic}) \quad (3)$$

Cu ball bond corrosion is most probably attributed to Cl^- attacking the edge of Cu ball bond region. Hydrolysis of

Fig. 7 The SEM image shows
that the Cu ball bond opens
after the UHAST test.
Arrows indicate the evidence
of possible hydrogen
embrittlement-induced
microcracking (between Cu to
CuAl IMC) which started from
the edge of the Cu ball bond

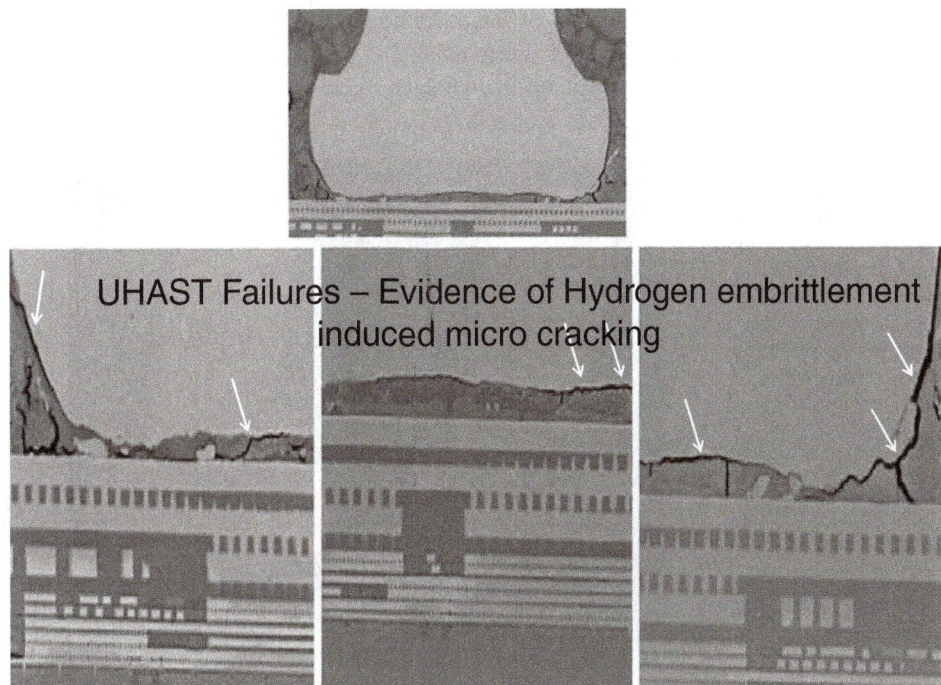

Fig. 8 The SEM image shows that in the microstructural analysis, the Au ball bond opens after the UHAST test. A thicker AuAl IMC is observed compared to CuAl IMC in the Cu ball bond

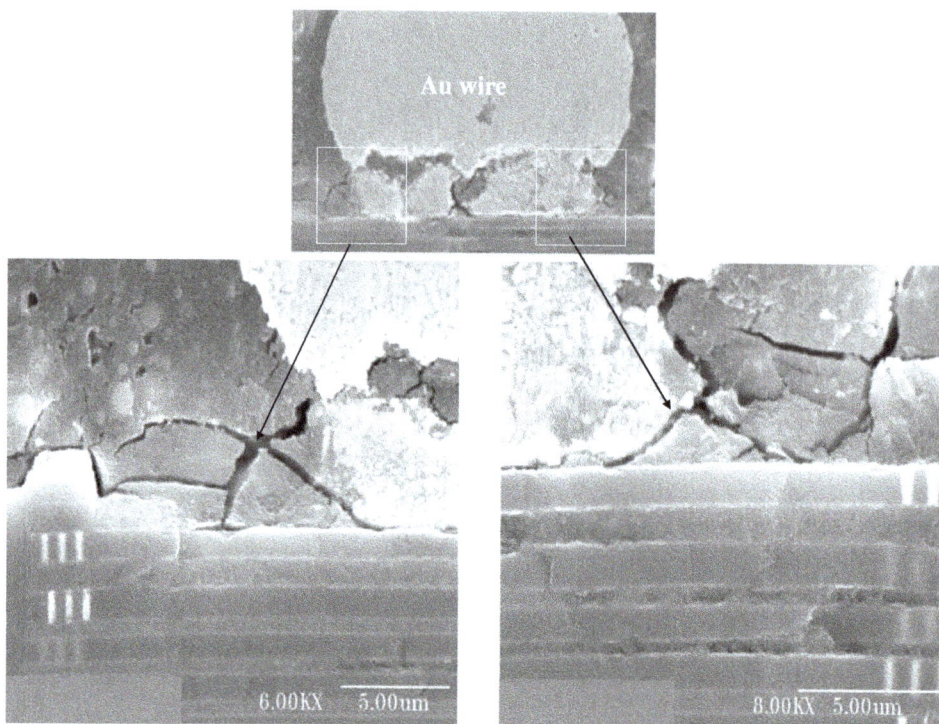

IMC and $AlCl_3$ (intermediate product) under moisture environment forms aluminum (III) oxide which is a resistive layer, and ionic Cl^- is usually found at the corroded ball bond [1, 10]. Equation 1 indicates the hydrolysis of Cu_9Al_4 into Al_2O_3 and out gassing. Cracking of the Al_2O_3 interface of Cu to the Cu IMC might be due to out gassing of H_2 during hydrolysis (as shown in Eqs. 2 and 3) in between the Cu IMC and Cu ball bonds. Cracking usually starts at the Cu ball bond periphery, and it will propagate toward the center of the Cu ball bond [9, 10, 13]. There is a possibility that under moist conditions, internal oxidation of intermetallic phases can result in oxidation of aluminum, precipitation of the noble metal (Au or Cu), and generation of hydrogen.

Possible reactions (not necessary correct) resulting in formation of aluminum oxide that might occur with gold and copper ball bonds are given in Eqs. 4 and 5 [11, 12]. Hydrogen gas evolution due to moisture in contact with intermetallics has been extensively documented and is one of the known causes of embrittlement [12].

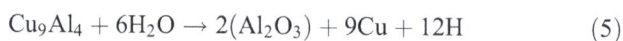

$$2Au_4Al + 3H_2O \rightarrow Al_2O_3 + 8Au + 6H \qquad (4)$$

$$Cu_9Al_4 + 6H_2O \rightarrow 2(Al_2O_3) + 9Cu + 12H \qquad (5)$$

Au ball bonds also exhibit similar Au oxidation and corrosion during UHAST stress. Figure 5 reveals the

Fig. 9 The optical image shows that the lifted Au ball bond opens after the UHAST test. The EDX analysis on the failed Au ball bond indicates the presence of O, Al, and Au elements. The presence of O element proves oxidation of AuAl IMC under moist UHAST environment

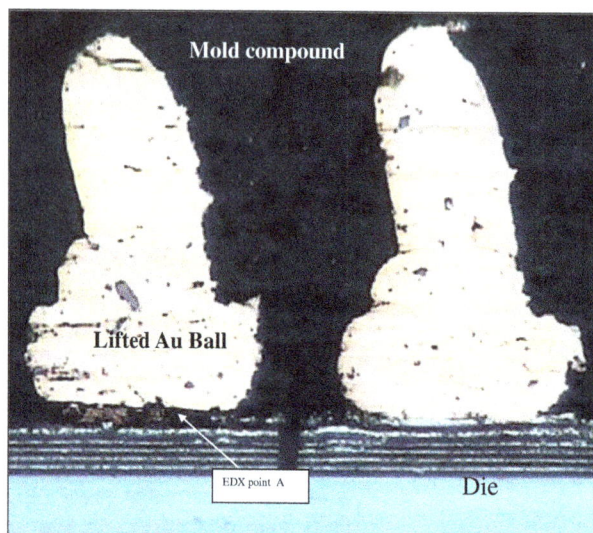

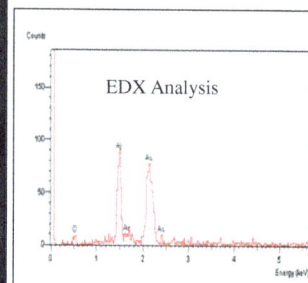

Elmt	Spect. Type	Inten. Corrn.	Std Corrn.	Element %	Sigma %	Atomic %
O K	ED	1.729	211.82	6.72	1.03	24.76
Al K	ED	1.680	242.89	12.67	1.19	43.15
Au M	ED	0.756	244.02	80.62	1.63	32.15
Total				100.00		100.00

Elmt	Spect. Type	Inten. Corrn.	Std Corrn.	Element %
O K	ED	0.838	0.52	6.85
Al K	ED	0.351	0.83	5.56
Si K	ED	0.428	0.91	1.68
Cl K	ED	0.605	0.94	0.24
Cu K	ED	0.970	1.00	79.41
Ta M	ED	0.367	0.77	6.27
Total				100.00

* = <2 Sigma

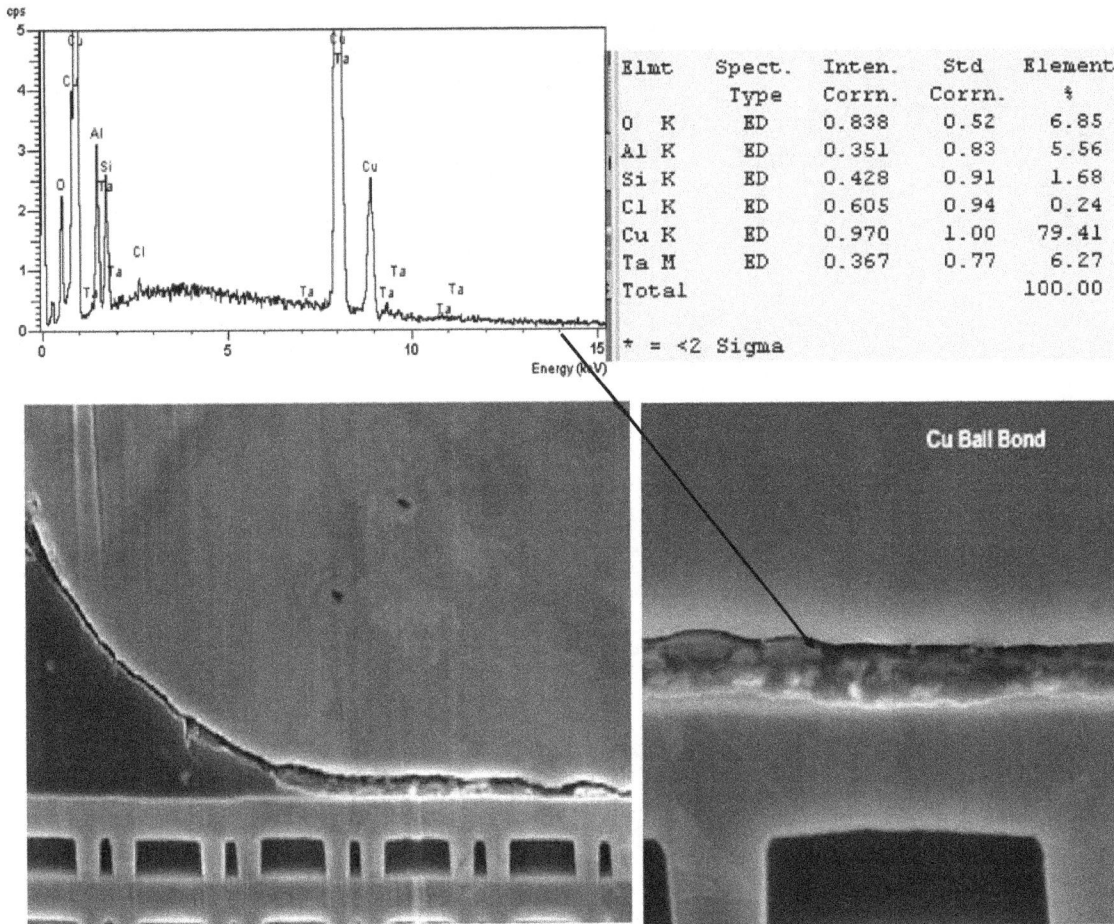

Fig. 10 Microstructural analysis of the Cu ball bond, which opens after the UHAST test. The EDX analysis on the failed Cu ball bond indicates the presence of O, Cu, Si, Ta, and Cl elements. The presence of O and Cl elements proves internal oxidation of Cu–Al IMC under moist UHAST environment and corrosion by Cl⁻ ion

representative SEM image of Au ball bonds and microcracking after UHAST 2,000 h. Au is well known for its corrosion resistance compared to Cu ball bonds and is being deployed in microelectronic packaging for more than 25 years.

Microstructural analysis of failed ball bonds

The Au and Cu ball sample which failed to open after UHAST test has been subjected to detailed microstructural analysis such as cross-section and EDX analyses. Figure 6 reveals that the microcrack starts from the edge of the Cu ball bond (with larger gapping of microcrack) and propagates toward the center of the Cu ball bond.

Table 5 EDX analysis of failed Cu and Au ball bonds after the UHAST test

Location	Element (atomic %)						
	Au	Cu	O	Al	Si	Ta	Cl
Au ball	32.15	–	24.76	43.15	–	–	–
Cu ball	–	79.41	6.85	5.56	1.68	6.27	0.24

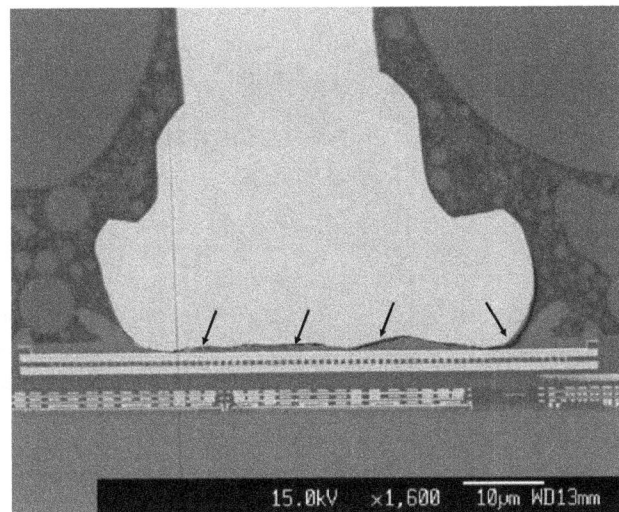

Fig. 11 Representative SEM image of Cu ball bond CuAl interfacial microcracking after TC 9,500 cycles

Fig. 12 Proposed TC failure mechanism of Cu ball bond

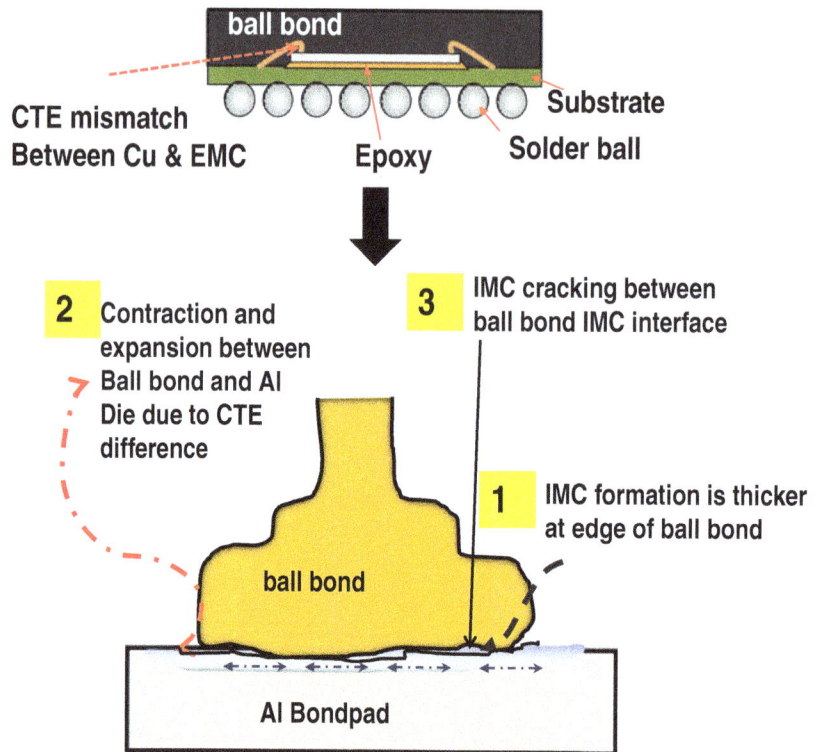

We have proposed the possible failure mechanism of Cu ball bond under moist UHAST conditions in which the hydrogen out gassing might possibly induce microcracking (hydrogen embrittlement-induced microcracking). Figure 7 shows the microstructure of failed Cu ball bonds with fine microcracking beneath Cu ball bonds.

Au ball bonds undergoing extended hours of UHAST test might fail owing to internal Au IMC oxidation, and microcracking will occur. Figure 8 indicates that the representative Au ball bond which is observed with thicker AuAl IMC failed UHAST and microcrack occurs beneath the Au ball bond.

EDX analysis has been performed on failed Au ball bonds (in Fig. 9) and Cu ball bonds (in Fig. 10),

respectively. A large amount of O peak is found for the failed Au ball bond, which is proven with the proposed failure mechanism of Au ball bond whereby a possible internal AuAl IMC oxidation occurred under the UHAST test and led to open failure.

Figure 10 shows the FIB cross-section and EDX analyses of the failed Cu ball bond. The EDX analysis on the microcrack at the edge of the Cu ball bond indicates the presence of O and Cl peaks. This proves that the hydrolysis of CuAl IMC under UHAST moist conditions and Cl peak originates from $AlCl_3$. The trace Cl^- is usually found in epoxy mold compound. Table 5 tabulates the summary of the EDX analysis of Au and Cu ball bonds.

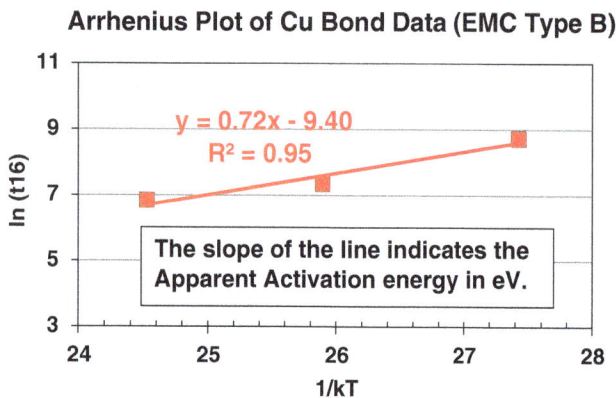

Fig. 13 Arrhenius plot of the Cu ball bonds in FBGA 64 package with EMC type B

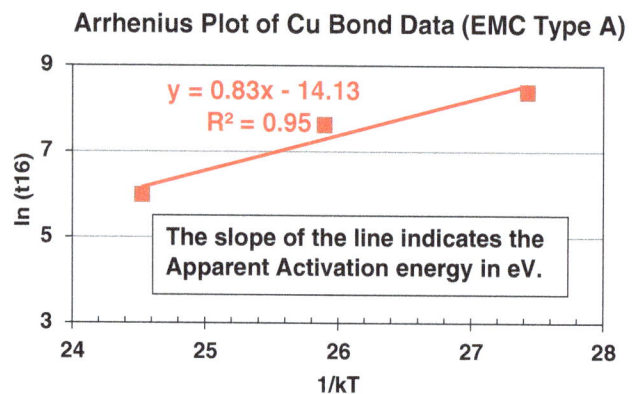

Fig. 14 Arrhenius plot of the Cu ball bonds in FBGA 64 package with EMC type A

Arrhenius Plot of Au Bond Data (EMC Type A)

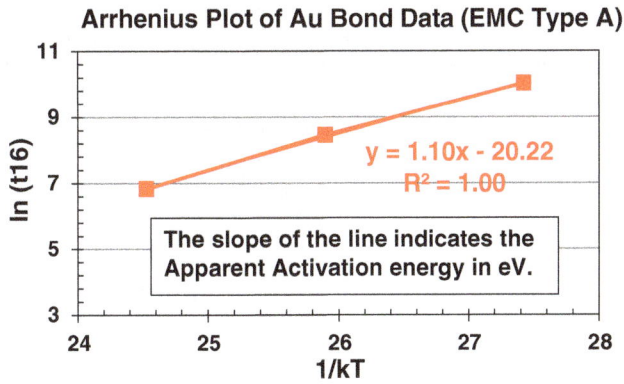

$y = 1.10x - 20.22$
$R^2 = 1.00$

The slope of the line indicates the Apparent Activation energy in eV.

Fig. 15 Arrhenius plot of Au ball bonds in FBGA 64 package with EMC type A

Temperature cycling

Temperature cycling is conducted on the four legs of Au and Cu ball bonds assembled with different EMCs to examine their thermomechanical reliability performance. The mismatch in CTE between the Cu (17 ppm/°C) and Au ball bonds (14 ppm/°C) to the silicon die (3.0 ppm/°C) induced different thermal expansions and contraction rates in the temperature cycling test. The CTE mismatch between the Au and Cu ball bonds with Al bondpad of silicon die will impose different thermal expansion rates during hot cycles (150 °C) and contraction rates during cold cycles (-40 °C). IMC formation started at the edge of the ball bond (due to the ball bond pressing force by capillary), and microcracking will be induced after long cycles of thermal cycling effects. The microcracking occurs in between ball bond IMCs (as shown in Fig. 11).

Figure 12 illustrates the typical package failures induced by TC stress due to CTE mismatch in Au and Cu ball bonds onto Al bondpad metallization.

Arrhenius Plot of Au Bond Data (EMC Type B)

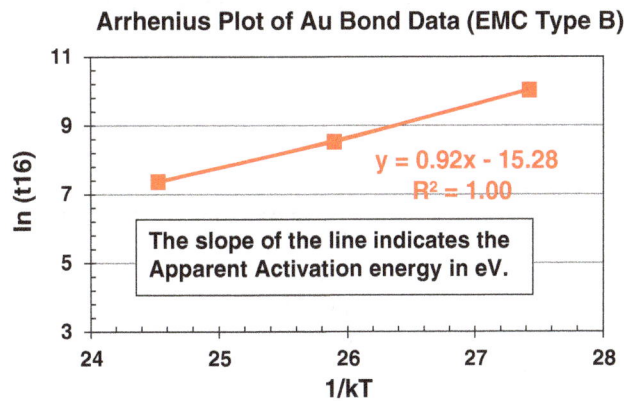

$y = 0.92x - 15.28$
$R^2 = 1.00$

The slope of the line indicates the Apparent Activation energy in eV.

Fig. 16 Arrhenius plot of Au ball bonds in FBGA 64 package with EMC type B

High-temperature storage life

The required activation energies (E_{aa}) of interdiffusion of Cu and Au atoms in Al were modeled by using Arrhenius model after the HTSL test. The fundamental basis of thermal activation is based on the probability of ascending a potential energy barrier due to Maxwell–Boltzmann energy distribution. This physical explanation was actually anticipated by Arrhenius work on chemical reaction rates, in which one would simply substitute the Rydberg gas constant for the Boltzmann constant and use different units. Thermally activated processes are modeled by Arrhenius equation, and it is given in Eqs. 6 and 7.

$$Rate = R_o \times \exp(-E_{aa}/kT) \qquad (6)$$

$$Or~Rate = R_o \times \exp(-E_{aa}/RT) \qquad (7)$$

where R_o is the rate constant characteristics of infinite temperature, E_{aa} refers to apparent activation energy in electronvolts per atom for physics units or kilocalories per mole for chemical units, k is the Boltzmann constant (8.62×10^{-5} eV/kelvin), R is the Rydberg gas constant (23,063 cal/mol kelvin), and T is the temperature in kelvin.

Using Eq. 7, the acceleration factor AF for T_1 versus T_2 is as follows (in Eq. 8):

$$AF = \exp[(E_{aa}/kT)(1/T_1 - 1/T_2)] \qquad (8)$$

It is noted that the acceleration factor is sensitive to the value of the apparent activation energy E_{aa} and the temperature difference. The apparent activation energy E_{aa}, which is temperature dependent, can be determined by plotting graph ln (lifetime of ball bonds) versus $1/kT$ as in Eq. 9. Graph *ln T* (lifetime) versus (1/*T*) can be plotted by using Eq. 10.

$$T = R_o \exp(-E_{aa}/kT) \qquad (9)$$

$$\ln T = -(E_{aa}/R)(1/kT) + \ln R_o \qquad (10)$$

where self-diffusion coefficient Ro is a constant, E_{aa} is the activation energy in electronvolts for the diffusion process, R is molar gas constant in joules per mole kelvin, and T is the lifetime of ball bonds. The apparent activation energy E_{aa} can be calculated from the gradient of the plot ln *T* versus 1/*kT*. HTSLs of Au and Cu ball bonds for EMCs A and B were conducted to understand and estimate its apparent activation energy (E_{aa}) after long duration of high-temperature bake. Previous studies show that the E_{aa} values of the Au ball bond range from 1.00 to 1.50 eV [15, 25], while the Cu ball bond is about 0.70 eV [14]. In our study, the E_{aa} values obtained are 0.72 eV (in Fig. 13) and 0.83 eV (in Fig. 14) for Cu wire assembled with EMC B and EMC

Table 6 Summary of E_{aa} and HTSL failure mechanisms from previous studies

Ball bond type	HTSL aging test conditions (°C)	E_{aa} (eV)	Failure mechanism	Reference
Cu	150, 175, 200	0.70	CuAl microcrack	[14]
Cu	N/A	0.75	CuAl microcrack	[25]
Cu	150, 175, 200	0.72~0.83	CuAl microcrack	This work
Au	150, 175, 200	1.00~1.50	Kirkendall void	[15]
Au	N/A	1.00~1.26	Kirkendall void	[25]
Au	150, 175, 200	0.92~1.10	Kirkendall void	This work

A, respectively. A lower E_{aa} is found on the Cu ball bonds compared to the Au ball bonds in the HTSL tests.

The Au wire exhibits higher E_{aa} values in the HTSL tests, with 1.10 eV for EMC A (in Fig. 15) and 0.92 eV for EMC B (in Fig. 16), respectively. The typical failure mechanism of AuAl IMC in the HTSL test is attributed to Kirkendall voiding-induced opens [4, 5, 25]. Cu ball bonds are known with slower CuAl IMC growth rate compared to AuAl IMC, and the failure mechanism in HTSL is slightly different from Au ball bonds in HTSL.

Table 6 tabulates the summary of E_{aa} obtained by previous researchers on Au and Cu ball bonds in HTSL test. The typical failure mechanism of Cu ball bond is CuAl IMC microcracking after long durations of high-temperature bake. Au ball bonds (with faster Au atomic diffusion rate into Al metallization) exhibit Kirkendall microvoiding in AuAl to Al bondpad interface and induce opens. Our study shows the similar observation and findings.

Ball bond lifetime analysis by using E_{aa} (apparent activation energy)

Table 7 shows the lifetime estimate of Cu wire bonding in FBGA package, derived from our experimental data herein, for a number of different market segments and usage models. The computations are based on a 0.1 % failure rate at 175 °C, which are approximately 408 h (for Cu ball bond with EMC A) and 302 h (for Cu ball bond with EMC B) from the data collected and shown in lognormal plot. The operating

conditions as well as typical life in the computations in Table 7 were taken from JESD94 [25]. The operating temperatures were obtained from the maximum ambient temperature conditions shown in JESD94. Storage conditions were chosen to be 30 °C uniformly across all applications. The data indicate that for all applications listed, there is more than sufficient reliability margin to meet all listed reliability requirements. Similarly, Au ball bond lifetime can be calculated based on the E_{aa} values obtained. We have calculated the lifetime of Au ball bonds (Au ball bond with EMC A and EMC B, respectively), and the data meet typical lifetime (years) as required in JESD94 standard [25].

Effects of molding compound on extended reliability

The EMCs used in this evaluation are from suppliers A and B. The important material characteristics of the mold compound datasheets are given in Table 4. The only difference between the mold compounds is that they are from different mold compound manufacturers. EMC A exhibits higher hours to failures in UHAST extended reliability for EMC A (Fig. 1) but lower cycles to failures in TC extended reliability tests (Fig. 2) compared to EMC B. In HTSL tests, EMC B shows no significant difference in the apparent activation energy (E_{aa}) value in Cu ball bonds while a much lower E_{aa} value in Au ball bonds (Fig. 13). This proves that the different types of epoxy mold compounds have a significant influence in the HTSL test of Au wire bonding. Au ball bonds are well known for its higher IMC growth rate and increased susceptibility to

Table 7 Lifetime estimations for various market segments (Cu wire bond FBGA package)

Market segment	Typical lifetime (years)	Operating condition		Storage conditions		# of lives (years) Cu EMC A	# of lives (years) Cu EMC B
		Time (h)	Temp. (°C)	Time (h)	Temp. (°C)		
Consumer desktop	5	13,000	30	30,800	30	264.61	51.82
High-end server	11	94,000	30	2,360	30	120.28	23.55
Avionics electronics	23	150,000	50	51,480	30	10.37	2.58
Telecom handheld	5	43,800	40	0	30	96.17	21.47
Telecom controlled	15	131,000	70	400	30	2.20	0.69
Automotive underdash	15	8,200	45	123,200	30	72.55	14.80
Automotive underhood	15	8,200	125	123,200	30	0.73	0.38

Kirkendall microvoiding after the HTSL test compared to slower CuAl IMC growth rate. Hence, we observed no significant E_{aa} values obtained in Cu ball bonds for EMC A and EMC B (Figs. 14 and 13). Both EMCs show promising extended reliability results which far exceed the typical 96 h of UHAST and 1,000 cycles of TC according to JEDEC standards. Hence, both EMC A and B are used in our flash memory BGA laminate.

Effects of wire types on extended reliability

Many previous studies reported Au wire bonding with higher reliability margins compared to Cu wire bonding. However, there are very few published data on extended reliability of Au and Cu ball bonds. In our study, Au ball bonds show higher UHAST reliability compared to Cu ball bond in FBGA package. This is notable as Au is much more stable and has higher corrosive resistance compared to Cu. Cu is easily oxidized and corroded under moist environments, especially in the UHAST or biased HAST tests. Our extended reliability study (Fig. 1) shows similar findings with Au wires in EMC A. Another factor affecting the first ball bond strength is Au wire bond shear or wire pull strength shows less variation compared to Cu ball bonds. This as-bonded stage strength value would influence the reliability of ball bonds in semiconductor packaging.

However, Cu ball bonds exhibit higher cycle to failure in TC test compared to Au ball bonds (Fig. 2) regardless of EMC types. The CTEs of Cu and Au are pretty similar in this case. This is an interesting finding since the reliability performance of TC stress is pretty much material CTE dependent with regard to Al bondpad. The mismatch in CTE between the Cu (17 ppm/°C) and Au ball bonds (14 ppm/°C) to the silicon die (3.0 ppm/°C) induced different thermal expansions and contraction rates in the temperature cycling test. The CTE mismatch between the Au and Cu ball bonds with Al bondpad of silicon die will impose different thermal expansion rates during hot cycles (150 °C) and contraction rates during cold cycles (-40 °C).

Future works and recommendation

Cu wire will be continuously developed to replace Au wire in higher pin counts of semiconductor packages, but transition is predicted to be less on power-device-based packages. Future engineering work should be focused on knowledge-based reliability testing and prediction to understand the initial failure point in semiconductor device packaging. Extended reliability concept would be used in this type of reliability studies. Further characterization should be carried out for Pd–Ag–Au or bare Ag wire bonding in nanoscale device packaging, especially for 45, 28, 22, or subnanoscale 10 nm below technology nodes.

Conclusions

In this research, we analyzed the effects of wire alloy on extended reliability of UHAST, TC, and HTSL stresses. Au ball bonds show a significant higher UHAST reliability compared to Cu ball bond in FBGA package with EMC A. This is notable as Au is much more stable and has higher corrosive resistance compared to Cu. Contrary results occur in TC, where Cu ball bond is more superior compared to Au ball bond. EMC B exhibits higher TC reliability margins compared to EMC A assembled with Au or Cu wires. However, both EMCs are far exceeding the minimum required 96 h of UHAST and 1,000 cycles of TC according to JEDEC standards. The E_{aa} values obtained for Au ball bonds range from 0.92 to 1.10 eV and 0.72 to 0.83 eV for Cu ball bonds. These values are close to previous HTSL studies conducted on Au and Cu ball bonds. Au wire bonding will still remain as a mainstay in microelectronic packaging, especially for more complicated semiconductor packages (with higher pin counts), while Cu wire bonding will equally gain some market shares in low-pin-count and power device packaging. Ag wire bonding would probably become an emerging technology as an option in microelectronic packaging. However, more engineering works should be carried out to understand the extended reliability performance as well as assembly yield monitoring before deployment for high-volume manufacturing. Future engineering work should be focused on knowledge-based reliability testing and prediction to understand the initial failure point in semiconductor device packaging. Extended reliability concept would be used in this type of reliability studies.

Acknowledgments The authors would like to take this opportunity to thank Spansion management (Gene Daszko, Tony Reyes, and Chong HL) for their management support for the paper publication.

References

1. Ellis TW, Bond W (2004) The future of gold in electronics introduction. Gold Bulletin 37:66–71
2. Appelt BK, Tseng A, Chen C-H, Lai Y-S (2011) Fine pitch copper wire bonding in high volume production. Microelectronics Reliability 51:13–20.
3. Breach CD (2010) What is the future of bonding wire? Will copper entirely replace gold? Gold Bulletin 43:150–168
4. Breach CD (2009) Intermetallic growth in gold ball bonds aged at 175c: Comparison between two 4 N wires of different chemistry. Gold Bulletin 42:92–105
5. Muller T, Schraplerl L, Altmann F et al (2006) Influence of intermetallic phases on reliability in thermosonic Au–Al wire

bonding. IEEE Electronics System Integration Technology Conference, Dresden, pp 1266–1273

6. Xu H, Liu C, Silberschmidt VV et al (2011) Intermetallics intermetallic phase transformations in AuAl wire bonds. Intermetallics 19:1808–1816.

7. Zulkifli MN, Abdullah S, Othman NK, Jalar A (2012) Some thoughts on bondability and strength of gold wire bonding. Gold Bulletin 45:115–125.

8. Gan CL, Toong TT, Lim CP, Ng CY (2010) Environmental friendly package development by using copper wire bonding, In Proceedings of 34th IEEE CPMT IEMT. Malacca 2010:1–5

9. Gan CL, Ng EK, Chan BL et al (2012) Technical barriers and development of Cu wirebonding in nanoelectronics device packaging. Journal of Nanomaterials 2012(173025):1–7.

10. Gan CL, Ng EK, Chan BL, Hashim U (2012) Reliability challenges of Cu wire deployment in flash memory packaging. IEEE Proceedings of International Microsystems, Packaging, Assembly and Circuit Technology Conference, Taipei, pp 498–501

11. Breach CD, Shen NH, Lee TK, Holliday R (2011) Failure of gold and copper ball bonds due to intermetallic oxidation and corrosion. 18th IEEE International Symposium on the Physical and Failure Analysis of Integrated Circuits (IPFA) 1–6

12. Breach CD, Lee TK (2011) Conjecture on the chemical stability and corrosion resistance of Cu–Al and Au–Al intermetallics in ball bonds. IEEE International Conference on Electronics Packaging Technology and High Density Packaging. 275–283

13. Gan CL, Ng EK, Chan BL, Kwuanjai T, Jakarin S, Hashim U (2012) Wearout reliability study of Cu and Au wires used in flash memory fine line BGA package. IEEE Proceedings of International Microsystems, Packaging, Assembly and Circuit Technology Conference, Taipei, pp 494–497

14. Classe F, Gaddamraja S (2011) Long term isothermal reliability of copper wire bonded to thin 6.5 μm aluminum. IEEE International Reliability Physics Symposium 685–689.

15. Blish RC, Li S, Kinoshita H et al (2007) Gold–aluminum intermetallic formation kinetics. IEEE Transactions on Device and Materials Reliability 7:51–63

16. Gan CL, Ng EK, Chan BL, Classe FC, Kwuanjai T, Hashim U (2013) Wearout reliability and intermetallic compound diffusion kinetics of Au and PdCu wires used in nanoscale device packaging. Journal of Nanomaterials 2013(486373): 1–9

17. Cho J, Yoo K, Hong S, et al. (2010) Pd effects on the reliability in the low cost Ag bonding wire. IEEE International Electronic Components and Technology Conference 1541–1546

18. Kai LJ, Hung LY, Wu LW, et al. (2012) Silver Alloy Wire Bonding. pp 1163–1168

19. Pagba A, Reynoso D, Thomas S, Toc HJ (2010) Cu wire and beyond—Ag wire an alternative to Cu? 2010 12th Electronics Packaging Technology Conference 591–596.

20. Tsai HH, Lee JD, Tsai CH, Wang HC, Chang CC, Chuang TH (2012) An innovative annealing twinned Au–Ag–Pd bonding wire for IC and LED packaging, IEEE International Microsystems Packaging Technology Conference 505–508

21. Long Z, Han L, Wu Y, Zhong J (2008) Study of temperature parameter in Au–Ag wire bonding. IEEE Transactions on Electronics Packaging Manufacturing 31:221–226

22. Chuang TH, Chang CC, Chuang CH, Lee JD, Tsai HH (2012) Formation and growth of intermetallics in an. IEEE Transactions on Components, Packaging and Manufacturing Technology 1–7

23. Chuang TH, Wang HC, Tsai CH et al (2012) Thermal stability of grain structure and material properties in an annealing-twinned Ag–8Au–3Pd alloy wire. Scripta Materialia 67:605–608.

24. JEDEC JEP 122 (2010) Failure mechanisms and models for semiconductor devices

25. JEDEC JESD94 (2008) Application specific qualification using knowledge based test methodology

Optical spectroscopy of functionalized gold nanoparticles assemblies as a function of the surface coverage

C. Humbert · O. Pluchery · E. Lacaze · A. Tadjeddine · B. Busson

Abstract Layers of thiophenol functionalized spherical gold nanoparticles grafted on Si(100) are probed by linear UV-vis, Fourier transform infrared and nonlinear infrared-visible vibrational sum/difference–frequency generation spectroscopies as a function of the nanoparticles surface coverage. Depending on the dipping time (5 min, 20 min, 1 h, and 24 h) in the colloidal solution, AFM imaging corroborates that the silicon surface coverage with gold nanoparticles increases, while the distance between neighbouring nanoparticles decreases, leading to their aggregation which dramatically impacts their optical properties. In the UV-vis reflectance spectra after the appearance of the 525-nm individual plasmonic band, a second broad band located at 660 nm and related to the gold nanoparticles aggregation on silicon rapidly dominates in intensity. Nonlinear vibrational spectroscopy is able to detect the specific vibration of the thiophenol molecules (3, 055 cm^{-1}) whatever the immersion time and at least down to 1 % of the substrate filling factor by the gold nanoparticles, overtaking the molecular sensitivity threshold of surface infrared and Raman spectroscopies on small gold nanostructures (17 nm) adsorbed on a semiconductor. Moreover, a quantitative analysis of the nonlinear vibrational fingerprint from 5 min to 24 h in the framework of the effective medium models of Maxwell-Garnett and Bruggeman illustrates the role played by the interband and the plasmonic properties of gold modulated by the silicon optical response. In this case, the sample reflectivity affects the molecular oscillator strength measured by nonlinear optical vibrational spectroscopy. For this latter technique, no coupling with the optical properties of aggregated AuNps is evidenced while the localized surface plasmon resonance excitation amplifies the molecular response.

Keywords Gold · Nanospheres · Silicon · Nonlinear optics · UV-visible spectroscopy · Atomic force microscopy

Introduction

The manufacturing of well-designed nanosensors is a key step to improve their sensitivity threshold in the probe-target scheme encountered in molecular recognition, especially when expecting the detection of materials traces to the single molecule level. Among the numerous methods of sensors design at the nanoscale, metal-based supports are routinely used taking advantage of their plasmonic properties [1]. In fact, their interest lies in the enhancement due to the surface plasmon resonance (SPR) of the nano-objects organized in networks with well-defined optical properties in the UV-visible (UV-vis) spectral range, depending on the nano-object size, shape and metal nature and on the lattice parameter. All these properties are related to the surface preparation. A silicon substrate is generally used as a transducer to take benefit of the particular optical and electronic properties of those systems. Nowadays, different ways of production are routinely used. Among them, one route consists in a physical method based on controlled deposition directly on the substrate by electron beam lithography [2]. Another well-known route is "wet" chemistry where metal nanoparticles are synthesized first, then grafted on the substrate through aminosilane molecular interlayer [3]. While the latter method seems easier to implement, its difficulty lies in preventing the aggregation of the nano-objects on the substrate and in their subsequent functionalization. These steps are crucial because

C. Humbert (✉) · A. Tadjeddine · B. Busson
Univ Paris-Sud, Laboratoire de Chimie Physique, CNRS,
Bâtiment 201 Porte 2, 91405 Orsay, France
e-mail: christophe.humbert@u-psud.fr

O. Pluchery · E. Lacaze
Univ Pierre et Marie Curie UPMC-CNRS, Institut des Nanosciences de Paris, 4 place Jussieu, 75005 Paris, France

it should not quench or shift significantly the SPR intensity or position, respectively, with respect to the SPR properties observed in the native colloidal solution. Indeed, spherical gold nanoparticles (AuNps) are commonly used to build nanosensors with a fixed SPR located in the visible light.

Nanosensors are based on a molecular recognition process evidenced by monitoring the SPR shift. For an efficient use of these sensors, a perfect knowledge of the chemistry of the process at play is mandatory, but plasmonics alone does not enlighten which molecular bonds are involved in the recognition of the target molecule by the probe molecule. To reach that goal, vibrational optical techniques are preferred such as infrared and Raman spectroscopic tools. PM-IRRAS and SERS are efficient for providing molecular recognition at the nanoscale level. Nevertheless, on silicon, the selection rules of PM-IRRAS prevent enhanced sensitivity. In SERS, the unequalled sensitivity can reach the single molecule level but is directly related to the presence of "hot spots", generally occurring on samples where nanoparticles are touching, which requires preparation of specific samples. An alternative method, whose sensitivity could be independent of the sample preparation, consists in taking advantage of the IR and Raman peculiar rules together: nonlinear optical sum (difference)–frequency generation spectroscopy (SFG/DFG) based on the mixing of two incident visible and infrared beams at the same point of the probed interface. This latter generates a SFG/DFG coherent light beam depending on the modulation of the afforded incident energy by IR molecular absorption and/or visible SPR enhancement (Fig. 1). Therefore, the main advantage of this technique for nanotechnological devices

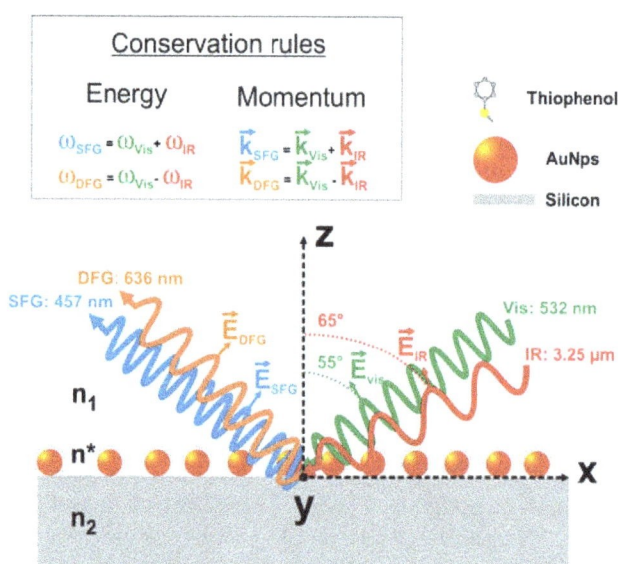

Fig. 1 Sketch of the SFG and DFG processes on the thiophenol/AuNps/Si(100) interface. All the beams are p-polarised, i.e. the wavevectors k are located in the *XZ* plane. The incident infrared and visible laser beams are always mixed at the same point of the probed interface whether the SFG or DFG configuration

characterization lies in its interface intrinsic molecular sensitivity at the sub-monolayer level and its potential coupling to the SPR resonance as recently demonstrated by considering samples based on solid substrate as platforms for nanosensors, an amplification factor of the molecular SFG signal was put in evidence for functionalized AuNps films with respect to a flat gold reference surface. AuNps were deposited either on silicon in external reflection configuration [4] or on glass substrates in total internal reflection configuration [5] to compare the sensitivity and molecular ordering of functionalized AuNp films as a function of the surface coverage on glass [6], to unravel the orientation of grafted molecules on AuNps and AgNps [7] and to extract and deduce the nature of the vibration modes thanks to the SPR amplification on functionalized AuNp films in the fingerprint spectral range of aromatic molecules with density functional theory (DFT) calculations [8]. In a general way, SFG has also proven its efficiency in biomolecular recognition on glass substrate [9], in DNA-based biosensors on Pt(111) single crystal [10] and in enlightening DNA hybridization on glass [11] and (100)-facetted gold films [12]. In summary, SFG/DFG spectroscopy is a promising tool in describing the chemistry occurring in biological recognition [13, 14] performed on nanostructured biosensors provided that we light on/off the SPR amplification in a controlled manner: AuNps nature, size, shape, and dispersion (surface coverage and lattice parameter) on the substrate, exciting visible wavelength. In those SFG/DFG works, no systematic analysis of the effect of an evolving plasmonic pattern on the level of the molecular sensitivity threshold, i.e. sub-monolayer sensitivity was achieved.

Therefore, we address in this paper the role of the plasmonic pattern on the sensitivity threshold for different nanostructured interfaces, as a function of the surface coverage by varying only one parameter: the surface coverage with AuNps on a silicon substrate that was previously functionalized with aminosilane monolayer. This AuNps coverage is controlled by the immersion time of the sample in colloidal AuNps suspension. The surface optical and chemical properties are therefore studied by grafting thiophenol molecules on the AuNps as a probe of the SPR enhancement and/or coupling to the vibrational activity and as a specific marker of the particle surface. To reach that goal, a careful pre-characterization step is necessary: atomic force microscope (AFM) imaging to deduce the surface coverage properties (AuNps diameter and dispersion on the silicon), UV-vis spectroscopy in reflection on the silicon substrate to obtain the plasmonic shape. Fourier transform infrared spectroscopy (FTIR) is used as a standard technique to give the chemical fingerprint of the sample and defined its sub-monolayer sensitivity threshold. Finally, SFG/DFG spectroscopy is performed to give the vibrational pattern of the interface in order to deduce the potential SPR coupling effect and its efficiency to improve the molecular sensitivity at the nanoscale.

Experimental details

Sample preparation

AuNps were synthesized according to the Turkevich method where 1 mL of 8.5×10^{-4} M trisodium citrate ($Na_3C_6H_5O_7$) was added to a boiling aqueous (Millipore water, resistivity= 18 MOhm cm) solution of 20 mL $HAuCl_4$ (2.5×10^{-4} M) under vigorous agitation. The resulting solution (pH=5.5) displays a UV-vis absorbance spectrum (see Fig. 5) with a single peak located at 520 nm [4, 15] corresponding to non-aggregated AuNps with diameter=17 ± 2 nm. Wafers of ultrasonically cleaned n-doped silicon (1×1 cm^2, Siltronix) were silanized in an absolute methanol solution containing 3-aminopropyl-triethoxysilane (APTES, $H_2N(CH_2)_3Si(OC_2H_5)_3$, 10 % vol.). The surface deposition of the four samples was achieved by dipping each silanized wafer in an equal amount of the native colloidal aqueous solution during 5 min, 20 min, 1 h and 24 h, respectively. No particle aggregation in the solutions was observed during and after the dipping as checked by UV-vis spectroscopy. After the AuNps grafting, samples were functionalized during 18 h in a 10^{-3} M thiophenol solution (C_6H_5SH) dissolved in dichloromethane (CH_2Cl_2). All chemicals were purchased from Sigma-Aldrich.

AFM microscopy

We make use of AFM imaging (Digital Instrument, DI3100) in tapping mode to analyze the surface density of the samples. The silicon tips have 130 KHz working frequency, with a curvature radius at the apex around 10 nm. It does not allow sufficient lateral resolution to image correctly isolated gold nanoparticles, but it is sufficient to identify each nanoparticle and to obtain their diameter, equal to their height with respect to the substrate: 17 ± 2 nm.

FTIR spectroscopy

FTIR measurements have been carried out to identify the vibrational modes of thiophenol in the range 2,500–3,400 cm^{-1}. In this range, silicon is transparent for the infrared beam and the measurement is performed in transmission geometry at 60° of incidence with a Bruker Tensor-27 spectrometer. The absorbance A=$-\log (I/I_0)$ is calculated by recording I_0 from the sample itself before thiophenol was deposited. In order to maximize the sensitivity, FTIR spectra were recorded by averaging over 1,000 scans with a spectral resolution of 4 cm^{-1}. The noise was as low as 3×10^{-5} absorbance units so that it was possible to identify surface species at sub-monolayer coverage.

Nonlinear optical vibrational spectroscopy

The sum/difference–frequency generation spectroscopic setup is described elsewhere [6]. Briefly, it is based on a 15 ps pulsed laser Nd:YAG source (1 μs train, repetition rate 25 Hz). After amplification, one part is used to pump an infrared optical parametric oscillator (OPO) built around a LiNBO$_3$ crystal giving access to the 2,500–4,000 cm^{-1} spectral range (10 μJ pulse energy, 3 cm^{-1} OPO bandwidth). The other part is used to obtain a green visible laser beam (5 μJ pulse energy and 532 nm wavelength) by frequency-doubling in a BBO crystal. The infrared and visible beams are then mixed at the same point of the probed surface for each sample with angles of incidence of 65° and 55° with respect to the surface normal, respectively. The infrared, visible and SFG/DFG beams are p-polarized. All the SFG/DFG data are normalized to the SFG/DFG signal of a ZnSe reference crystal in order to compensate for eventual laser fluctuations or atmospheric absorption. The geometrical configuration depicted in Fig. 1 is the same for each sample. The only difference resides in the detection scheme for SFG and DFG. In fact, SFG and DFG signals have a different emission direction. In these conditions, to switch from SFG to DFG configuration, only one mirror is tilted in the lateral direction to send either SFG or DFG photons in the same direction for their detection after spatial and spectral filtering through a monochromator. To compensate for eventual misalignment of the SFG detection direction, the baseline of the SFG spectra is recovered by a linear fit of the experimental data. In order to compare quantitatively the SFG data with respect to the DFG data, an experimental scaling factor set to 1.55 has to be applied to the DFG vibration mode amplitudes reported in Tables 1 and 2 of "SFG/DFG measurements". It is easily explained by considering Fig. 1 because of the non-uniform response characteristics over the visible wavelength range of the detection chain (gratings and photomultipliers) for SFG and DFG energies. The SFG photons are detected in the blue while the DFG photons are detected in the red.

Within this experimental configuration, SFG and DFG spectroscopies respect the rules of the energy ($\hbar\omega$) and momentum (k) conservation parallel to the interface (i.e. the processes are phase-matched). Both spectroscopies sum up two contributions: one coherent (phase-matched SFG/DFG) and one scattered [16]. Sum–frequency scattering contribution has been measured from the bulk colloidal solutions of bigger particles [16, 17]. Nevertheless, it is negligible in our experiments due to the small size of the particles and to the very small thickness of the monolayer. In addition, the scattering angle distribution completely differs from the phase-matching angle, and we have checked experimentally that the SFG/DFG

Table 1 Evolution of the AuNps surface density N_s, filling factor f and fitting parameters of the SFG/DFG spectra

Optical probe	SFG				DFG			
Dipping time	5 min	20 min	1 h	24 h	5 min	20 min	1 h	24 h
N_S (10^{10}/cm^2)	1.05±0.12	2.48±0.48	6.2±1.28	9.48±0.96	1.05±0.12	2.48±0.48	6.2±1.28	9.48±0.96
f (%)	1.57±0.17	3.80±0.72	9.18±1.93	14.95±1.49	1.57±0.17	3.80±0.72	9.18±1.93	14.95±1.49
$|C_{FG}|$	0.074±0.008	0.096±0.019	0.098±0.02	0.11±0.011	0.023±2.5e^{-4}	0.054±0.011	0.098±0.021	0.005±5e^{-4}
ϕ_{FG} (°)	91.61	88.5	57.49	85.05	26.67	18.45	−17.61	23.78
$|a_0|$	0.076±0.008	0.14±0.028	0.53±0.111	0.64±0.064	0.11±0.012	0.23±0.046	0.56±0.118	0.48±0.062
ω_0 (cm^{-1})	3,056.59	3,049.76	3,057.95	3,050.58	3,060.8	3,058.17	3,055.56	3,053.94

signals shown in this paper propagate along the phase-matched direction.

UV-visible spectroscopy

UV-visible (UV-vis) experiments are carried out to obtain the optical signature of the AuNps monolayer deposited on the silicon substrate. In the UV-vis range, this signature is mostly influenced by the plasmon resonance of the nanoparticles. Since the silicon substrate is a reflecting material in the UV-vis range, the spectra were recorded in reflection geometry with an incidence angle of 10° and with a Cary 5 spectrophotometer (Varian). However, in order to discriminate the optical signature of the particles from the typical reflectivity spectrum of bare silicon, a differential method is performed. The reflectivity R_0 of silicon with APTES is recorded and used as a reference. After the reflectivity R of the sample with AuNps is measured, the reflectance is processed: Reflectance=−log (R/R_0). It requires a precise alignment of the reference and the sample within the same measurement procedure in order to avoid spectral features related to different optical paths. In these conditions, we checked that the measurement reproducibility is ensured at a level better than 1×10^{-3} absorbance units.

Results and discussion

AFM measurements

We show the representative and typical AFM pictures of our four samples differing by the AuNps surface coverage (monitored by the dipping time in the colloidal solution) in Fig. 2 (from left to right: 5 min–20 min–1 h–24 h). In order to ensure the accuracy of the AFM measurements and have a correct count of the AuNps to deduce their surface density (N_s), multiple scans are performed at different scales (10×10 μm^2–1×1 μm^2–500×500 nm^2) on several areas of the silicon wafers. As shown on Fig. 2, each individual nanoparticle can be identified which allows to obtain from the AFM images of 1 μm^2 and 500×500 nm^2 the local surfacic

density of gold nanoparticles. For each sample, we have averaged on at least five different areas on the sample, the density variation for each sample being indicated in Table 1. From these measurements, in addition to the AuNps height (i.e. the diameter), we can deduce that the AuNps surface density (N_s) reported in Table 1 ("SFG/DFG measurements") is multiplied by a factor of ~10 when increasing the dipping time from 5 min to 24 h. It is worth noting that it does not evolve significantly on the timescale between 1 and 24 h. From N_s and by considering gold spheres (~17 nm diameter), we can deduce the volumic filling factor f of the gold inclusions grafted on the silicon substrate with respect to an equivalent surface of bare silicon. Moreover, by performing profile measurement in the lateral direction, it is possible to evaluate the average distance between the centres of two AuNps. We observe that they can be in close contact for 24 h dipping time (d_{np-np}~20 nm). In these conditions, it is expected that the optical properties will be drastically modified as checked by UV-visible measurements.

FTIR measurements

The differential spectra acquired with FTIR spectroscopy are aimed at detecting when the thiophenol vibrational bands in the 2,500–3,400 cm^{-1} can be detected. Figure 3 clearly shows the CH stretching vibration modes of the thiophenol aromatic core at 3,057 cm^{-1} from 1 h immersion time. We checked that when a sample without AuNps is dipped into the thiophenol solution, no band is detected at 3,057 cm^{-1}. This demonstrates that thiophenol exclusively interacts with gold. Therefore, the stretching mode is detected for $f > 9$ %. Nevertheless, in the case of a very low amount of AuNps on the surface such as f between 1 and 4 %, FTIR spectroscopy is not sensitive enough. Moreover, the IR peak is fairly broad which does not allow identifying clearly all these features. It should be noted that the correct assignment of this vibration mode requires being careful because it is a rich spectral range influenced by the molecular adsorption on gold atoms [18]. Whatever the precise nature of the vibration mode, it should have a sufficient infrared activity to be detected in FTIR.

Table 2 Evolution of the interface Fresnel factors (F) and thiophenol molecular amplitude (a)

Optical probe	SFG				DFG			
Dipping time	5 min	20 min	1 h	24 h	5 min	20 min	1 h	24 h
F_{zzz} (MG)	1.2498±0.0137	0.9833±0.1967	0.5573±0.117	0.3119±0.0312	1.1958±0.1315	0.9121±0.18242	0.4818±0.1012	0.2503±0.025
a_{zzz} (MG)	0.06057±0.0066	0.1383±0.0277	0.959±0.201	2.052±0.205	0.0953±0.0104	0.2528±0.0506	1.1659±0.2448	1.9335±0.0193
F_{zzz} (BM)	1.2503±0.1375	0.9978±0.1995	0.6454±0.1355	0.4285±0.0428	1.1866±0.1305	0.8837±0.1767	0.5222±0.1097	0.3509±0.0351
a_{zzz} (BM)	0.0605±0.0066	0.1362±0.0272	0.8281±0.1739	1.4936±0.1494	0.0960±0.0105	0.2609±0.0005	1.0755±0.2258	1.3791±0.1379

Moreover, from the SFG/DFG principles, it should also have simultaneously an important Raman activity to be active. The interested reader will find in the Electronic Supplementary Information some discussion based on literature and DFT calculations on this peculiar point.

SFG/DFG measurements

We present the corresponding SFG and DFG measurements of the four samples in Fig. 4 left and right, respectively. In both cases, we observe the presence of a dip (SFG) or a peak (DFG) at the frequency $\omega_0 = 3,055 \pm 6$ cm^{-1} interfering with a strong (SFG baseline) or weak (DFG baseline) background depending on the immersion time in the colloidal solution. This vibrational feature corresponds to the stretching vibration mode of the CH groups of the aromatic core of the thiophenol [4] that we already mentioned in the FTIR measurements. Whatever f, we always observe the vibration mode from 1 to 15 % contrary to FTIR spectroscopy which is a first proof of an exaltation of the molecular signal due to the gold LSPR.

For both nonlinear configurations, we observe that the vibration mode intensity increases with the immersion time in the AuNps solution, which is evidently related to the increasing N_s as deduced from AFM measurements. We want to know how much SFG/DFG could be quantitative and how it is possible to assess the quantity of adsorbed thiophenol on the AuNps. To that end, the SFG/DFG intensity in reflection mode (coherent emission) is modelled by the following equation [8]:

$$I_{FG}(\omega_{IR}) = \left| \chi_{eff}^{(2)} \right|^2 = \left| C_{FG} e^{i\phi_{FG}} + \frac{a_0}{\omega_{IR} - \omega_0 \pm i\Gamma_0} \right|^2 .$$

We then apply the SFG/DFG data fitting developed in a procedure detailed elsewhere [8]. Moreover, to allow a direct comparison of a_0 as a function of the dipping time, the bandwidth of the vibration mode (Γ_0 damping constant) is set to 7.5 cm^{-1} in the equation of the intensity I_{FG} where the subscript FG stands either for SFG ($+i\Gamma_0$) or DFG ($-i\Gamma_0$). $\chi_{eff}^{(2)}$ is the nonlinear second order effective susceptibility of the interface, which includes the surface reflectivity (Fresnel factors). In order to obtain a more quantitative description of the molecular vibration, we thus need to correct a_0 using Fresnel factors, leading to the so-called absolute molecular amplitude a_{zzz}. C_{FG} and ϕ_{FG} are the amplitude and phase of the nonresonant (NR) contribution to the nonlinear signals. The evolution of the relevant parameters as a function of N_s obtained from the fitting procedure is given in Table 1. In our case, the NR contribution, which appears as a strong (SFG) or weak (DFG) baseline, depends on the AuNps electronic properties and more precisely to s–d interband electronic transition of gold as well-established for single crystals and colloidal

5min	20min	1h	24h

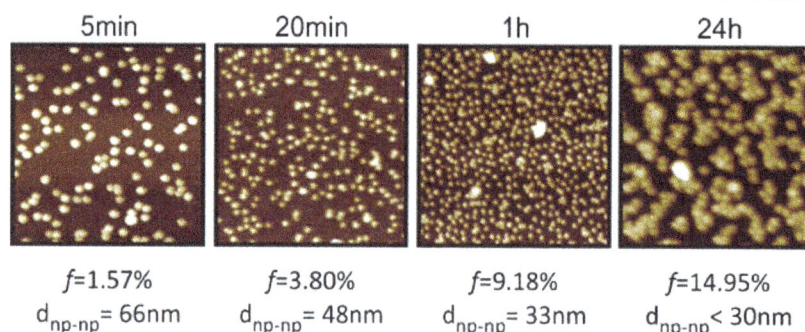

f=1.57%	f=3.80%	f=9.18%	f=14.95%
d_{np-np}= 66nm	d_{np-np}= 48nm	d_{np-np}= 33nm	d_{np-np}< 30nm

Fig. 2 AFM pictures (width scale: 1×1 μm^2 for 5 min to 1 h, $500\times$ $500nm^2$ for 24 h; height scale: 30 nm from black to white) of the four thiophenol/AuNps/APTES/Si(100) interfaces as a function of the immersion time in the colloidal solution. Relevant parameters are equally given: filling factor (f) and AuNps separation (d_{np-np})

materials [4, 8]. This point will be discussed further in "Influence of the interband transition character of gold nanospheres on the SFG/DFG spectra".

We equally observe a discrepancy of the ϕ_{FG} parameter for the sample immersed 1 h in the AuNps solution when looking at the SFG/DFG spectra. A different interference pattern is observed. We have no definitive explanation for that effect. It suggests that another chemical component may interact with the thiophenol (a strong OH band due to the presence of water molecules) for this specific preparation conditions.

In our experimental configuration, a direct comparison between the different a_o as a function of N_s makes sense provided that the surface reflectivity is taken into account via the Fresnel factors (F_{zzz}) contribution to the SFG/DFG signals. In fact, as the polarization scheme of the SFG/DFG experiment is ppp (for SFG/DFG, visible and IR beams, respectively) as depicted in Fig. 1, only the F_{zzz} contribution is significant [8]. To address this point, we consider a three-

layer model, N_s being therefore included in an effective medium model [19] where the wavelengths of light (450–4, 000 nm) are much greater than the surface roughness (~17 nm). The appropriate effective medium model has to be selected depending on the filling factor f describing the gold inclusions in a host matrix. If we develop two extreme cases, the Maxwell-Garnett (MG) model for low f (1–10 %) and the Bruggeman (BM) model for high f (>30 %), we can deduce and compare the absolute molecular amplitude (a_{zzz})

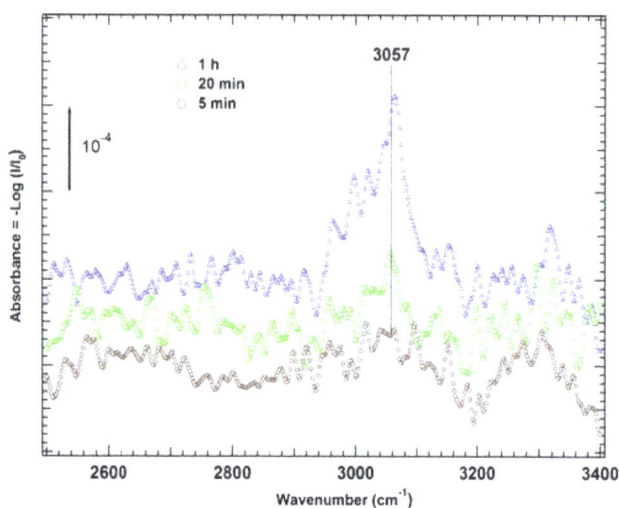

Fig. 3 FTIR differential spectra showing the CH stretching mode of thiophenol adsorbed on the AuNps. The reference spectrum corresponds to the same samples but before adsorption of thiophenol. No clear thiophenol vibration can be detected on the spectra taken with a low amount of AuNps (5–20 min dipping time). With a greater amount of AuNps (1 h dipping time), a broad vibration mode clearly emerges from the noise

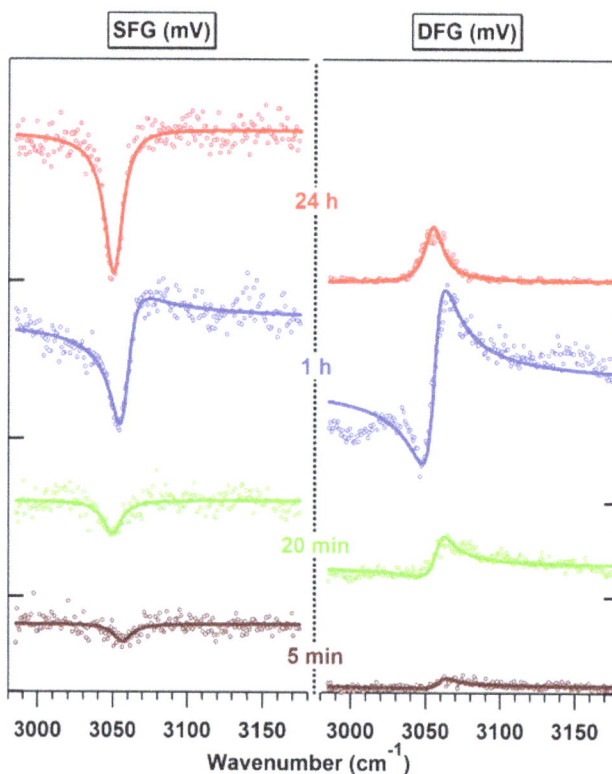

Fig. 4 SFG (*left panel*) and DFG (*right panel*) spectra of the four thiophenol/AuNps/APTES/Si(100) interfaces as a function of the increasing immersion time in the colloidal solution (5 min–20 min–1 h–24 h). SFG and DFG intensity scales are identical (min, 0 mV; max, 18 mV) to facilitate direct comparison between SFG and DFG data whatever the immersion time. The zero of each curve is indicated by *dashes on the left* (SFG) *and right* (DFG) axes

of the SFG and DFG signals for the four samples as reported in Table 2. In this manner, we will be able to investigate the possible importance of the sample reflectivity on the results.

The difference between MG and BM lies in the role attributed to the dielectric constant ε^* of the composite layer of refractive index n^* located between the silicon substrate (n_2) and the ambient air (n_1) in the three-layer model as depicted in Fig. 1. In the MG model, the composite layer is defined as a host matrix of air ($\varepsilon_h = 1$) with a low concentration of gold spherical inclusions. In the BM model, because of the high concentration of gold inclusions, the matrix host is considered as the composite layer ($\varepsilon_h = \varepsilon^*$), i.e. each of its component is considered as an inclusion. The interested reader will find in the Electronic Supplementary Material the details of the procedure used to deduce a_{zzz} from the SFG and DFG data in the framework of both effective medium models as well as the Fresnel factors calculations. We finally find, as discussed further on Fig. 6, that the vibrational amplitude is essentially proportional to the actual molecular coverage N_s for both SFG/DFG measurements when considering the appropriate effective medium model.

UV-vis measurements

We present the corresponding UV-vis reflectance curves of the four samples in Fig. 5. The optical features are strongly modified with the immersion time in the colloidal solution. The localized surface plasmon resonance (LSPR) of AuNps is expected between 505 and 520 nm depending on the immediate molecular surrounding of the particles. The absorbance of AuNps exhibits a positive peak localized at 520 nm in solution and at 515 nm when deposited on glass and measured

in transmission [20]. However, on silicon in the reflection geometry, the results seem to be very different and counter-intuitive. The LSPR does not show up in any positive peak as presented in Fig. 5.

After 5 min, no clear evidence of the LSPR is detected. After 20 min, a single negative feature appears at 525 nm. After 1 h, with a denser AuNps coverage, the dip at 525 nm is confirmed and a second stronger negative band appears at 660 nm. At 24 h, this latter band dominates the spectrum. Very often, the appearance of shoulder at ca. 650 nm is indicative of aggregated nanoparticles in a solution. However, in the present case, such a predominant aggregation is confirmed by AFM for $f_{max} = 15\%$ only. Moreover, these negative features are unusual. All these peculiarities can be explained when studying more closely the analytical expression of the reflectance in the case of AuNps deposited on silicon. The electric field reflected from a surface can be calculated analytically with the Fresnel Formula. In the case of a thin layer on top of the silicon surfaces, the three-layer model already mentioned in SFG/DFG measurements can be used and given that the thin layer is much smaller than the wavelength, the variation of the reflected intensity due to the thin layer is given by:

$$\frac{\Delta R}{R} = \frac{8\pi}{\lambda}\sqrt{\varepsilon_{vac}}d \; Im\left(\frac{\varepsilon_{Si} - \varepsilon_{MG}}{\varepsilon_{Si} - \varepsilon_{vac}}\right).$$

If divided by Ln(10)=2.3, this formula describes the reflectance spectra actually measured in normal incidence (no distinction anymore between s and p polarization). The input parameters are the complex dielectric functions $\varepsilon = \varepsilon' + i\varepsilon''$ of the silicon substrate of the surrounding medium (assumed

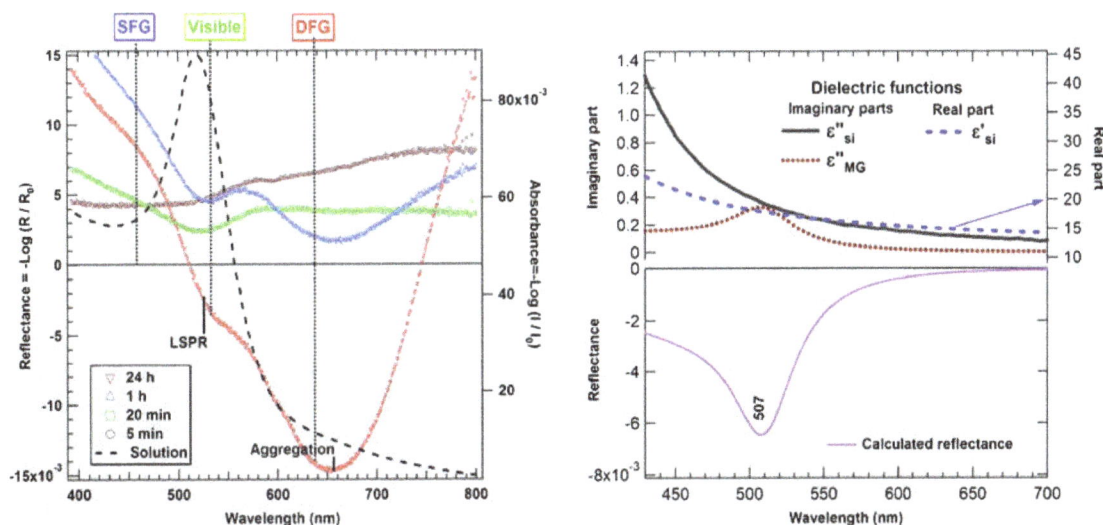

Fig. 5 *Left*: UV-vis absorbance curve of the AuNps native colloidal solution and reflectance curves of the four thiophenol/AuNps/APTES/ Si(100) interfaces as a function of the increasing immersion time in the colloidal solution (5 min–20 min–1 h–24 h). The corresponding SFG,

visible and DFG beams wavelengths are given to facilitate comparison between linear and nonlinear optical properties of the samples. *Right*: calculation of the theoretical reflectance spectrum of AuNps on silicon for 1 h immersion time within the MG model. See text for explanation

here to be vacuum for simplification). ε_{MG} is the MG effective medium model approximation of gold spheres in an air matrix as used above. This model satisfactorily captures the plasmon resonance of AuNps (in term of plasmon position and intensity). The analysis of the formula is essential to understand the main features that give the shape to the spectra of Fig. 5 (left panel). With some simplifications, ε_{Si} can be considered as mostly real ($\varepsilon_{Si}=18+0.4i$ whereas $\varepsilon_{vac}=1$). As a result, the imaginary part applies to the numerator $\varepsilon_{Si}-\varepsilon_{MG}$ only and it simplifies into:

$$\frac{\Delta R}{R}\bigg|_{approx} = \frac{8\pi}{\lambda}\sqrt{\varepsilon_{vac}}d\frac{\varepsilon''_{Si}-\varepsilon''_{MG}}{17}.$$

Therefore, the shape of optical spectrum mostly depends on $\varepsilon_{Si}''-\varepsilon_{MG}''$ where ε'' is the imaginary part of the dielectric functions. It is extremely instructive to plot the values of ε'' as a function of wavelength as done in Fig. 5 (right panel) in the case of the MG effective medium model approximation. The calculation reproduces the AuNps density corresponding to 1 h dipping time. The reflectance spectrum will have the shape of the difference between the two curves of the imaginary parts of the dielectric constants of Fig. 5 (right panel), and therefore, it clearly explains the negative appearance of the LSPR peak and is satisfactorily reproduced. Indeed, the calculated spectrum exhibits a negative LSPR peak at 507 nm instead of 525 nm. This discrepancy is mostly due to the presence of the thiophenol molecules around the AuNps whereas the model considers the AuNps are in vacuum. Moreover, Fig. 5 (right panel) also gives indication about what happens at 660 nm. Since the two plots of the imaginary parts of the dielectric constants ε_{Si}'' and ε_{MG}'' are almost parallel, just a slight tendency for aggregation will display a shallow shoulder in the LSPR peak that will be readily amplified when the difference $\varepsilon_{Si}''-\varepsilon_{MG}''$ is calculated. Therefore, the aggregation peak appears exaggerated in the reflection spectra on silicon.

As a temporary conclusion, the UV-vis spectra of AuNps deposited on silicon confirm the presence of the main plasmon peak at 525 nm even if it shows up negatively in the reflectance spectra. The negative broad contribution at 660 nm is due to aggregated AuNps, but it appears strongly exaggerated and does not lead to the conclusion that most of the AuNps are aggregated in agreement with AFM data. Therefore, the amplification of the local field due to the LSPR is still expected at 525 nm with all these samples.

Discussion

FTIR versus SFG: molecular sensitivity threshold

An interesting feature observed by comparison of the FTIR and SFG/DFG measurements is related to the different sensitivity threshold of the molecular fingerprint on small gold nanospheres. We show, in our experimental conditions on the same samples, that SFG/DFG is more sensitive than FTIR. In the latter case (Fig. 3), for f below 9 %, no thiophenol is observed. This lack of sensitivity can be explained by the specificity of each technique. In SFG/DFG, it is a process intrinsically sensitive to the symmetry breaking at the surface of AuNps which produces photons in the visible range easily detected by photomultipliers. Furthermore, the incident visible laser beam is located in the green and exalts the LSPR of the sample (Fig. 5), increasing dramatically the amplitude of the local electric fields around the AuNps and therefore the SFG/DFG response, which is not the case in FTIR. In this way, we have an *absolute* measurement of the thiophenol signal, being amplified by the LSPR excitation of the AuNps. Both thiophenol and AuNps constitute the probed interface. Moreover, silicium has no SFG/DFG activity in the involved energies and therefore does not disturb directly the nonlinear process generated by the interface. Nevertheless, we have to remember that the interface reflectivity modulates the nonlinear response through the Fresnel factors contribution. For FTIR, we perform *differential* measurements with respect to a reference sample without thiophenol. In this way, in the division procedure of the spectra, the sensitivity is related to the IR detector sensitivity. Moreover, we have to compensate for the broad and intense OH (water molecules) contribution to extract the weak CH vibration mode of the thiophenol, which becomes tricky for low f. The IR sensitivity also depends on surface reflectivity and is well adapted to metal surfaces not for silicon which is dominant for low N_s. As we have no FTIR signal in these conditions, it is more convenient to use the SFG/DFG data for a quantitative analysis of low surfacic molecular coverages. For the present paper, they allow to test the accuracy of the effective medium models developed previously and thus to understand how to take into account interface reflectivity in surfacic spectroscopic tools.

Influence of the interband transition character of gold nanospheres on the SFG/DFG spectra

By considering the SFG/DFG spectra (Fig. 4) and the parameter C_{FG} (Table 1), we see that the baseline intensity of the spectra increases with N_s, with the notable exception of the DFG curve for 24 h immersion time. This point will be addressed further. Nevertheless, in both cases, it confirms that the contribution of gold nanospheres has a qualitative and quantitative strong impact on the vibrational fingerprint of the spectra. However, the most specific feature of SFG/DFG spectroscopy lies in the interference pattern observed in the spectra. We have the thiophenol vibration mode appearing as a dip for SFG and a peak for DFG. It is related to the s–d interband electronic transition of gold as mentioned earlier and extensively detailed in previous references [4, 8]. The experimental marker of this electronic effect is the phase shift ϕ_{FG}. It is the strongest when ϕ_{FG} equals 90°. In the present work, the SFG wavelength coincides with the maximum of the s–d interband transition located in the blue (460 nm; $\phi_{FG}\sim$ 90°); therefore, the gold contribution to the nonlinear response is stronger than in the DFG case, whose wavelength is in the red (640 nm; $\phi_{FG}\sim$20°). This resonance effect of the SFG beam with the s–d interband electronic transition explains why we have a destructive interference pattern strongly marked in SFG and a constructive one in DFG. Moreover, it explains a big difference with a previous study [8] where ϕ_{FG} was close to 140° for SFG and 70° for DFG. This is because the SFG and DFG wavelengths were at 500 and 550 nm (IR spectral range centred on 10 μm), i.e. well beyond the maximum of the s–d interband transition. In summary, in our SFG/DFG data, due to the considered wavelengths, the SFG spectra are strongly influenced by the interband electronic transition of gold while DFG is only weakly affected. This property is very interesting because it proves that the vibration detected at 3,057 cm^{-1} is due to thiophenol molecules which are adsorbed on gold. In the case these molecules would have been adsorbed on other substrates, the interference pattern would have been different.

Influence of the reflectivity of gold nanospheres on the SFG/DFG spectra

As calculated in "SFG/DFG measurements" for SFG/DFG and observed in "UV-vis measurements" for UV-vis, the surface coverage of AuNps (N_s) plays a crucial role on the optical response of the sample. Depending on the dipping time in the AuNps solution, the molecular amplitude a_{zzz} is expected to scale linearly with N_s [8, 21] because our samples

produce SFG/DFG photons in a coherent way (cf. "Nonlinear optical vibrational spectroscopy"). To illustrate the influence of the sample reflectivity, we report the evolution of a_{zzz} in Fig. 6. In this manner, it is possible to compare and discuss the accuracy or discrepancy between the molecular amplitudes within the framework of the effective medium models developed above. The error bars are related to the measurement uncertainty of N_s and f extracted by AFM (Tables 1 and 2).

In spite of the small number of points per curve, we checked the correctness of the linear fit hypothesis by performing power fits of a_{zzz} as a function N_s with a free exponent n. The mean values for n are 1.57 for SFG and 1.31 for DFG. In addition, this analysis shows that linearity in the BM model is better than in the MG frame, as the n values are closer to 1 (1.49 and 1.24 for SFG and DFG, respectively). However, these two effective medium models are not sufficient to fully account for the linear hypothesis as shown by the slight deviations from the ideal case. A quick inspection of Fig. 5 (left) shows that, from the energy point of view, DFG is not influenced in the same way as SFG. The former coincides with the broad band in the red related to AuNps aggregation (ca. 650–660 nm) at high N_s (1–24 h dipping time). The latter does not undergo any local electronic effect, but the s–d interband transition as explained before. At low N_s (5–20 min dipping time), it should be noted that only the LSPR of isolated AuNps may impact the reflectivity through its excitation by the incident visible beam in the nonlinear processes. For both cases, it proves that the LSPR contribution is not correctly accounted for by MG nor BM as shown in "UV-vis measurements" where the LSPR is calculated at 507 nm instead of the expected value of 525 nm. It is a clear indication that the local fields at the visible SFG and DFG wavelengths play a significant role. The fact that the vibration amplitude linearly increases with N_s for both SFG and DFG is also consistent with the previous observation of an exaltation factor for the detection of molecules adsorbed on AuNps. Indeed,

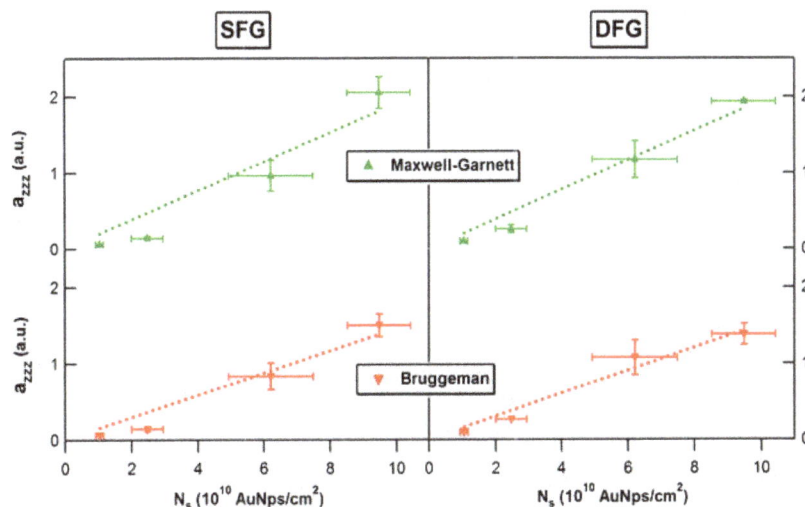

Fig. 6 Evolution of the molecular amplitude a_{zzz} as a function of the surface coverage of AuNps on silicon (N_s) for SFG (*left*) and DFG (*right*) within the Maxwell-Garnett and Bruggeman effective medium models. *Error bars* depend on the precision of N_s deduced by AFM measurements (Table 1). The *dashed curves* are the best linear fits to data

we had previously estimated this factor to be around 20 to explain the differences observed with a flat gold surface covered by a full monolayer of thiophenol [4]. However, this latter result implied that the vibrational amplitude increased linearly with N_s, which is confirmed in the present work.

Conclusions

In summary, we have studied and compared the sub-monolayer sensitivity threshold of thiophenol adsorbed on AuNps as small as 17 nm and grafted on silicon for different filling factors f checked by AFM imaging. We showed that, on silicon, SFG/DFG spectroscopy is intrinsically more sensitive than conventional FTIR spectroscopy especially for f below 9 %. Moreover, we have shown that SFG/DFG spectroscopy could take profit of the LSPR amplification by the incident visible laser beam to enhance the sensitivity without the need of hot spots as mandatory required for instance in SERS spectroscopy of small nanoparticles with similar diameters. We have quantitatively related the nonlinear SFG/DFG molecular amplitudes to the samples reflectance measured by UV-vis measurements by developing the three-layer effective medium models of MG and BM and showing that the BM approach was more appropriate to explain the SFG and DFG data when AuNps aggregation occurs. As a consequence, our results show that no coupling of nonlinear optical spectroscopy with the optical properties of aggregated AuNps occurs. For small AuNps, only the LSPR amplification of isolated AuNps can be coupled to nonlinear optics to boost the molecular sensitivity for f ranging from 1 to 4 %. Finally, we illustrated that the effective medium models of MG and BM were not sufficient to precisely take into account the LSPR coupling with the adsorbed molecules. Neglecting the molecular contribution was not possible to properly quantify the coupling between plasmonics and nonlinear optical vibrational spectroscopy. To go further quantitatively and describe the role of the local fields, the next steps should include a continuous tuning of the incident visible laser beam on several samples of different AuNps densities. In this perspective, the fine tuning of the LSPR with the visible, SFG or DFG beams would allow to increase the molecular sensitivity required in plasmonic plateforms used as biosensors where the precise targeting of specific chemical bonds between molecular probes and targets is of crucial importance to characterize the biomolecular recognition process at play in those systems.

Acknowledgments   Research leading to these results has been supported by the Région Île-de-France in the framework of the funding program C'Nano IdF under grant agreement CREMOSOFT. The authors acknowledge the BQR financial support of the Université Paris-Sud. The authors also acknowledge C. Six and A. Gayral for technical assistance on the SFG/DFG experimental setup and L. Dalstein for fruitful discussion on theoretical aspects of SFG and SFS spectroscopy of nanoparticles.

References

1. Morel AL, Boujday S, Méthivier C, Krafft JM, Pradier CM (2011) Biosensors elaborated on gold nanoparticles: a PM-IRRAS characterisation of the IgG binding efficiency. Talanta 85:35–42
2. Barbillon G, Bijeon JL, Plain J, de la Chapelle ML, Adam PM, Royer P (2007) Electron beam lithography designed chemical nanosensors based on localized surface plasmon resonance. Surf Sci 601:5057–5061
3. Enders D, Nagao T, Pucci A, Nakayam T (2006) Reversible adsorption of Au nanoparticles on SiO₂/Si: an in situ ATR-IR study. Surf Sci 600:L71–L75
4. Pluchery O, Humbert C, Valamanesh M, Lacaze E, Busson B (2009) Enhanced detection of thiophenol adsorbed on gold nanoparticles by SFG and DFG nonlinear optical spectroscopy. Phys Chem Chem Phys 11:7729–7737
5. Tourillon G, Dreesen L, Volcke C, Sartenaer Y, Thiry PA, Peremans A (2007) Total internal reflection sum–frequency generation and dense gold nanoparticles monolayer: a route for probing adsorbed molecules. Nanotechnology 18:415301, 1–7
6. Humbert C, Busson B, Abid JP, Six C, Girault HH, Tadjeddine A (2005) Self-assembled organic monolayers on gold nanoparticles: a study by sum-frequency generation combined with UV-vis spectroscopy. Electrochim Acta 50:3101–3110
7. Bordenyuk AN, Weereman C, Yatawara A, Jayathilake HD, Stiopkin I, Liu Y, Benderskii AV (2007) Vibrational sum frequency generation spectroscopy of dodecanethiol on metal nanoparticles. J Phys Chem C 111:8925–8933
8. Humbert C, Pluchery O, Lacaze E, Tadjeddine A, Busson B (2012) A multiscale description of molecular adsorption on gold nanoparticles by nonlinear optical spectroscopy. Phys Chem Chem Phys 14:280–289
9. Dreesen L, Sartenaer Y, Humbert C, Mani AA, Méthivier C, Pradier CM, Thiry PA, Peremans A (2004) Probing ligand-protein recognition with sum-frequency generation spectroscopy: the avidin-biocytin case. Chem Phys Chem 5:1719–1725
10. Sartenaer Y, Tourillon G, Dreesen L, Lis D, Mani AA, Thiry PA, Peremans A (2007) Sum–frequency generation spectroscopy of DNA monolayers. Biosens Bioelectron 22:2179–2183
11. Walter SR, Geiger FM (2010) DNA on stage: showcasing oligonucleotides at surfaces and interfaces with second harmonic and vibrational sum frequency generation. J Phys Chem Lett 1:9–15
12. Howell C, Zhao J, Koelsh P, Zharnikov M (2011) Hybridization in ssDNA films: a multi-technique spectroscopy study. Phys Chem Chem Phys 13:15512–15522
13. Tourillon G, Dreesen L, Volcke C, Sartenaer Y, Thiry PA, Peremans A (2009) Close-packed array of gold nanoparticles and sum frequency generation spectroscopy in total internal reflection: a platform for studying biomolecules and biosensors. J Mater Sci 44:6805–6810
14. Humbert C, Busson B (2011) SFG spectroscopy of biointerfaces. In: Pradier CM, Chabal YJ (eds) Biointerface characterization by advanced IR spectroscopy. Elsevier, Amsterdam, pp 279–321
15. Ji XH, Song XN, Li J, Bai YB, Yang WS, Peng XG (2007) Size control of gold nanocrystals in citrate reduction: the third role of citrate. J Am Chem Soc 129:13939–13948

16. Roke S, Gonella G (2012) Nonlinear light scattering and spectroscopy of particles and droplets in liquids. Annu Rev Phys Chem 63:353–378

17. de Aguiar HB, Scheu R, Jena KC, de Beer AGF, Roke S (2012) Comparison of scattering and reflection SFG: a question of phase-matching. Phys Chem Chem Phys 14:6826–6832

18. Feugmo CGT, Liégeois V (2013) Analyzing the vibrational signatures of thiophenol adsorbed on small gold clusters by DFT calculations. ChemPhysChem 14:1633–1645

19. Aspnes DE, Theeten JB (1979) Investigation of effective-medium models of microscopic surface roughness by spectroscopic ellipsometry. Phys Rev B 20:3292–3302

20. Pluchery O, Lacaze E, Simion M, Miu M, Bragaru A, Radoi A (2010) Optical characterization of supported gold nanoparticles for plasmonic biosensors. Semiconductor Conference (CAS), International Sinaia, IEEE Electron Devices Society, Romania, pp 159–162

21. Zhuang X, Miranda PB, Kim D, Shen YR (1999) Mapping molecular orientation and conformation at interfaces by surface nonlinear optics. Phys Rev B 59:12632–12640

Permissions

List of Contributors

Masataka Hakamada
Department of Energy Science and Technology, Graduate School of Energy Science, Kyoto University, Yoshida Honmachi, Sakyo, Kyoto 606-8501, Japan

Masaki Takahashi
Department of Energy Science and Technology, Graduate School of Energy Science, Kyoto University, Yoshida Honmachi, Sakyo, Kyoto 606-8501, Japan

Mamoru Mabuchi
Department of Energy Science and Technology, Graduate School of Energy Science, Kyoto University, Yoshida Honmachi, Sakyo, Kyoto 606-8501, Japan

Alfred Z. Msezane
Department of Physics and Center for Theoretical Studies of Physical Systems, Clark Atlanta University, Atlanta, GA 30314, USA

Zineb Felfli
Department of Physics and Center for Theoretical Studies of Physical Systems, Clark Atlanta University, Atlanta, GA 30314, USA

Kelvin Suggs
Department of Chemistry, Clark Atlanta University, Atlanta, GA 30314, USA

Aron Tesfamichael
Department of Chemistry, Clark Atlanta University, Atlanta, GA 30314, USA

Xiao-Qian Wang
Department of Physics and Center for Theoretical Studies of Physical Systems, Clark Atlanta University, Atlanta, GA 30314, USA

Muhammad Nubli Zulkifli
Institute of Microengineering and Nanoelectronics (IMEN), Universiti Kebangsaan Malaysia, 43600 UKM, Selangor, Malaysia

Shahrum Abdullah
Department of Mechanical and Materials Engineering, Universiti Kebangsaan Malaysia, 43600 UKM, Selangor, Malaysia

Norinsan Kamil Othman
School of Applied Physics, Faculty of Science and Technology, Universiti Kebangsaan Malaysia, 43600 UKM, Selangor, Malaysia

Azman Jalar
Institute of Microengineering and Nanoelectronics (IMEN), Universiti Kebangsaan Malaysia, 43600 UKM, Selangor, Malaysia

Shuo-Hong Wang
Department of Materials Science and Engineering, National Tsing Hua University, No. 101, Section 2, Kuang-Fu Road, Hsinchu 30013, Taiwan

Tsung-Shune Chin
Department of Materials Science and Engineering, National Tsing Hua University, No. 101, Section 2, Kuang-Fu Road, Hsinchu 30013, Taiwan
Center for Nanotechnology, Materials Science and Microsystems, National Tsing Hua University, No. 101, Sec-2, Kuang-Fu Road, Hsinchu 30013, Taiwan
Department of Materials Science and Engineering, Feng Chia University, No. 100, Wenhwa Road, Seatwen District, Taichung 40724, Taiwan

Shinji Sakai
Division of Chemical Engineering, Department of Materials Science and Engineering, Graduate School of Engineering Science, Osaka University, 1-3 Machikaneyama-cho, Toyonaka, Osaka 560-8531, Japan

Shogo Kawa
Division of Chemical Engineering, Department of Materials Science and Engineering, Graduate School of Engineering Science, Osaka University, 1-3 Machikaneyama-cho, Toyonaka, Osaka 560-8531, Japan

Koichi Sawada
Division of Chemical Engineering, Department of Materials Science and Engineering, Graduate School of Engineering Science, Osaka University, 1-3 Machikaneyama-cho, Toyonaka, Osaka 560-8531, Japan

Masahito Taya
Division of Chemical Engineering, Department of Materials Science and Engineering, Graduate School of Engineering Science, Osaka University, 1-3 Machikaneyama-cho, Toyonaka, Osaka 560-8531, Japan

L. P. Ward
School of Civil, Environmental and Chemical Engineering, RMIT University, Melbourne, VIC, Australia

D. Chen
Department of Chemistry and Chemical Engineering, Foshan University, Foshan, Guangdong, China

A. P. O'Mullane
School of Applied Sciences, RMIT University, Melbourne, VIC, Australia

Weiwei Xu
Key Laboratory for Special Functional Materials of Ministryof Education, Henan University, Kaifeng 475004, People's Republic of China
Life Science Division, Graduate School at Shenzhen, Tsinghua University, Shenzhen 518055, People's Republic of China

Jinzhong Niu
Department of Mathematical and Physical Sciences, Henan Institute of Engineering, Zhengzhou 451191, People's Republic of China

Hangying Shang
Key Laboratory for Special Functional Materials of Ministryof Education, Henan University, Kaifeng 475004, People's Republic of China

Huaibin Shen
Key Laboratory for Special Functional Materials of Ministryof Education, Henan University, Kaifeng 475004, People's Republic of China

Lan Ma
Life Science Division, Graduate School at Shenzhen, Tsinghua University, Shenzhen 518055, People's Republic of China

Lin Song Li
Key Laboratory for Special Functional Materials of Ministryof Education, Henan University, Kaifeng 475004, People's Republic of China

S. A. Nikolaev
Department of Chemistry, M.V. Lomonosov Moscow State University, 1 Leninskie Gory, 119991, Moscow, Russia

D. A. Pichugina
Department of Chemistry, M.V. Lomonosov Moscow State University, 1 Leninskie Gory, 119991, Moscow, Russia

D. F. Mukhamedzyanova
Department of Chemistry, M.V. Lomonosov Moscow State University, 1 Leninskie Gory, 119991, Moscow, Russia

B. Henriques
Center for Mechanical and Materials Technologies, Universidade do Minho, Campus de Azurém, 4800-058 Guimarães, Portugal

P. Pinto
Center for Mechanical and Materials Technologies, Universidade do Minho, Campus de Azurém, 4800-058 Guimarães, Portugal

J. Souza
Center for Mechanical and Materials Technologies, Universidade do Minho, Campus de Azurém, 4800-058 Guimarães, Portugal

J. C. Teixeira
Center for Mechanical and Materials Technologies, Universidade do Minho, Campus de Azurém, 4800-058 Guimarães, Portugal

D. Soares
Center for Mechanical and Materials Technologies, Universidade do Minho, Campus de Azurém, 4800-058 Guimarães, Portugal

F. S. Silva
Center for Mechanical and Materials Technologies, Universidade do Minho, Campus de Azurém, 4800-058 Guimarães, Portugal

Ngac An Bang
VNU-University of Science, 334 Nguyen Trai, Thanh Xuan, Hanoi, Vietnam

Phung Thi Thom
VNU-University of Science, 334 Nguyen Trai, Thanh Xuan, Hanoi, Vietnam

Hoang Nam Nhat
VNU-University of Engineering and Technology, 144 Xuan Thuy, Cau Giay, Hanoi, Vietnam

Guozhen Chen
Key Laboratory of Applied Surface and Colloid Chemistry, Ministry of Education, Key Laboratory of Analytical Chemistry for Life Science of Shaanxi Province, School of Chemistry and Chemical Engineering, Shaanxi Normal University, Xi'an 710062, China
China State Key Laboratory of Chemo/Biosensing and Chemometrics, Hunan University, Changsha 410082, People's Republic of China

Yan Jin
Key Laboratory of Applied Surface and Colloid Chemistry, Ministry of Education, Key Laboratory of Analytical Chemistry for Life Science of Shaanxi Province, School of Chemistry and Chemical Engineering, Shaanxi Normal University, Xi'an 710062, China
China State Key Laboratory of Chemo/Biosensing and Chemometrics, Hunan University, Changsha 410082, People's Republic of China

Wenhong Wang
Key Laboratory of Applied Surface and Colloid Chemistry, Ministry of Education, Key Laboratory of Analytical Chemistry for Life Science of Shaanxi Province, School of Chemistry and Chemical Engineering, Shaanxi Normal University, Xi'an 710062, China

China State Key Laboratory of Chemo/Biosensing and Chemometrics, Hunan University, Changsha 410082, People's Republic of China

Yina Zhao
Key Laboratory of Applied Surface and Colloid Chemistry, Ministry of Education, Key Laboratory of Analytical Chemistry for Life Science of Shaanxi Province, School of Chemistry and Chemical Engineering, Shaanxi Normal University, Xi'an 710062, China
China State Key Laboratory of Chemo/Biosensing and Chemometrics, Hunan University, Changsha 410082, People's Republic of China

Jie Shen
Institut des Sciences Moléculaires d'Orsay, ISMO, Centre National de la Recherche Scientifique (CNRS), Université Paris Sud, UMR 8214, Bâtiment 351, 91405 Orsay, France

Juanjuan Jia
Institut des Sciences Moléculaires d'Orsay, ISMO, Centre National de la Recherche Scientifique (CNRS), Université Paris Sud, UMR 8214, Bâtiment 351, 91405 Orsay, France

Kirill Bobrov
Institut des Sciences Moléculaires d'Orsay, ISMO, Centre National de la Recherche Scientifique (CNRS), Université Paris Sud, UMR 8214, Bâtiment 351, 91405 Orsay, France

Laurent Guillemot
Institut des Sciences Moléculaires d'Orsay, ISMO, Centre National de la Recherche Scientifique (CNRS), Université Paris Sud, UMR 8214, Bâtiment 351, 91405 Orsay, France

Vladimir A. Esaulov
Institut des Sciences Moléculaires d'Orsay, ISMO, Centre National de la Recherche Scientifique (CNRS), Université Paris Sud, UMR 8214, Bâtiment 351, 91405 Orsay, France

Pandian Lakshmanan
Laboratoire de Réactivité de Surface, UMR 7197 CNRS, Université Pierre etMarie Curie-UMPC, 4 place Jussieu, 75252 Paris Cedex 05, France
CNRS UMR 7285, Institut de Chimie des Milieux et Matériaux de Poitiers (IC2MP), University of Poitiers, 4 rue Michel Brunet, 86022 Poitiers Cedex, France

Frédéric Averseng
Laboratoire de Réactivité de Surface, UMR 7197 CNRS, Université Pierre etMarie Curie-UMPC, 4 place Jussieu, 75252 Paris Cedex 05, France

Nicolas Bion
CNRS UMR 7285, Institut de Chimie des Milieux et Matériaux de Poitiers (IC2MP), University of Poitiers, 4 rue Michel Brunet, 86022 Poitiers Cedex, France

Laurent Delannoy
Laboratoire de Réactivité de Surface, UMR 7197 CNRS, Université Pierre etMarie Curie-UMPC, 4 place Jussieu, 75252 Paris Cedex 05, France

Jean-Michel Tatibouët
CNRS UMR 7285, Institut de Chimie des Milieux et Matériaux de Poitiers (IC2MP), University of Poitiers, 4 rue Michel Brunet, 86022 Poitiers Cedex, France

Catherine Louis
Laboratoire de Réactivité de Surface, UMR 7197 CNRS, Université Pierre etMarie Curie-UMPC, 4 place Jussieu, 75252 Paris Cedex 05, France

Qiang Zhang
College of Chemistry, Chemical Engineering and Materials Science, Soochow University, Suzhou 215123, People's Republic of China

Ruirui Yue
College of Chemistry, Chemical Engineering and Materials Science, Soochow University, Suzhou 215123, People's Republic of China

Fengxing Jiang
College of Chemistry, Chemical Engineering and Materials Science, Soochow University, Suzhou 215123, People's Republic of China

Huiwen Wang
College of Chemistry, Chemical Engineering and Materials Science, Soochow University, Suzhou 215123, People's Republic of China

Chunyang Zhai
College of Chemistry, Chemical Engineering and Materials Science, Soochow University, Suzhou 215123, People's Republic of China

Ping Yang
College of Chemistry, Chemical Engineering and Materials Science, Soochow University, Suzhou 215123, People's Republic of China

Yukou Du
College of Chemistry, Chemical Engineering and Materials Science, Soochow University, Suzhou 215123, People's Republic of China

Hazar Guesmi
CNRS—Laboratoire de Réactivité de Surface, Université Pierre et Marie Curie (UMR 7197), 3 rue Galilée, 94200 Ivry, France
CNRS—Institut Charles Gerhardt-équipe MACS, Ecole Nationale de Chimie de Montpellier (UMR 5253), 8 rue de l'Ecole Normale, 34296 Montpellier, France

Georg Steinhauser
Atominstitut, Vienna University of Technology, Stadionallee 2, 1020 Vienna, Austria

Stefan Merz
Atominstitut, Vienna University of Technology, Stadionallee 2, 1020 Vienna, Austria

Franziska Stadlbauer
Atominstitut, Vienna University of Technology, Stadionallee 2, 1020 Vienna, Austria

Peter Kregsamer
Atominstitut, Vienna University of Technology, Stadionallee 2, 1020 Vienna, Austria

Christina Streli
Atominstitut, Vienna University of Technology, Stadionallee 2, 1020 Vienna, Austria

Mario Villa
Atominstitut, Vienna University of Technology, Stadionallee 2, 1020 Vienna, Austria

Ji-In Jeong
Department of Dental Materials, Institute of Translational Dental Science, School of Dentistry, Pusan National University, Beomeo-Ri, Mulgeum-Eup, Yangsan-Si, Gyeongsangnam-Do 626-814, South Korea

Hyung-Il Kim
Department of Dental Materials, Institute of Translational Dental Science, School of Dentistry, Pusan National University, Beomeo-Ri, Mulgeum-Eup, Yangsan-Si, Gyeongsangnam-Do 626-814, South Korea

Gwang-Young Lee
Department of Dental Materials, Institute of Translational Dental Science, School of Dentistry, Pusan National University, Beomeo-Ri, Mulgeum-Eup, Yangsan-Si, Gyeongsangnam-Do 626-814, South Korea

Yong Hoon Kwon
Department of Dental Materials, Institute of Translational Dental Science, School of Dentistry, Pusan National University, Beomeo-Ri, Mulgeum-Eup, Yangsan-Si, Gyeongsangnam-Do 626-814, South Korea

Hyo-Joung Seol
Department of Dental Materials, Institute of Translational Dental Science, School of Dentistry, Pusan National University, Beomeo-Ri, Mulgeum-Eup, Yangsan-Si, Gyeongsangnam-Do 626-814, South Korea

Fengqun Lang
Advanced Power Electronics Research Center, National Institute of Advanced Industrial Science and Technology (AIST), Tsukuba 305-8568, Japan

R & D Partnership for Future Power Electronics Technology, c/o, AIST, Tsukuba, Japan

Hiroshi Nakagawa
Advanced Power Electronics Research Center, National Institute of Advanced Industrial Science and Technology (AIST), Tsukuba 305-8568, Japan

Hiroshi Yamaguchi
Advanced Power Electronics Research Center, National Institute of Advanced Industrial Science and Technology (AIST), Tsukuba 305-8568, Japan

Sung-Min Kim
Department of Dental Materials, Institute of Translational Dental Science, School of Dentistry, Pusan National University, Beomeo-Ri, Mulgeum-Eup, Yangsan-Si, Gyeongsangnam-Do 626-814, South Korea

Hyung-Il Kim
Department of Dental Materials, Institute of Translational Dental Science, School of Dentistry, Pusan National University, Beomeo-Ri, Mulgeum-Eup, Yangsan-Si, Gyeongsangnam-Do 626-814, South Korea

Byung-Wook Jeon
Department of Dental Materials, Institute of Translational Dental Science, School of Dentistry, Pusan National University, Beomeo-Ri, Mulgeum-Eup, Yangsan-Si, Gyeongsangnam-Do 626-814, South Korea

Yong Hoon Kwon
Department of Dental Materials, Institute of Translational Dental Science, School of Dentistry, Pusan National University, Beomeo-Ri, Mulgeum-Eup, Yangsan-Si, Gyeongsangnam-Do 626-814, South Korea

Hyo-Joung Seol
Department of Dental Materials, Institute of Translational Dental Science, School of Dentistry, Pusan National University, Beomeo-Ri, Mulgeum-Eup, Yangsan-Si, Gyeongsangnam-Do 626-814, South Korea

A. Bouvrée
Institut des Matériaux Jean Rouxel (IMN), CNRS, Université de Nantes, 2 rue de la Houssinière, BP 32229, 44322 Nantes cedex 3, France

A. D_Orlando
Institut des Matériaux Jean Rouxel (IMN), CNRS, Université de Nantes, 2 rue de la Houssinière, BP 32229, 44322 Nantes cedex 3, France

T. Makiabadi
Institut des Matériaux Jean Rouxel (IMN), CNRS, Université de Nantes, 2 rue de la Houssinière, BP 32229, 44322 Nantes cedex 3, France

S. Martin
Institut des Matériaux Jean Rouxel (IMN), CNRS, Université de Nantes, 2 rue de la Houssinière, BP 32229, 44322 Nantes cedex 3, France

G. Louarn
Institut des Matériaux Jean Rouxel (IMN), CNRS, Université de Nantes, 2 rue de la Houssinière, BP 32229, 44322 Nantes cedex 3, France

J. Y. Mevellec
Institut des Matériaux Jean Rouxel (IMN), CNRS, Université de Nantes, 2 rue de la Houssinière, BP 32229, 44322 Nantes cedex 3, France

B. Humbert
Institut des Matériaux Jean Rouxel (IMN), CNRS, Université de Nantes, 2 rue de la Houssinière, BP 32229, 44322 Nantes cedex 3, France

Jyun Lin Li
Department of Materials and Optoelectronic Science, National Sun Yat-Sen University, 70, Lien-Hai Road, Kaohsiung 804, Taiwan

Pei Jen Lo
Department of Materials and Optoelectronic Science, National Sun Yat-Sen University, 70, Lien-Hai Road, Kaohsiung 804, Taiwan

Ming Chi Ho
Department of Materials and Optoelectronic Science, National Sun Yat-Sen University, 70, Lien-Hai Road, Kaohsiung 804, Taiwan

Ker-Chang Hsieh
Department of Materials and Optoelectronic Science, National Sun Yat-Sen University, 70, Lien-Hai Road, Kaohsiung 804, Taiwan

Byung-Wook Jeon
Department of Dental Materials, Institute of Translational Dental Sciences, School of Dentistry, Pusan National University, Beomeo-Ri, Mulgeum-Eup, Yangsan-Si, Gyeongsangnam-Do 626-814, South Korea

Sung-Min Kim
Department of Dental Materials, Institute of Translational Dental Sciences, School of Dentistry, Pusan National University, Beomeo-Ri, Mulgeum-Eup, Yangsan-Si, Gyeongsangnam-Do 626-814, South Korea

Hyung-Il Kim
Department of Dental Materials, Institute of Translational Dental Sciences, School of Dentistry, Pusan National University, Beomeo-Ri, Mulgeum-Eup, Yangsan-Si, Gyeongsangnam-Do 626-814, South Korea

Yong Hoon Kwon
Department of Dental Materials, Institute of Translational Dental Sciences, School of Dentistry, Pusan National University, Beomeo-Ri, Mulgeum-Eup, Yangsan-Si, Gyeongsangnam-Do 626-814, South Korea

Hyo-Joung Seol
Department of Dental Materials, Institute of Translational Dental Sciences, School of Dentistry, Pusan National University, Beomeo-Ri, Mulgeum-Eup, Yangsan-Si, Gyeongsangnam-Do 626-814, South Korea

Yong-Ryeol Lee
Department of Dental Materials, MRC for Biomineralization Disorders, School of Dentistry, Chonnam National University, Gwangju 500-757, Korea

Mi-Kyung Han
Department of Dental Materials, MRC for Biomineralization Disorders, School of Dentistry, Chonnam National University, Gwangju 500-757, Korea

Min-Kang Kim
Department of Dental Materials, MRC for Biomineralization Disorders, School of Dentistry, Chonnam National University, Gwangju 500-757, Korea

Won-Jin Moon
Korea Basic Science Institute, Gwang-Ju Center, 300Yongbong-dong, Buk-gu, Gwangju 500-757, Korea

Ho-Jun Song
Department of Dental Materials, MRC for Biomineralization Disorders, School of Dentistry, Chonnam National University, Gwangju 500-757, Korea

Yeong-Joon Park
Department of Dental Materials, MRC for Biomineralization Disorders, School of Dentistry, Chonnam National University, Gwangju 500-757, Korea

Chong Leong Gan
Institute of Nano Electronic Engineering (INEE), Universiti Malaysia Perlis, 01000 Kangar, Perlis, Malaysia
Spansion (Penang) Sdn. Bhd., Bayan Lepas, 11900 Penang, Malaysia

Classe Francis
Spansion (Penang) Sdn. Bhd., Bayan Lepas, 11900 Penang, Malaysia

Bak Lee Chan
Spansion (Penang) Sdn. Bhd., Bayan Lepas, 11900 Penang, Malaysia

Uda Hashim
Institute of Nano Electronic Engineering (INEE), Universiti Malaysia Perlis, 01000 Kangar, Perlis, Malaysia

C. Humbert
Univ Paris-Sud, Laboratoire de Chimie Physique, CNRS,
Bâtiment 201 Porte 2, 91405 Orsay, France

O. Pluchery
Univ Pierre et Marie Curie UPMC-CNRS, Institut des
Nanosciences de Paris, 4 place Jussieu, 75005 Paris, France

E. Lacaze
Univ Pierre et Marie Curie UPMC-CNRS, Institut des
Nanosciences de Paris, 4 place Jussieu, 75005 Paris, France

A. Tadjeddine
Univ Paris-Sud, Laboratoire de Chimie Physique, CNRS,
Bâtiment 201 Porte 2, 91405 Orsay, France

B. Busson
Univ Paris-Sud, Laboratoire de Chimie Physique, CNRS,
Bâtiment 201 Porte 2, 91405 Orsay, France